Reinventing Biology

Race, Gender, and Science

Anne Fausto-Sterling, *General Editor*

Feminism and Science (1989)
Nancy Tuana, Editor

The "Racial" Economy of Science: Toward a Democratic Future (1993)
Sandra Harding, Editor

The Less Noble Sex: Scientific, Religious, and Philosophical Conceptions of Woman's Nature (1993)
Nancy Tuana

Love, Power and Knowledge: Toward a Feminist Transformation of the Sciences (1994)
Hilary Rose

Women's Health—Missing from U.S. Medicine (1994)
Sue V. Rosser

Deviant Bodies (1995)
Jennifer Terry and Jacqueline Urla, Editors

Im/partial Visions of Life: Gender Ideology in Molecular Biology (1995)
Bonnie B. Spanier

Reinventing Biology
Respect for Life and the
Creation of Knowledge

Edited by Lynda Birke and Ruth Hubbard

Indiana University Press

Bloomington and Indianapolis

© 1995 by Indiana University Press

The paper used in this publication meets the minimum require-
ments of American National Standard for Information Sciences—
Permanence of Paper for Printed Library Materials, ANSI
Z39.48-1984.

Manufactured in the United States of America

Library of Congress Cataloging-in-Publication Data

Reinventing biology : respect for life and the creation of knowledge /
    edited by Lynda Birke and Ruth Hubbard.
        p.   cm. — (Race, gender, and science)
    Includes bibliographical references and index.
        ISBN 0-253-32909-4 (cl : alk. paper). — ISBN 0-253-20981-1 (pa :
alk. paper)
        1. Biology—Philosophy. 2. Biology—Research. 3. Animal
experimentation—Moral and ethical aspects. 4. Human-animal
relationships. I. Birke, Lynda I. A. II. Hubbard, Ruth, date.
III. Series.
QH331.R425 1995
574'.01—dc20   95-1443

1 2 3 4 5   00 99 98 97 96 95

So the real presence of an animal in a laboratory—that is, an animal perceived by the experimenting scientist not as an object, . . . but as a subject in the philosophical/grammatical sense of a sentient existence of the same order as the scientist's existence—so such presence and perception in a laboratory where experiments are performed upon animals would profoundly change the nature, and probably the results, of the experiments.

—Ursula K. LeGuin, Introduction to "Schrödinger's Cat," in *Buffalo Gals and Other Animal Presences*, Penguin Books, 1987, p. 187

There are times in life when the question of knowing if one can think differently than one thinks, and perceive differently than one sees, is absolutely necessary if one is to go on looking and reflecting at all. People will say, perhaps, that these games with oneself would better be left backstage; or, at best, that they might properly form part of those preliminary exercises that are forgotten once they have served their purpose. But, then, . . . in what does [philosophy] consist, if not in the endeavor to know how and to what extent it might be possible to think differently, instead of legitimating what is already known?

—Michel Foucault, *The Use of Pleasure* (volume 2 of *The History of Sexuality*), Vintage Books, 1986, pp. 8, 9.

Contents

Part IV. Border Crossings: Human/Animal, Live/Inanimate

Introduction

Lynda Birke and Ruth Hubbard

Two decades ago, when the women's liberation and health movements were coming into their own, the two of us, working within them in Great Britain and the United States, began to raise questions about science. In particular, we were concerned to explore what effects the absence of women scientists had had on the theoretical frameworks and the practices of biology and medicine. This led to our participation in the editorial collectives that produced, at much the same time, on one side of the Atlantic, *Alice through the Microscope*, and on the other, *Women Look at Biology Looking at Women*.[1] Clearly, though we did not know each other, we were traveling on similar tracks.

Now, two decades later, and this time together, we are setting out to explore the further question of how the fact that biologists look on humans as radically different from, yet fundamentally similar to, the objects of their research—be they animals, plants, bacteria, or humans—has shaped biological thinking and practice. As before, we have approached our question by asking colleagues and friends to join us, though this time geographical distances have made the process less of a collective venture.

The product before you is a group of essays linked less by their specific subject matter than by a common point of view, which can be encapsulated in the question: how do biologists conceptualize the nature of the organisms they work with, and what alternative outcomes would we expect to see if the conceptual framework or the rules of practice were different? And not just different, but, specifically, if scientific objectivity were defined not as an attitude of separation and detachment between scientific actors and the passive objects they manipulate but as a cooperative venture in which scientists and their research subjects are partners. Put another way, what would science look like if it respected the living organisms it studies as individuals with their own histories and integrities? What would it look like if scientists thought of other organisms as rational and capable of intelligent thought?

We have asked a range of people to reflect on the issues these questions raise for them, based on their perspectives as historians, sociologists, biologists, or anthropologists and as thoughtful participants in various political struggles for a better world. In pursuing these issues, the authors have not felt limited by their academic training but have taken the opportunity to probe what for all of us represents relatively uncharted terrain.

There are areas of biology that we have not covered, though we would have liked to do so. Agricultural research, for example, poses many problems about respect for other creatures that are not explicitly touched on here. Yet the contributors to this volume have ranged widely enough to bring up a good many important issues and to give some idea of how science might look if detachment and a narrowly defined objectivity were not the order of the day. The result is a fairly eclectic mix of essays that reflect the authors' different disciplinary and personal commitments, linked by a common interest in describing the different ways boundaries are drawn both among different groups of people and between people and other organisms.

Though traditional scientists usually explain these boundaries in naturalistic terms, clearly, economic and social interests are involved. This is most apparent in the discussions of the use of people as subjects for scientific experimentation. Like the fictional accounts about Victor Frankenstein and Dr. Moreau, these descriptions of contemporary scientific practices raise in a particularly stark form the question of the relationship between biologists and the organisms they study. What would biology look like if biologists and biomedical scientists stopped thinking of themselves as apart from their subjects? By implication, how would they look on the subjects on whom they experiment if they granted that these are capable of both suffering and rational thought? It seems inevitable that a science that objectifies nonhuman organisms will, in some circumstances, use similar justifications to objectify people.

The solution does not lie in abandoning all forms of biological experimentation. Rather, we see an urgent need for including concerns about the potential welfare and suffering of experimental subjects in all evaluations of research questions and design. Potential benefits to humans, and indeed usually only to specific subgroups of humans, must not be used to justify subjecting people or other organisms to experimental situations we would not accept for ourselves or our loved ones.

One of the most controversial areas, and one in which there has been the most public criticism of science, concerns the way scientists use animals. Our knowledge of anatomy and physiology inevitably has relied on both dissection of human cadavers and the use of living animals; both have been controversial. One of the earliest records of the use of animals for experiments came from Greece, when Galen used living pigs for anatomical studies. The body of medical knowledge to which Galen contributed was little challenged for over a thousand years, partly because dissection was proscribed by the Christian church and later by Islam.

The use of bodies for anatomical dissection was revived with the rise of modern science in Europe. Although no longer so controversial in religious terms, the practice met considerable opposition by the end of the eighteenth century. At that time, public fears centered on growing anxiety about what doctors might be doing with bodies and how they might acquire them: respect for people's wishes about their bodily fate after death was often discarded in the pursuit of science.[2]

Meanwhile, living animals were also used in the name of science. Although there have been dissident voices opposed to the practice since the beginning of the scientific revolution in the seventeenth century,[3] organized opposition began only toward the end of the nineteenth century, as Hilary Rose notes in her chapter. A source of that earlier opposition was the concern that lack of respect itself might alter what we come to know about nature. Thus the microscopist Robert Hooke expressed doubt in 1665, asking whether by "dissecting and mangling creatures whils't there is life yet within them, we find [nature] indeed at work, but put into such disorder by the violence offer'd".[4]

An organized antivivisectionist movement emerged in the nineteenth century as part of a growing concern with the welfare of animals. But it was also an expression of mounting fears about science and what scientists might be doing. If they are torturing animals now, people felt, might they start on us next?[5]

After the end of the First World War, the antivivisectionist cause lost prominence. It appeared again after 1975, when Peter Singer published *Animal Liberation*,[6] and has been gaining strength since. Whatever else it reflects, modern concern with animal rights is part of a widespread rejection of the values that many people associate with science. A lack of respect for living organisms is part of that suspect value system.

To antivivisectionist organizations, nothing less than the total abolition of all animal use in science will do, while most scientists think that such a move would totally cripple scientific inquiry and medical progress. The two sides seem irreconcilable, each speaking a different language. Scientists sometimes accuse their opponents of putting animals before people, thus disrespecting humans; the opposition accuses scientists of putting self-interest first.

At the heart of the controversy are ethical questions. Scientists justify their use of animals by reference to potential medical benefits—and who among us would not like to see cancer or AIDS cured, if not prevented altogether? On this line of thought, it would be unethical *not* to use animals, if doing so advances medical knowledge. But to those concerned with animal rights, it is ethically unacceptable to prioritize humans in this way: animals suffer in experiments, they argue, and such suffering cannot ever be justified by nebulous claims about possible benefits in the future—especially since it is largely humans who are said to benefit.

Yet these polarizations gloss over many other issues. Part of the problem is that the oppositional stance of the controversy helps to perpetuate a rigid demarcation between humans and animals. But is it that simple? Science is hardly respectful of humans in its practice. As Karen Messing and Donna Mergler point out in their chapter, scientists treat some groups of humans with blatant disregard. Epidemiological studies of occupational health often pay little heed to the needs of the workers, for example. One important difference, though, is summed up in the title of Karen Messing and Donna Mergler's chapter: that rats cannot speak about how the researchers treat them, but humans can. And loud enough voices may begin to change the way scientists behave.

Thus the voices of the animal liberation movement have contributed to changes in the law as regards the use of animals. Public outcries in Britain in the late 1970s contributed to parliamentary debate about animal experimentation, culminating in a new law, the 1986 Animals (Scientific Procedures) Act, governing animal use by scientists. Such changes are of no account to antivivisectionists, to whom they appear to be merely a sop. Still, the new law requires that scientists spend time thinking in advance about how best to minimize the pain and suffering specific experiments are likely to produce. In that sense, the act is part of a sea change in the way scientists must think and behave.[7] Although scientists might still justify their use of animals in terms of medical benefits, they are perhaps more likely now than they were a few years ago to pay attention to the welfare of the animals. A small step, perhaps—and certainly insignificant to those opposed to all animal experimentation. But it is likely to be a big step for those individual animals who are the subjects of experimental work.

At the moment, scientists justify their work on animals by recourse to a kind of utilitarian balancing act, in which the costs (largely to the animals) are weighed against the potential benefits (usually to people) and also against the quality of the science. This balancing act was made explicit in the 1986 act in Britain, which requires that scientists make statements about the balance before being granted licenses by the Home Office.

Similar balances are more implicit in other countries. Laws in the United States, for instance, cover primarily the housing and care of laboratory animals rather than scientific procedures, although some federal granting agencies, such as the National Institutes of Health, may require a statement about balance in animal experiments. In addition, research practices are sometimes monitored by institutional ethics committees, which, at least in principle, serve a similar "balancing" function.

In practice, of course, decision making, whether by ethics committees or by scientists, in conjunction with government representatives, is likely to be influenced also by economic and professional concerns. That would be true whether the decisions related to proposals for animal research or for research using human subjects.

Justification nonetheless assumes that present use of animals can be weighed against possible medical benefits in the future. Indeed, with the drive for scientists publicly to justify their research, there is even greater incentive for them to *claim* possible clinical benefit, even where there is little chance of that ever being realized. But what would happen if such health justification were to be removed? What other criteria might be applied to decision making, short of a complete ban on animal use? And if direct health criteria were removed from the balance sheet, what criteria would be appropriate for research on human subjects?

While the emphasis on justificatory strategies is important, it also has its drawbacks. Requiring that experiments be justified in terms of human health begins to pull basic questions about animal or plant biology into the realm of medicine. What place does that leave for research into animal behavior or botanical taxonomy? What is likely to happen is that scientific medicine accrues more power, as all forms of biological research become redefined in relation to it. It then becomes less possible to ask what we would need or want to know about animals or plants in their own right. Increasingly, biologists must seek to justify their work not in terms of the increase in knowledge about the organisms they are studying but in terms of how useful that knowledge might be *to people*. Learning about how animals behave for its own sake (or theirs) becomes a luxury that we can do without, while botany becomes an adjunct to agriculture or pharmacology.

If the health justification were removed, what would happen to scientists' justificatory strategies? Given that funding for basic science is increasingly being squeezed, biologists would have to provide other kinds of justification if they are going to use living organisms. But doing so would be much harder, in view of the fact that many people question (and with good reason) some of the arcane experiments scientists do. The need for justification helps to make science more accountable to the general public, but only in limited ways. What lessens the accountability of science is its authority and its refusal to take seriously other voices, other knowledges. Thus scientific medicine and physiology ignore, and sometimes ridicule, the voices of alternative health practitioners; researchers in agricultural genetics pay little heed to the knowledge of indigenous peoples about the dynamics of their own ecosystems; and the science of animal behavior pays little attention to the experience of farmers or animal trainers. And, of course, science fails consistently to take seriously the voices and knowledges of nonhuman organisms themselves.

In the title of a recent book, Sandra Harding asks the question *Whose Science? Whose Knowledge?*.[8] She points out how the rise of modern science in the West has meant the validation of only some knowledges. These are not likely to be the sciences or knowledges of those who lack power. But there is a stronger point. Science, Harding reminds us, claims unto itself supreme objectivity, the separation of knower from known that we explore in this book. Yet a science that fails to take seriously other knowledge claims can only be limited in its objectivity.

Recently there has been a growth in intellectual questioning, not only about animal rights and related philosophical issues but also about the political and social significance of separating the living world into two categories, human beings and "other animals." Western culture in particular has developed a rigid demarcation of humanity from the rest of nature.[9] This bifurcation is coming under increasing scrutiny, partly from the environmental movement but also from observers writing from a variety of other intellectual perspectives.[10] In the face of global environmental damage, it becomes essential to rethink the barriers between humans and the rest of nature.

To the extent that those barriers are products of Western culture, Western science helps to perpetuate them. As Vandana Shiva points out in her chapter, many other cultures experience themselves as being *in* nature and do not have this kind of exploitative relationship to nature. It is precisely that kind of relationship that lends itself to exploiting human "others" as part of nature. Separating humans from the rest of nature has enabled the West to justify controlling and exploiting all those who are cast as "other"—be they non-Western peoples, women, people with disabilities, or nonhumans. An important part of this conceptualization is that "others" are categorized as having less intelligence, rationality, or whatever characteristics the dominant culture deems most typically "human." Within Western culture itself, some of those groups of others have challenged that categorization; one effect of feminism, for instance, has been that at least some women are now admitted to the terrain of intelligent life.

Not so animals. Implicit in the power to classify and name ourselves and other organisms is the power to define what capabilities different animals have and their significance. Thus a few kinds of animals are beginning to be given special status across the divide and to be granted some level of intelligence or self-consciousness. The great apes, for example, are increasingly thought of as sufficiently similar to humans that we can acknowledge their intelligence, and therefore many more people are challenging their use in scientific research.[11]

Other kinds of animals may be admitted to the circle of intelligent beings because they live within a primarily *human* social world—domestic dogs, for instance.[12] Yet this reluctant granting of intelligence or rationality poses its own problems for those of us who question the human–animal divide. In the first place, it relies explicitly on a categorization imposed by humans and defined by us; we tend to recognize as "intelligent" only those species whose intelligence takes forms similar to our own. Second, it privileges rationality over other ways of knowing or being.[13]

Nonetheless, boundaries are being breached on all sides. Ironically, it is from within science itself that some of the biggest challenges to perceived bodily boundaries are coming, as Emily Martin and Stuart A. Newman note in their chapters. Crossing those boundaries is not new in itself; hybrid creatures have long existed both in nature and in cultural imagination.[14] But new technologies, such as organ transplantation and genetic engineering, promise further possibilities of boundary transgressions. Although Martin would concur with Donna Haraway, who sees creative possibilities in these transgressions,[15] she reminds us that the power to decide who can breach boundaries, what boundaries can be breached, and in what way will remain with those who are already dominant.

Given the authors' diverse backgrounds, each one has addressed the questions posed by the editors in her or his own way. Their reflections in turn pose further questions, partly about the relationship between humans and animals or between humans and other humans, and partly about the nature of science as we know it. Despite their diversity, the various chapters can be grouped into four parts, each united by common themes and introduced by a brief statement by the editors.

The thread linking the three chapters in part I is the various ways in which the practice of science, its distancing *from* the subjects of inquiry, leads to disrespect or even abuse. Animals, of course, are used by science and are clearly "other" to the experimenter. Hilary Rose examines this theme in her analysis of the connections between the Victorian antivivisectionist movement and the animal rights movement of today. But many people can be "others," too. Vandana Shiva describes the imperialist tendencies of Western science, which exploits not only the peoples of other countries but also their indigenous flora and fauna. Also in Western countries, people may be the exploited objects of science, and Karen Messing and Donna Mergler explore this theme in their discussion of inhumanity to workers in occupational health research.

The essays in part II use personal testimony to illustrate some of these issues. The three chapters here, from biologists Betty J. Wall, Ruth Hubbard, and Lynda Birke, describe personal histories. Learning to do science, these scientists concur, meant undergoing a process of desensitization and learning not to empathize with experimental animals. And alongside their initial enthusiasm for what they were learning, there grew for each of these scientists an unease arising from their increasing respect for nonhuman organisms.

Like part I, the third part is concerned specifically with the theory and practice of biology. The chapters by Anne Fausto-Sterling and Marianne van den Wijngaard analyze the book's themes through examples drawn from studies of animal sexuality. Fausto-Sterling uses primate studies in particular to show the inadequacies of much of the research and to ask why we would want to know this. Van den Wijngaard similarly focuses on the inadequacies and assumptions underlying much research, particularly research on sex hormones and behavior of rodents. Both seek to move away from the reductionist thinking that dominates modern biology, and toward accounts that are more respectful of both human and animal lives.

Lesley J. Rogers examines scientific claims about the minds and brains of nonhuman animals and questions the alleged intellectual inferiority of these animals. She points out that inferring a lack of intelligence may be a consequence of using animals who live under highly constrained and impoverished conditions. What would happen, she asks, if scientists posed their questions in more respectful ways, if they asked their questions of animals living their own lives, and if the questions furthermore reflected the animals' ways of living rather than our own?

Judith C. Masters's concern is with current evolutionary theory and its emphasis on the role of competition for static resources among organisms who differ in only minor ways. By casting organisms as active and their environments as passive, preformed niches, theorists have tended to ignore the dynamic mutual relationships that must underlie evolutionary change.

Part IV explores various ways in which the boundaries between humans and animals have been, or are being, breached. Like Masters, Stuart Newman critiques the rhetoric of limitless change implied by much of evolutionary theorizing and, more recently, by genetic engineering. He traces its origins in Western religious thought and philosophy.

Concentrating on the human–animal divide, Arnold Arluke and Boria Sax examine its curious construction in Nazi ideology, which cast some humans as "worse than animals." Their chilling chapter reminds us not only that boundaries can be constructed in different and contradictory ways but that doing so can have disastrous consequences.

Taking a different approach to the social consequences of the way boundaries get drawn, Emily Martin, in the final chapter, looks at recent attempts to redefine boundaries between humans and animals and between humans and machines. She shows that welcome though such activity is, it tends to obscure the fact that the border crossings themselves are constructed within traditional hierarchies of power. Boundaries can never be unproblematic, since they can, by definition, only be crossed by those with power.

Notes

1. Brighton Women and Science Group, *Alice through the Microscope* (London: Virago, 1980), and Ruth Hubbard, Mary Sue Henifin, and Barbara Fried (eds.), *Women Look at Biology Looking at Women* (Cambridge, Mass.: Schenkman, 1979).

2. See Ruth Richardson, *Death, Dissection and the Destitute* (London: Pelican, 1989).

3. Anita Guerrini, "The Ethics of Animal Experimentation in Seventeenth Century England," *Journal of the History of Ideas*, 50 (1989), 391–408.

4. Guerrini, p. 401.

5. See Coral Lansbury, *The Old Brown Dog: Women, Workers and Vivisection in Edwardian England* (Madison: University of Wisconsin Press, 1985).

6. Peter Singer, *Animal Liberation* (London: Jonathan Cape, 1976).

7. Mike Michael and Lynda Birke, "Diatribes and Dialogues: Scientific Uncertainty, Public Unease and Animal Experiments," *Science, Technology and Human Values*, 19 (1994), 189–204.

8. Sandra Harding, *Whose Science? Whose Knowledge?* (Buckingham: Open University Press, 1991).

9. See, for example, Keith Thomas, *Man and the Natural World* (Harmondsworth: Penguin, 1983).

10. These include Ted Benton, *Natural Relations: Ecology, Animal Rights and Social Justice* (London: Verso, 1993); Barbara Noske, *Humans and Other Animals* (London: Pluto, 1989); James Sheehan and Morton Sosna (eds.), *The Boundaries of Humanity: Humans, Animals, Machines* (Berkeley: University of California Press, 1991); Josephine Donovan, "Animal Rights and Feminist Theory," *Signs*, 15 (1990), 350–375; Keith Tester, *Animals and Society: The Humanity of Animal Rights* (London: Routledge, 1991).

11. See Paola Cavallieri and Peter Singer (eds.), *The Great Ape Project: Equality beyond Humanity* (New York: St. Martin's Press, 1993).

12. See Benton, *Natural Relations*, and Elizabeth Marshall Thomas, *The Hidden Life of Dogs* (Boston: Houghton Mifflin, 1993).

13. Feminists particularly have criticized the dominance of rationality in patriarchal Western discourses. See, for instance, Genevieve Lloyd, *The Man of Reason* (London: Methuen, 1984).

14. Hybrids between humans and other creatures pepper the mythology of Western culture, for example, and form part of the iconography of "monsters." See, e.g., A. Davidson, "The Horror of Monsters," in Sheehan and Sosna. It is, moreover, debatable whether women have ever experienced bodily boundaries as impenetrable.

15. Donna Haraway, *Simians, Cyborgs, and Women* (London: Free Association Books, 1991); see esp. the essays "The Cyborg Manifesto" and "The Biopolitics of Postmodern Bodies."

PART I

Exploitation of the "Other"

IN SEEKING TO move beyond the stance that assumes a distance between scientists and the organisms they study, we need to remind ourselves of some of the consequences within the recent history of science of that distancing and of the pursuit of "objectivity." Separating ourselves from the individuals studied has been central to the belief in scientific methods and in the validity and predictability of the claims made. Yet it has often led to inhumanity, for what is human or humane about an interaction that denies one actor agency while giving power to the other?

The chapters in this part illustrate, in quite diverse ways, how the distancing stance of science can contribute to disrespect or abuse. Animals, for example, are used extensively in biomedical research; their use is justified biomedically on grounds of their physiological similarities to us—they are "animal models." Yet their use is justified ethically by claiming difference from us—they are of less ethical value. Recognizing that animals have often suffered pain and abuse at the hands of scientists has been central to the recent resurgence of the antivivisectionist movement. Yet not only animals are treated as "others" in the scientific enterprise. Science treats as others those who are not part of the dominant social groups, including women, working-class people, people from non-Western cultures.

Hilary Rose begins this exploration, looking at what we can learn from the antivivisection movements of the late nineteenth century and their opposition to the growing power of physiology. The current wave of concern about animal rights shares much in common with that earlier movement. Both echo wider public unease about the status of science. Understanding the similarities and differences is important to the project of "reinventing biology," she argues.

It is not, of course, only animals who suffer from the distancing stances of science. As Vandana Shiva points out, science has gone hand in hand with Western imperialism to exploit the peoples, flora, and fauna of the less industrialized world. More than that, it threatens to reduce biological and cultural diversity, and ultimately, it threatens the survival of us all. The imperializing drive of modern science and its refusal to hear the voices of others are themes taken up in a different context by Karen Messing and Donna Mergler. Their chapter focuses on the way certain groups of people—in this case, primarily workers in

North America—are treated inhumanely by scientists. Their concern is with the practice of occupational health research, which pays little heed to the specific experiences of workers or to their need to avoid, or recover from, occupationally derived illness. Science, on the contrary, insists on extensive proof that the industrial processes themselves are responsible for causing illness.

1 | Learning from the New Priesthood and the Shrieking Sisterhood

Debating the Life Sciences in Victorian England

Hilary Rose

THE CHALLENGE to begin "reinventing biology" stems from the widespread unease at the kind of Western biology which is emerging as dominant at the end of the twentieth century. A century ago the writer Ouida denounced the new experimental physiologists as the new priesthood. Today the molecular biologists are poised to take over the physiologists' priestly role, claiming, in their turn, to be the newest scientific revolutionaries offering immense theoretical and therapeutic powers. In reaction many turn to alternative medicine and to the development of other new sciences more responsible to people and nature alike. Currently physiologists find themselves resisting the claims of the upstart discipline of molecular biology, which threatens that position won within the hierarchies of knowledge during those crucial battles during the nineteenth century. The foreword written by the physiologist and Nobel laureate Sir James Black to a recent set of essays seeking to defend the claims of physiology against those offered by molecular biology argues: "Hopes of realising the optimistic forecasts about the benefits that molecular biology will bring to pharmacology are likely, I believe, to be circumscribed by the current state of physiological knowledge, models and concepts. In the current hegemony of molecular biology, with splitters having a field day, lumpers like me often get gloomy about the future."[1]

While the debates between competing experimental life sciences at the end of the nineteenth and the twentieth centuries are not identical, there are some interesting echoes between past and present. To Sir James's questioning of the therapeutic potential of molecular biology has to be added the careful reevaluation of physiology and its claims by historians of medicine. In an overview of research on the laboratory in medicine historians Andrew Cunningham and Perry Williams conclude:

> Taking all these contributions together, we can see certain common themes.
> One is that the claims made on behalf of the laboratory—both the cognitive

claim of representing nature, more accurately and authentically in an unmediated way, and the practical claims of delivering clinical benefits—were not of themselves self-evident or naturally compelling.[2]

The revisionist historical research of the last three decades has added complexity to the Whiggish and unilinear view of progress offered by the biologists and their medical allies. Instead, what comes into perspective is that the bearers of the rising discipline were also members of a rising professional class and within that a dominant gender. The discourses of knowledge were also discourses of power and as such sought proximity to the state, since legitimacy, recognition, research institutes, posts, and financially secure careers were also its prizes.

For both the new priesthood of the second half of the nineteenth century and the knight errants of molecular biology searching for the Holy Grail of the human genome[3] at the close of the twentieth, there is a central tenet of faith—that for medicine to advance it must become "scientific." Rhetorically there is a straight lineal descent between the claims of the French physiologist Claude Bernard and those of the ideologues of the new genetics. What Bernard did was to propose nothing less than a new epistemological foundation for medicine. Where "hospital medicine" had succeeded "bedside medicine,"[4] the *Introduction to the Study of Experimental Medicine* proclaimed the future:

> I consider hospitals only as the entrance to scientific medicine: they are the first field which the physician enters: but the true sanctuary of medical science is the laboratory; only there can he seek explanations of life in the normal and pathological states by means of experimental analysis.[5]

In this much quoted apologia for the new physiology, Bernard goes on to insist that the physician must enter the laboratory and that "there, by experiments on animals, he will seek to account for what he has observed in his patients, whether about the action of drugs or about the origins of morbid lesions in organs or tissues." While medicine has historically turned to the study of animals to augment the study of the patient, Bernard is arguing for a new knowledge and with it an entirely new sort of training for doctors. They have to learn by being in the laboratory, by directly carrying out experimentation on living animals, and they have to enter a conceptual world where the animal is to become the model for Man.

But while there is a rhetorical continuity which insists on commonality between the life sciences, there is also difference between them, and in a way that is masked by the rhetoric, "scientific" is not some immutable and essential quality. What is and what is not science is continuously under cultural and political negotiation, whether in the nineteenth century or today.[6] When the physicist Rutherford said "there is only one science, physics; all the rest is stamp collecting" or the molecular biologist and initial director of the Human Genome Project Jim Watson more recently declared that "there is only one science, physics; all the rest is social work," what they were doing is seeking to establish hierarchies and to police the boundaries of knowledge.

Thus thinking about "reinventing biology," with its suggestion of different, other biologies from the growing hegemony of molecular biology or the new genetics, pulls me in two ways. First into the present with the disturbing and triumphalist claims of molecular biology and the multistranded resistance of those within and without biology, from the social movements of ecology and feminism and from the multiplicity of struggles of Third World people to protect biodiversity simply in order to survive.[7] The second pull is back into the nineteenth century and in particular to a reconsideration of the fierce ethical, political, and scientific debates over the competing schools for the prize of being *the* discipline which was to provide the scientific foundations for medicine. Perhaps because the training of natural scientists encourages them to look only at today's conflicts within and about science, as a social scientist I find I am drawn to that central belief of historians that by understanding the recentish past better we can see ourselves and maybe act more effectively.

Thus to turn back to the struggle surrounding nineteenth-century physiology is to reinterrogate a history in which the triumph of that new experimental biology was by no means guaranteed; nor was it, in the context of a rich proliferation of thought systems claiming knowledge of nature, the only possible "biology." In this chapter I explore the intellectual and ethical debates about what kind of biology was to become preeminent, setting these struggles within their connected contexts of other and related forms of power. In particular I want to consider the intense opposition from feminists to vivisection, as it speaks not simply of the struggle against cruelty to animals which formed a general preoccupation of bourgeois femininity but also of the specific association of masculine science with cruelty to, above all, domestic animals. In the defense of the pet dog against the merciless constructors of knowledge, how far were Victorian feminists also symbolically defending those "other" domesticated animals who could be beaten, raped, and robbed of their money—entirely legally—by those who ideologically were cast as their protectors? I speak of course of bourgeois women themselves, who despite their class privilege over and above working-class women, nonetheless were nearer to the slave, the unfree laborer of Engels's analysis[8] when it came to their relationships with husbands and fathers.

A Physiologist Delirious with Cruelty?

Despite the best efforts of both scientists and the mainstream historians of science, those discourses of sexuality and of sexual difference which pervade Victorian culture are not entirely excludable from the representations of laboratory life, not least because recent feminist scholarship has scrutinized the mediation of gender silenced within the mainstream. The antivivisection feminists, with their unceasing attack on what they saw as the sensuous cruelty of the biological laboratory, thus disrupt a discourse which symbolically represented women, men, and nature in such a way as to sustain power relations. Feminist historian Judith Walkowitz's discussion of the serial killer called Jack the Ripper[9] cites a letter to the *Times* by the feminist Frances Power Cobbe, who was both a leading figure within the antivivisection movement and deeply involved in the struggle against violent husbands. Cobbe disruptively (and with some wit) writes:

> Should it fall out that the demon of Whitechapel prove really to be, as Mr. Baxter seems to suspect, a physiologist delirious with cruelty, and should the hounds be the means of his capture, poetic justice will be complete.[10]

Other feminist historians of the vivisection debate, such as Coral Lansbury[11] and Mary Ann Elston,[12] have demonstrated that a gender analysis is much more fruitful when it comes to examining the representations of nature within what are either dismissed as the "partisan" accounts written by the antivivisectionists[13] or talked up as the more "neutral" accounts of the professional historians.[14] One common distinguishing feature of such gender-blind accounts regardless of their ostensible partisanship or neutrality is reflected in their treatment of the figure of Frances Power Cobbe. Typically, both pay tribute to her political significance as a leading antivivisectionist while ignoring her involvement in other feminist campaigns against the violence of husbands[15] or the institutionalized violence against the entire class of prostitutes. The thought that her antivivisection was grounded in a coherent world view is outside the comprehension of a disciplined androcentric mind.

By contrast today, at the end of the twentieth century, we are in a slightly curious place, I hope transitional, in the history of the history of science. Lansbury's pioneering book, which has a nuanced discussion on the parallel between the cruel complexity of horse bridles, gynecological straps, and the devices of the brothels—but, in my view, rather weakly documents her optimistic interpretation that working-class and feminist movements shared a common opposition to vivisection—seems to be cited only by feminists.[16] Elston's chapter is included in the important collection of Nicolaas Rupke; nonetheless the ideas it contains are still sealed by an impermeable membrane and the other contributors within the same book continue to produce androcentric knowledge. My own experience of contributing to a book on the laboratory revolution within medicine was much the same. At the Cambridge conference which preceded the book, one of the men historians present jokingly observed while considering the limitless possibilities which the biological laboratory offered the scientists: "You can do anything you like in a laboratory or in a whorehouse."

While the utterer was extraordinarily embarrassed by his words, a sexualized and violent language is not in itself new in or about science. The words which had just slipped out juxtaposed those two places where animals and women are the slaves of men and indicated that both offered contexts where there were no restraints other than those fettered by the masculine imagination. It went without saying that the "you" in question had the resources to enter both the laboratory and the whorehouse. But even after this event, which caused a small perturbation if not quite a scandal, the men historians present remained unable to consider the gender dimension of their analyses. The subversive link between the lapse and the feminist arguments of either the past or the present remained too dangerous to make. The women's movement of the late twentieth century has inserted feminist voices within the academy, but that does not mean that those fluent within the dominant discourses will choose to listen.

Choice between different approaches to nature was extraordinarily rich, however, in the early nineteenth century. Nicholas Jardine quotes an 1840 list of competing approaches:

> Metaphysicians, Idealists, Iatromechanics, Iatrochemists, Experimental Physi-
> ologists, Natural Philosophers, Mystics, Magnetizers, Exorcisers, Galenists,
> modern Paracelsian Homunculi, Stahlinaists, humano Pathologists, Gastri-
> cists, Infarct-men, Broussaisists, Contrastimulists, Natural Historians, Physia-
> tricists, Ideal-Pathologists, German Christian Theosophists, Schoelleinian Epi-
> gones, Homeobiotics, Homeopathists, Isopathists, Homeopathic Allopathists,
> Psorists and Scorists, Hydropathists, Electricity-men, Physiologists after Ham-
> burger, Heinrothians, Sachsians, Kieserians, Hegelians, Morisonians, Phre-
> nologists, Iatrostasticians.[17]

By the end of the nineteenth century it was Experimental Physiology, with the laboratory as the locus for the production of knowledge, which had surpassed all these other methods of inquiry and therapeutic approaches. In the early years of the century these diverse approaches swam around together in the cultural soup of the time with a certain egalitarian equivalence, so that those which were eventually to become designated as "science" and those to be designated as "nonscience" were equally likely to command interest among the relatively small intellectual elite. Just to take a few examples: the Brontë sisters were passionately interested in phrenology and went from their Haworth parsonage to nearby Keighley to listen to lectures on the new materialist science of the mind, in which specific areas of the head were thought to be associated with specific capacities. The physiognomies of the characters within Charlotte's *Jane Eyre* are indelibly part of the images fashioned within a phrenological reading of psychology. Harriet Martineau, committed feminist, writer, and positivist (she translated Auguste Comte) and part of the Darwin circle, regarded mesmerism—hypnotism—as having been critical in the restoration of her health. The novelist George Du Maurier was also interested in mesmerism but portrayed it rather more hostilely in his figure of Svengali, the sinister mesmerist lover of

Trilby. George Eliot was to portray all the hopes for the new scientific medicine in her figure of Dr. Lydgate in *Middlemarch,* who in the abandonment of his commitment fails both morally and intellectually.

But the novel's account of Lydgate's personal defeat was not the prevailing story of the century. By its end most of the earlier approaches' claims to speak about living nature had been successfully marginalized and named as "quackery." The experimentalists had triumphed and successfully placed themselves at the apex of knowledge with its corollary that the future of the clinic was to be founded on a laboratory-based science. "Hands on" learning in the laboratory as part of medical education was going to transform the mentality of the doctors, and indeed do nothing less than civilize them. At the same time in their successful emphasis on the experiment, the physiologists and their allies the bacteriologists had reduced the mere observational sciences—whether of the bedside or of green nature—to subordinate intellectual status.

A Platform for the Life Sciences

Within Britain it was the British Association[18] which had become the venue for the great scientific and cultural debates of the century, above all that between the Bishop Wilberforce and Darwin's friend and fellow evolutionist T. H. Huxley in 1860.[19] As a platform the British Association and in particular its presidency was to become both the means of dissemination and the prize of recognition. It was thus a mark of the success of the new physiology that its most famous protagonist within Britain, John Burdon-Sanderson, was elected president in 1893. A hero to the experimentalists, Burdon-Sanderson was also the embodiment of cruelty for the antivivisection movement. Less than two decades previously, this movement, uniquely powerful in British cultural life, had nearly blocked animal experimentation. In no other country had the antivivisection movement come so close to success.

While the physiologists' scientific and political victory was considerable, the opposition was still vocal and had influential allies, and the occasion therefore was of some delicacy.[20] The presidential address of the British Association was something of a state-of-the-science speech, to be delivered magisterially but preferably not too controversially, serving both to reassure the scientific community of its value and to inform the general public of its achievements. Even today the organization and its presidential address carry something of this double task.

Burdon-Sanderson claimed that a radical new approach to the study of living nature had come about since the early nineteenth century. This claim and program, for it was both, was set within a particular origins story of the emergence of the new subject. First he reminded his audience that the term *biology* as the philosophy of living nature had been proposed by Treviranus in the early

nineteenth century, and (while other tellers of the history were to credit Lamarck) that the term now commanded among students of living nature both
widespread usage and substantial agreement that the focus of biology was the
organism's responsiveness to both internal and external conditions. Thus for
those attracted to the study of life itself, the concept of "adaptation" with or
without a commitment to evolutionary theory was central.[21]

Because he was still something of a vitalist—that is, someone who believes
in a separate set of laws which explain life itself—Burdon-Sanderson was well
placed, both to insist on the distinctive features of this new science of life and
to demonstrate his hostility to the idea that biological explanation could theoretically be reduced to mere physics and chemistry. Instead he underlined the
crucial importance of the experimental approach which had been developed by
the physical sciences, but he sought the method, not the physicochemical explanation. Like many nineteenth-century thinkers, he drew on a stages theory of
the development of knowledge and in consequence saw botany, anatomy, and
zoology as necessary precursors which opened the way to a conception of biology as a new experimental science with explanatory powers. "Botanizing" to
use Burdon-Sanderson's diminishing term, was simply the necessary completion of the task of careful observation and classification begun by Linnaeus,
while zoologists were to do much the same for the animal world. Observational
knowledge was thus to serve as a handmaiden to the master discourse, which
could only be generated through an interventionist experimental science. (Not
by chance was botanizing, as a handmaiden science, widely seen as a suitable
intellectual activity for ladies—both healthful and feminine.) Physiology by
contrast was an unquestionably masculine activity. How, asked one physiologist, could one have dinner with the ladies "smelling of dog"?

The address both surveyed the achievements of the new experimental approach and set out a research program which by looking into the organism
would be able to answer the "how" questions of living nature. While Burdon-
Sanderson prioritized looking within, he also acknowledged Haeckel's very different theoretical project, "oecology," of looking at the relationship among organisms and between organism and environment. But acknowledgment of this
holistic approach to biology was not to be confused with giving support. Ecology as a major social and cultural movement fusing a new approach to nature
intellectually and practically was going to have to wait until Rachel Carson's
Silent Spring, for both intellectual space and the research resources were already
commanded.

Instead Burdon-Sanderson used a magisterial but restrained language to
advocate physiology. The aesthetics of the experiment which mattered to him
led him to describe particular pieces of research as "beautiful" or "elegant"; at
no point does he reflect on what others found objectionable either aesthetically
or ethically. There is no hint of the bloodiness or the stench of the procedures,

nor of the pain. For Burdon-Sanderson the perspectives of alternative[22] knowledge and therapies had no place within a survey of biology; alternatives were as thoroughly excluded from the authorized discourse of nature as they are today.[23]

Burdon-Sanderson's strategy of discreet silence as a method of handling the controversial issue of the "method of cruelty" made dramatic contrast with the frank recognition of the appalling price of the new knowledge by Claude Bernard. In Bernard's often quoted words,

> my idea of the science of life, I should say that it is a superb and dramatically lighted hall which may be reached only by passing through a long and ghastly kitchen.[24]

This was more than a cultural matter of Gallic lushness and English understatement, for neither of these leading physiologists takes any account of animal suffering in their classical textbooks,[25] nor it would seem in their scientific practice, until (in Burdon-Sanderson's case) animal experimentation and the use of anesthetics were regulated in Britain by the 1876 Cruelty to Animals Act. But the presence of such legislation has long varied from country to country, so that the United States, for example, is still controlled only by convention, whereas the British are regulated by law, and, after much pressure from the animal welfare and antivivisection lobby, by inspection.[26]

The relationship between Man and Beast had long been a matter of cultural contestation,[27] but the experimentalists, as a precondition of their work, necessarily set aside the question of animal suffering—for them it was as if the power of the machine metaphor actually transformed animals into "mechanisms" for whom the concept of suffering was simply irrelevant. Power Cobbe's close textual reading of Burdon-Sanderson's *Handbook* drew attention to his moral insensitivity to the animals' distress.[28] When a hundred years later the historian and physiologist Stuart Richard examined Burdon-Sanderson's laboratory notebooks, his work serves to confirm Power Cobbe's conclusions, though without acknowledging her prior claim. Richard shows that the physiologist routinely experimented on dogs to investigate the circulatory effects of asphyxia, only using curare (which stilled the animal but did not inhibit pain) or morphine when the animals struggled. Even then it is the struggling—which interferes with the experiment—not the pain which concerns the experimenter.[29]

Both secular and religious cultural traditions sustained the physiologists' project. From a Cartesian perspective the animal lacked the essentially human capacity for thought and was on a par with an inanimate object; thus removing the leg from a dog or a table was, apart from this matter of struggle, not an issue.[30] From a Judeo-Christian perspective, Man had been made in the image of God and given dominion over Nature. A number of physiologists drew strongly on this argument to justify their actions. The new physiology may have been based at University College London, that home of dissenters, but they were untouched by Jeremy Bentham's more generous impulses toward both beasts and other men which had led him to conclude that human behavior toward animals should be guided not by whether an animal has the capacity for thought but by whether it can suffer.[31] But while the physiologists were unscathed by such reasoning, it found considerable resonance in Victorian England, where the opposition to cruelty against animals, above all the domestic animal, the noble horse and the loyal dog, was embedded within bourgeois and particularly feminine bourgeois culture. The Royal Society for the Prevention of Cruelty to Animals sent out lady educators—the Bands of Mercy—to teach children to be gentler to animals and birds. Nor were such animal welfare initiatives entirely fruitless. Flora Thompson's *Larkrise at Candleford* tells of how, after the Band had visited, the village children gave up their cruel habits of bird nesting. The novel, whether directed toward child or adult readers, was a vehicle for cultural and political messages both for social and for animal welfare issues. So were Dickens's exposé of the Yorkshire schools and Eliot's defense of Jews; Anna Sewell's *Black Beauty* defended the horse. Within this bourgeois literature the chief agents of cruelty to animals were cast as working-class men. It took the socialist Henry Salt to observe that the brutal butcher was kept in place not by his essentially violent nature but by the demand of middle-class England for meat.[32]

Biology Enters the Factory

The protagonists of the new experimental biology needed both institutional support and resources, not least physical structures. While Charles Darwin was able to do research at Down as part of the craft era of science still practicable for plant experiment, most animal experimenters needed space to house the animals and laboratories for the research itself. Although it was almost a hundred years before biology, particularly molecular biology with its robotic production, entered Big Science, during this century biological research begins to leave the domestic space and enter the public space of the institute and the university laboratory. As feminists have discussed for all sorts of production, whether of things—such as brewing or optics—or knowledge itself, the move from domestic production to factory production has served to exclude women.[33]

Success crowned the efforts of the experimenters here too, and in 1893 the new British Institute of Preventative Medicine building was moving to successful completion. Its supporters extended a warm welcome to the new research laboratory, both because it was inspired by what they understood as the success of the Institut Pasteur in defeating hydrophobia (rabies) and because it had managed to overcome the not inconsiderable resistance mounted by the antivivisection movement. This was seen in distinctly negative and gendered terms. The

Pall Mall Gazette affirmed that the new institute "should be congratulated on shaking off the shrieking sisterhood of anti-vivisectionist sensationalists."[34]

Such a triumph was all the sweeter to the supporters of the new biology, for the shrieking sisterhood and their allies had, in the 1870s, moved with considerable political skill against the method of cruelty of the new experimental biology. When Power Cobbe, recognizing the need to establish the existence of animal sentience and the capacity to suffer, published an article in 1872 in the influential *Quarterly Review*[35] which drew on and amplified a growing public discourse against cruelty to domestic animals, the antivivisectionists had successfully seized the initiative and had turned to the House of Lords to legislate against this moral evil. They shared and fostered a widespread public concern at the extension of what were seen as barbaric Continental physiological practices (with the names of Müller, Magendie, and Bernard as a cruel trinity) into British science. Self regulation—offered by the British Association in 1871—demonstrably failed. Members of the movement saw every sign of a British physiology developing all too quickly and unethically. They resisted the claims of the vivisectors to be improving science, and they rejected the animal model[36] as the basis for understanding human disease. Against Eliot's Dr. Lydgate, who is offered as more civilized than the traditional doctor, the antivivisectors saw doctors as morally coarsened by the fearful things required of them in the physiological, and for that matter the bacteriological, laboratories. The evidence of those who had witnessed the practices of Bernard's laboratory was used in a nationwide campaign to alert the public to this social evil.[37]

At this historical juncture the balance of cultural politics favored the antivivisectionists, and the list of the great and the good who were opposed to knowledge gained through experimentation on living animals included such figures as Carlyle, Tennyson, Manning, Jowett, and, as her letter to Disraeli made clear, the Queen herself. The class structure within the politics of social reform meant that the leadership of all the key societies was necessarily male, even though the organizing and indeed the strategic thinking seems to have been frequently carried out by women.

The danger posed by the antivivisectionists to the development of experimental biology was widely recognized among the biologists. Thus while neither Thomas Huxley nor Charles Darwin carried out animal experimentation themselves, and indeed Darwin had given up medicine on the grounds he could not bear watching or inflicting pain, both were part of an intellectual network committed to experiment which included the physiologists.[38] Darwin was initially a confusing figure in the debate because his work combined the most exquisite sensibility in observational science (not least in his classificatory activities from the Galapagos material) with work on human expression and feeling. Power Cobbe had corresponded with Darwin, and had indeed received an advance copy of the *Origins*, and saw him as a gentle person—perhaps a potential ally—

speaking of him as "a man who would not allow a fly to bite his pony's neck."[39] Thus although Darwin was an avid experimenter on plants, the constructions of cruelty of the time simply excluded plants from the category of victims. His violence against the plants as he manipulated variables was thus intellectually as one with that of his fellow animal experimenters, yet nonetheless placed his work outside the cruelty debate. Controversy around Darwin was limited to the not inconsiderable matter of evolutionary theory. However, as the grandees of biology, Darwin and his friend and ally Huxley were well placed to understand this new threat to the developing discipline. They moved to protect it. Darwin's discreet and high-level lobbying successfully neutralized the proposed legislation, which appeared as the 1876 act, and persuaded the government of the relevance of that flexible instrument central to British political life for managing complex social or cultural reform pressures, the Royal Commission.

But even while this successful countermove was being planned from Down, part of that success lay as much in defining the problem as in how it was to be managed. The countermove had to fashion a culturally and politically manageable issue out of those multistranded linkages between sex and sensibility which were integral to the construction of Victorian bourgeois masculinity and femininity. What is more, the move had to work in both the public and private arena.

To give one example of the delicacy and complexity of this project: although Darwin had managed to dissuade his daughter Henrietta from actively supporting the antivivisectionists, and indeed had recruited her lawyer husband to help draft proresearch legislation, nonetheless he was aware of the potentially dangerous sensibilities of the Down ladies. The increasing mobilization of all kinds and conditions of women against the Contagious Diseases Act was, although the Down House ladies were themselves unscathed, an ever-present warning. Darwin's instructions to his brother scientists that not a word concerning the defense of vivisection was to be breathed in the presence of the ladies was thus a sensible domestic political precaution.

It was not necessary for the great biologist to echo Bacon's characterization of the new experimental knowledge as consciously "masculine"; instead the sensibilities of nineteenth-century gentlemen transmuted the aggressive claims of a blunter epoch into the mostly silent and taken for granted association between bourgeois masculinity and the organized production of knowledge of nature.

Supposing the Antivivisectors Had Won?

Behind this chapter lies a speculation of what biology and indeed medicine and culture might have been like if the antivivisection movement had been successful. Would it, as Power Cobbe, that brilliant campaigner who was responsi-

ble for Lord Henke's bill, suggested, simply redeploy that evident scientific creativity toward the construction of a different, noncruel science? And although
she and her allies offered little more as an alternative than the resurrection of
bedside medicine (the scandal of the death rates at the Chelsea Hospital for
Women became at one stage fused by the antivivisectionists, who linked patient
experimentation with animal experimentation), would there have been a cultural space for a more ecological approach to biology? Might this have produced
a biology to sustain health rather than a biology directed toward treating sick-
ness? Would stopping the new hands from being trained in vivisection have
made for gentler doctors, more concerned with their patients and less concerned
with their publications? Would success have reduced the prevailing cultural en-
dorsement to men, particularly professional men, that they can do anything
they like to women and to animals, not just in the ghastly kitchen, in the charity
hospitals, and in their examinations for contagious diseases, but everywhere?

For many of the nineteenth-century antivivisectionists, the "animals" they
sought to defend, if not confined to domestic pets, at the least took them as their
point of departure. Thus the horse, much used by the bacteriologists to supply
serum, was also the subject of an extensive middle-class discourse seeking to
prohibit cruelty, which was widely seen as a practice associated with rough
working-class men. Very few acknowledged that men paid poor wages drove
their beasts cruelly so that they and their dependents might live, for that would
have raised questions about the self-regarding genteel strata.[40] To infer that bac-
teriology threatened the noble horse was to cast doubts about the gentility of its
practitioners. The physiologists by contrast threatened the pet cat and the pet
dog. These were represented as at particular risk of being stolen by the murder-
ous suppliers to the laboratories. Antivivisectionists feared an even more vicious variant of Burke and Hare, no longer content to snatch human corpses
from the grave for medical dissection but now snatching the furry living bodies
of their pets from the streets, exposing them to those hideous torturers the
physiologists and their almost equally iniquitous allies the bacteriologists.

Despite the apparent solidity of a post-Darwinian cultural universe in
which human and other animals are all understood as animals and in which
the combination of evolution and natural selection no longer left God much part
to play, then and now it is clear that some animals are more equal than others.
The delicate political fudging so that it was entirely possible for the writer of
The Origin to be buried in Westminster Abbey also enabled science to develop
an hierarchical model entirely compatible with Judeo-Christian thought, in
which because God had made man in the image of himself, it automatically followed that man had dominion over the other creatures with the misfortune not
to be based on the divine. For that matter, science did not challenge but sustained the hierarchy between "races." It was not by chance that Victorian representations of Jesus depict a fair-haired man with blue eyes, for these traits, as

stereotypically English, were manifestly closest to the divine. Correspondingly, animals were classified according to their closeness to being human, whether biologically, which prioritizes primates as near kin, or socially, which prioritizes the dogs, cats, and horses integral to domestic life. Drawing the line between animals, between the deer and the horse, the fox and dog, the magpie and the eagle, constructing hierarchies of regard and thus determining their treatment, is in itself a fluid historical and cultural matter.

Following Bentham, Peter Singer, the philosopher of animal rights, argues that the chief criterion is sentience and suggests that the line may be drawn between the shrimp and the oyster. While neurobiologists have been heard to suggest that his solidarity with the shrimp owes more to his anthropomorphic impulses than his knowledge of neural systems, the philosopher Mary Midgley points out that it is precisely anthropomorphism which enables "us" to interpret animals' behavior at all. In everyday dietary practices we see the rich abundance of line drawing in the distinction between vegans, vegetarians who eat eggs and dairy produce, and vegetarians who eat fish but not other meat. Even while vegetarians and even fish-eating vegetarians will explain their practices through ethical and political discourse, there is a sense in which those who do the least violence to others, whether the laboratory rat, the beef cow, the oyster, or the beetle, once the issue of violence is raised, secure the moral high ground. However, vegetarianism is not only about human relationships to green nature, for it also touches issues of class and gender. Middle-class feminists who advocate vegetarianism need to remember that not eating meat was historically not a matter of choice for working-class women. Regardless of who actually did the most work, men were seen as having a stronger claim for the best food. The poverty literature records that where there was only enough meat for one, it went to the man; the woman and her children did without.

But also then and now the definition of those animals who are to be defended and to whom rights or welfare are to be extended is a matter of contestation.[41] Even today furry and preferably young animals with cute faces are likely to find more defenders than the slimy and the shelled. Thus at the recent international meeting of geneticists at Birmingham, on Public Understanding of Science Day, a Regents Park Zoo scientist eager to refurbish the image of the zoo as no longer the imprisoner of animals but as the protector of endangered species, used the picture of a baby marmoset of a rare and diminishing kind to illustrate the change.

The struggle around animal rights and animal welfare tells a modernist story about a changing concept of the relationships between human and other animals. The antivivisectionists of the past saw humans as stewards who were morally responsible for their treatment of other animals, which would ultimately make humans more humane. Today the discourse of animal rights defines the other animals as equal citizens along with human animals. But the point that I want to make most strongly is that it is the human animals who have constructed the definitions, even as citizens, of the nonhuman. The animals, despite the claims that primates such as chimpanzees and human beings can share language and communicate, have played rather little part in setting the agenda.

Nowhere is this seen more bizarrely than in the claim for animal liberation, which, in Peter Singer's original formulation in 1975, was set alongside the struggles for women's liberation and for black liberation. Where the women's movement and the struggle of black women and men have been identity movements, made by those people declared Other by those with power, animal liberation was to be carried out by human beings on behalf of others. Arguably the silence of the oppressed was a significant part of their attraction; the liberators could speak on their behalf unfettered by any voices which said "that is not how I feel, that is not how I am, how dare you presume to speak for me."

Precisely at a time when these immense movements of feminism and black resistance to racism were speaking in their own newly found voices and were resisting the claims of expert white males to speak for them, a human movement has spoken and acted for and on behalf of animal liberation. While within the multistranded world of animal welfare and animal rights there are many women, and there are both gender values and feminist values within these movements, the loudest voices speaking on behalf of animal liberation and against "speciesism" have been those of white men: among the new social movements, Peter Singer and Tom Regan in animal liberation and Arne Naess and Alan Tourraine in the ecology movement. Here in the defence of Nature it has been possible for white bourgeois men to play those leading roles as theoreticians and strategists they occupied in the old mass working-class movements of the European past.

While feminist approaches to ecology have developed, I also have real difficulty with those forms of ecofeminism which by reproducing essentialist claims about women and nature simply reverse but do not oppose the patriarchal representations. While I don't want to share Janet Biehl's conclusion that the concept of ecofeminism has been lost to the essentialists and mystics, I can see that building a feminist ecology, as Vandana Shiva has attempted to do, is a formidable intellectual and political task. Not by chance do the movements Shiva discusses name themselves as women's movements, rather than feminist. Western feminists have to reflect on the differences in the struggle for survival, so while the Western academic theorists deconstruct the category of "woman," Third World women engaged in a mass movement to survive cannot afford that theoretical or political move. For that matter I am unconvinced that the arguments about a nonviolent relationship to nature can begin to thrive in the West, except among those people whose lives are not surrounded by violence and who can be sure of both shelter and food.

Rebuilding the Life Sciences?

Even while I write these skeptical sentences I also remember the warmth of my own response, watching a television interview during the late sixties with Herbert Marcuse, and must insist that while skepticism is often an ally of radical social movements, cynicism is not. Thus after discussing his ideas of human liberation and the student movements sweeping the Western world, the interviewer asked—and the question seemed entirely possible in such a utopian time—"And what, Professor Marcuse, will you do, after you have achieved human liberation?" Marcuse reflected for a moment, then smiled happily, "Why, free the animals of course."

The increased pressure against animal experimentation, whether it is the refusal of the children in the school biology class to do dissections or the parcel bomb (however much we may dislike the tactic) sent to the experimenter, undoubtedly compels scientists to think a little harder about the ethics of animal experimentation and the legitimacy of the animal model, both issues of a hundred years ago. Yet we have to be clear that animal experimentation is decreasing from the high peak of the seventies, at least as much because of technical changes in biology itself as from this pressure. Molecular biology simply does not use such quantities of animals, yet it is the new genetics which seeks hegemony in explaining living nature. Opposing animal cruelty as today's big issue probably is not going to hurt that bid. For that matter, because the Victorian antivivisection movement did not have a compelling alternative science to offer, it could only call for a restoration of the past. Raging against the cruelty of science as part of modernism, large sections of the movement took for granted the premodern cruelty of the hunt. As we know, most recently from British prime minister John Major's back-to-basics campaign, political demands to return to mythical golden ages rarely work.

Reflecting on this history of the struggle against the cruelty of physiology leads me to the conclusion that the challenge to reinvent biology has to take up the intellectual as well as the political task of developing and strengthening sciences gentler to people and nature alike. The historically questionable claims of the physiologists that their discipline would enhance medical practice should increase our caution about the current claims made on behalf of molecular biology that therapies will flow. Instead we should look to ecology as a major potential candidate, even though today it gets rather similar treatment to that accorded its forerunner oecology. It is spoken of but dismissed or transmuted into sociobiology by the prevailing new priesthood. In consequence "reinventing biology" is not just an intellectual matter. Changing the approach of the life sciences is going to require engagement in all those institution-building activities which foster the development of new knowledge.

Let me conclude with a hint as to the direction of action. Bruno Latour rightly observed that Bernard's "ghastly kitchen" was also a "costly ghastly kitchen."[42] Hence reinventing biology is also about recasting the science budget. Such a move is both about contesting particular priorities among the life sciences (not least the overly large slice taken by the Human Genome Project) and about questioning the overall direction of the research system, crucially the balance between civil and defense research. Excessive spending on defense research helps fashion economies overly preoccupied with the production and sale of arms, some of whose implications we can see in the globalization of conflicts where highly sophisticated military technology is deployed to deadly effect, often in already desperately poor countries. Debates about reinventing biology and the increasingly costly (and sometimes ghastly) kitchen of research are inseparable from debates about the nature of our culture and society.

Notes

I wish to acknowledge the Fawcett librarians for their interest and assistance; Anne Scott, my Ph.D. student, for pleasurable and illuminating discussions; Steven Rose for letting me test out my ideas and text on him; and Ruth Hubbard for incisive editorial help.

1. Sir James Black, foreword to C. A. R. Boyd and D. Noble (eds.), *The Logic of Life: The Challenge of Integrative Physiology* (Oxford: Oxford University Press, 1993), p. vii.

2. Andrew Cunningham and Perry Williams (eds.), *The Laboratory Revolution in Medicine* (Cambridge: Cambridge University Press, 1992), p. 10.

3. The metaphor of the Holy Grail is widely shared by the elite molecular biologists who lead the U.S. Human Genome Project; a number of Genome critics have commented on this mobilization of an icon for medieval Christian knights to put the mystery back into a distinctly worldly project. See Richard Lewontin, "The Dream of the Human Genome," *New York Review of Books*, May 28, 1992, and Hilary Rose, chap. 8, "The Genetic Turn," *Love, Power and Knowledge: Towards a Feminist Transformation of the Sciences* (Bloomington: Indiana University Press, 1994).

4. See Michel Foucault, *The Birth of the Clinic: An Archaeology of Medical Perception* (New York: Pantheon, 1973), and N. D. Jewson, "The Disappearance of the Sick Man from Medical Cosmology: 1770–1870," *Sociology*, 10 (1978), pp. 225–244.

5. Claude Bernard, *An Introduction to the Study of Experimental Medicine* [1865], trans. Henry Copley Greene (New York: H. Schumann, 1949), pp. 146–47.

6. A point made early and elegantly by M. F. D. Young (ed.), *Knowledge and Control* (London: Collier Macmillan, 1971).

7. Vandana Shiva, *Staying Alive: Women, Ecology and Development* (London: Zed Press, 1989).

8. The attack on the Contagious Diseases Act was much broader and more generous, the solidarity between women extended across the class barriers as bourgeois women sought to defend the rights of working-class women and prostitutes.

9. Judith Walkowitz, "Jack the Ripper," *Feminist Studies*, 7 (1982), pp. 543–574.

10. Frances Power Cobbe (FPC) letter to the *Times*, October 11, 1888.

11. Coral Lansbury, *The Old Brown Dog: Women Workers and Anti-vivisection in Edwardian England* (Madison: University of Wisconsin Press, 1985).

12. Mary Ann Elston, "Women and Anti-vivisection in Victorian England," in Nicolaas Rupke (ed.), *Vivisection in Historical Perspective* (London: Routledge, 1990).

13. E. Westacott, *A Century of Vivisection and Anti-vivisection* (Rochford, Essex: Daniel, 1949).

14. R. French, *Anti-vivisection and Medical Science in Victorian England* (Princeton, N.J.: Princeton University Press, 1975).

15. FPC, "Wife Torture in England," *Contemporary Review*, 32 (1878), pp. 55–87.

16. There is, however, a long and well-founded suspicion concerning experimentation on working-class patients. Doctors are more or less open about this. "For the past twenty years, surgeons have been banging material into patients, who are after all the cheapest experimental animals," says experimental surgeon Richard Coombs in describing new work on resurfacing cartilage at Hammersmith Hospital; *New Scientist*, August 21, 1993, p. 17.

17. While I cannot claim to know what all of these were, the list gives a marvelous feeling of the diversity of approach. However, as Jardine goes on to note, of the list, only Experimental Physiology required a laboratory. Nicholas Jardine, "The Laboratory Revolution in Medicine as Rhetorical and Aesthetic Accomplishment," in Cunningham and Williams, *Laboratory Revolution*.

18. That the biologists used the British Association as their launch pad has to be understood in the context of the collapse of the Royal Society into a mere fashionable club, particularly during the forty-two-year presidency of Sir Joseph Banks. When reform of the society had been blocked in 1832 with the election of the Duke of Sussex, reformers set about establishing the British Association for the Advancement of Science as a rival platform for science. The crucial difference was that the control of the British Association was to be solely in the hands of scientists. In large measure the reformer's strategy worked. Even though the nineteenth century saw the gradual restoration of the Royal Society to its former status as the preeminent scientific institution, the new experimental biology had been able to use the British Association very successfully and the physiologists as the new priesthood were to become handsomely represented in both.

19. It was here too in 1859 that Prince Albert, that single representative of the British monarchy over the past two hundred years with a serious interest in science and technology, who recognized its importance for the state, spoke of the connections between a thriving scientific and technical culture and economic and social well-being.

20. However the story was not uncontested, for it was also at a British Association meeting that a demonstration of the new life science so enraged the anti-vivisectionists, that they brought one of the only four prosecutions over the last hundred years against an animal experimentalist.

21. Physiologists still construct an alliance between their discipline and that of evolutionary theory; thus Boyd and Nobles's strategic text (see n. 1) includes an essay by Steven Jay Gould and a proposal for an "Evolutionary Physiology" by Jared Diamond.

22. To use the concept of "alternative" is to use today's categories on a past where the boundaries were not so sharply drawn, but it will have to serve.

23. The zeal of the Imperial Cancer Research Fund to denounce the alternative approach of the Bristol Cancer Center, a denunciation which outstripped the fund's statistical competence, is a recent example of this ontological and epistemological intolerance. There are exceptions. Thus the neurophysiologist Patrick Wall, well known for his work on pain, has a more complex view on alternative medicine in an attempt to produce a nondualistic account of therapeutic intervention and the placebo.

24. Bernard, *Experimental Medicine*, p. 15.

25. Bernard, *Experimental Medicine*; John Burdon-Sanderson, *Handbook for the Physiological Laboratory* (London, 1873).

26. The British Research Defence Society makes much of the regulation and inspection but is quiet about the modest rate of prosecutions.

27. See Keith Tester, *Animals and Society: The Humanity of Animal Rights* (London: Routledge, 1991).

28. FPC.

29. Stuart Richards, "Anaesthetic, Ethics and Aesthetics," in Cunningham and Williams, *Laboratory Revolution*, pp. 151–152.

30. Peter Carruthers's book is a recent expression of this Cartesian strong boundary between Man and other animals: *The Animals Issue—Moral Theory in Practice* (Cambridge: Cambridge University Press, 1992).

31. Nor it would seem were they reminded of Bentham's views by his mummified presence in the college.

32. Henry Salt, *Animal's Rights Considered in Relation to Social Progress* [1892] (London: Centaur, 1980). Salt is discussed by Tester in *Animals and Society*.

33. Rose, *Love, Power and Knowledge*.

34. *The Pall Mall Gazette*, Sept. 21, 1894.

35. FPC, "The Consciousness of Dogs," *Quarterly Review*, 133 (1872), p. 281.

36. Anna Kingsford, *Unscientific Science* (Edinburgh, 1883).

37. The Fawcett Library has a rich collection of newspaper clippings drawn from all over England. The movement included many vigorous and well-informed letter writers. The *English Women's Review* also carried antivivisection as a standard item.

38. George Romanes, who worked with Darwin, had studied with Burdon-Sanderson, who himself had assisted Darwin in a study of insectivorous plants.

39. FPC, *Life of Frances Power Cobbe: As Told by Herself* (London, 1904), pp. 469 and 491.

40. The socialist Henry Salt was the exception, observing in *Animal Rights* that it was the enthusiasm of the middle classes for meat eating which sustained the bloody trade of the butcher, not the butcher's natural savagery.

41. While the focus of the antivivisection movement was on the domestic animals, so that their leaders, such as the Duke of Portland, might hunt or the titled and other ladies attend the meetings covered with fur and feather, this did not pass entirely unscathed by criticism. Other sections of the feminist movement, notably the socialistic feminists surrounding the magazine *Chart* and G. B. Shaw, deplored the hunting and fur-and-feathers brigade among their "allies."

42. Bruno Latour, "The Costly Ghastly Kitchen" in Cunningham and Williams, *Laboratory Revolution*.

2 | "The Rat Couldn't Speak, But We Can"

Inhumanity in Occupational Health Research

Karen Messing and
Donna Mergler

In the introduction and the preceding chapter, we have learned that animals used in biological research are treated badly and that this inhumanity is justified by reference to an abstract set of values called "science." Keeping animals in deprived, regimented conditions is conceived to be necessary to get reproducible results, that is, results which could be obtained elsewhere under similar deprived, regimented conditions. However, many results obtained with animals raised in unnatural circumstances are unlikely to apply to humans or animals in their natural environment.

In our experience as biologists doing research in occupational health, we have often seen "science" used as a higher value to justify cruelty to humans. Workers are exposed to cold temperatures, poisonous chemicals, and dizzying work speeds in part because of the requirements of obscure scientific tests. And, similar to the situation with animals, we have seen that results obtained by applying standards of scientific practice are unlikely to contribute to the promotion of occupational health. In this chapter we will explain how this comes about; in particular, we can see that the further the workers are from scientists in social class, race, and sex, the more easily their needs can be ignored. Many occupational health scientists appear to us to feel as distant from their subjects as they would if they were working on rats.

One recent experience comes to mind. In South America we were introduced to a full professor and researcher at a major medical school. To persuade us that collaboration would be desirable, this man proudly described a recent research project on the toxic effects of pesticides. He had compared spinal fluid taken from one hundred pesticide sprayers with fluid taken from one hundred worker controls and had found an excess of pesticide among the sprayers. He explained that differences had previously been found in blood and urine but that this was the very first time that spinal fluids had been compared. When asked whether he had had trouble gaining cooperation from the unexposed controls, he replied that once the importance of the project had been explained, there had been no problem.

In fact, this project was horrifyingly dangerous, since spinal taps can cause infections, paralysis, even death. The same information could have been gained by other means, and there was little possibility that any results of this research would benefit workers. That the workers collaborated in allowing this very painful sampling procedure can only be due to the power of the scientists to impress the workers.[1]

Lest Northern Hemisphere scientists invoke the superiority of our ethics committees, we should remind them of the experiment described in admiring terms in the high-prestige journal *Science* in 1980.[2] Workers in a Texas factory were known to be exposed to the suspected carcinogen 1,1,1-trichloroethane (TCE). Geneticists jumped at the opportunity to test a new method. Blood cells of the TCE-exposed workers would be isolated and treated with identical doses of mutagenic chemicals, and the amount of resulting genetic damage would be recorded for each worker. The results of the test would be stored, and workers would be followed to see whether their cells' propensity to develop mutations would predict the likelihood that they would contract TCE-induced cancer. *Science* refused to publish a letter we wrote suggesting that it would be better for scientists to put their energies toward decreasing exposure to known carcinogens than to amuse themselves by telling cancer-stricken workers, "Ha, ha, I thought all along you would get cancer."[3] This journal kept its accolade for the experimenters, whose results could potentially be used for pre-employment screening of susceptible workers; it stated that "validated tests could be used by industries to screen populations exposed to carcinogens, such as workers at a chemical plant. Those most susceptible to cancer might be denied employment. . . . "[4]

How is it that occupational health scientists can be so cruel and heartless? And how has it come about that indifference to workers' health is passed over in the pages of *Science*? In this chapter, we will show that these phenomena are not accidents, but part of the structure of academia in occupational health. We will explain how the rules and standards of occupational health research act to support inhumane treatment of workers, particularly women workers, and most particularly nonwhite women workers. We suggest that the following mechanisms hold: (1) Occupational health research is done in a context which opposes the interests of workers to those of employers and governments but in which the workers have many fewer resources. (2) In order to avoid naming these issues, scientists are encouraged to attribute an "absolute" value to their studies which abstracts them from the situation to which they apply. (3) Standard practices for the conduct of occupational health research and rules for the determination of scientific quality contain hidden biases against recognition of occupational health hazards, particularly those affecting women. Emphasis is on well-controlled studies done in situations which bear little resemblance to real life. (4) Judgment of the value of scientific research takes place in the dark—anonymously, with no confrontation of judge and judged and with no recourse. (5) As with animal studies, this forced exclusion of human feeling in regard to the experimental situation can lead to cruelty. (6) Myths of excellence, relevance, rigor, and responsibility are used to justify the type of research that is being done and the people who are allowed to do it.

We question the use of such standard techniques as controlling for confounding variables, choosing unexposed reference groups, testing for statistical significance, and identifying pathological conditions. We suggest that a paradigm shift is necessary in occupational health research and that organizing for change is vital for obtaining accurate information as well as for prevention of suffering.

We also describe some existing mechanisms for cutting loose from this vicious circle and some ways workers have succeeded in challenging it. We draw on fifteen years of biological research in occupational health, in a context which makes it somewhat easier than usual to listen to the voices of workers.

Employers' vs. Employees' Interests

In North America, research and practice in occupational health have been conditioned by the workers' compensation system. Accidents and illnesses covered by workers' compensation are not subject to other recourse.[5] The compensation system is an insurancelike setup paid into by employers. Employers' payments, like other insurance premiums, are affected by their workers' rates of compensation, as well as by the overall level of compensation paid. Thus employers have a collective and individual interest in limiting the number of compensable conditions.[6]

This situation is further complicated in the United States where compensation and litigation around toxic substances involve very large amounts of money.[7] Lawyers representing companies or workers hire scientists to give expert testimony on behalf of their clients. Employers have more money and are more likely to hire scientists. Furthermore, major companies sit on the boards of many universities, provide much of the support for scientific associations and scientific meetings, and are influential in government-appointed commissions; they can make difficulties for scientists who take the part of workers.[8]

Research in occupational health has been directed by the need to establish causality, since problems which do not or could not result in compensation are not considered to be occupational health problems. So, for example, effects of women's unpaid work are not considered to be occupational health problems,[9] and occupational AIDS research has concentrated on health care workers rather than on the much more heavily exposed sex workers, who are not covered by workers' compensation.

The link to compensation is well illustrated by the situation in Québec, where the research money available from the occupational health granting agency is explicitly tied to the concept of "priority sectors": those occupational sectors where workers have been most heavily compensated in the past, such as construction and mining. (Only 13 percent of workers in the priority sectors are women.) Thus the emphasis is away from identifying new compensable conditions and toward finding low-cost engineering solutions to known problems such as noise and lifting heavy weights. This method of determining research priorities ordinarily results in low interest in women workers: since risks in women's work have been little studied, women's working conditions have not been integrated into the standards for compensation.[10,11] During the first six years after its creation, 73 percent of the projects funded by the Québec Institute for Research in Occupational Health and Safety involved no women workers.[12]

Because of their low rate of compensation,[13] mental health problems resulting from poor work organization and neurotoxic exposures have received little attention; only 3 percent of the 1985-1989 budget of the Québec granting agency went for studies including mental health parameters. This neglect is particularly unfortunate since women workers emphasize stress in their identification of priorities in occupational health research. In fact, women's stress-based claims for compensation for physical problems (e.g., heart attacks) caused by stress are refused four times as often as men's.[14] However, mental health claims are rising,[15] and recent calls for projects have included a reference to work organization and psychological factors.

Thus occupational health research is primarily driven not by the necessity to provide information leading to the protection of workers' health but by a financial incentive: to lower compensation costs for employers and governments. That science is not supposed to be influenced by such pecuniary motives leads granting agencies to wrap their criteria in some very fancy language.

Criteria for Receiving Research Funds

Occupational health research grants which are not obtained directly from employers can be requested from government, private, or university funding agencies. The latter sources employ "peer review" (a panel of scientific experts) to determine whose research will be funded. Peer review is also used to judge the quality of research when a scientific paper is presented for publication.

We once examined determinants of our own success in getting grants and found that for a given researcher, grants were more likely to be accepted if the project involved nonhumans or human cells in vitro rather than live humans, if the work was done in a laboratory rather than in the field, and if there was no visible worker input at any level of the project.[16] Projects which have been refused while others from the same researcher were accepted by the same granting agency included a questionnaire-based study of reproductive problems of health care workers (a study of their cells was accepted); a study of neurotoxic effects suffered by metal-exposed workers (a comparison of different neurotoxic tests of unexposed people was accepted); a study of ergonomic and social difficulties suffered by women in nontraditional jobs (a study of musculoskeletal problems of men and women in traditional jobs was accepted). Thus studies were funded in inverse proportion to their likelihood of supporting compensation or social change.

In particular, no studies dealing with problems specific to women were funded during that time. Since then, one granting agency asked us to remove the word *women* from the project summary so as to maximize our chances of getting funds, then cut the part of the project dealing specifically with women.

We are not alone. Although women are now almost half the workforce, research in women's occupational health is strikingly underfunded. In 1989 we were asked by the Health and Welfare and Labor ministries to prepare a critical review of women's occupational health studies in Canada[17] and found that our federal granting agencies did not have much to offer. According to Health and Welfare Canada, sixty-four projects on women's health were supported, of which only three addressed women's occupational health in any way. Only 3.1 percent of the small amount of money spent in that year on women's health went for research on women's occupational health. Many projects supported by Health and Welfare which should have included the effects of women's jobs did not appear to do so: premenstrual syndrome, factors associated with outcomes of surgery, risks for development of hypertension during pregnancy, women and health in the middle years, etc. The situation has not changed dramatically since then.[18]

How does this work? How can supposedly objective committees act with such unanimity to prevent improvement of working conditions?

Peer review takes place almost entirely in the dark. Grant applications are sent for review to experts who know the name of the applicant but whose names are concealed from both applicants and other committee members. Judgments are irreversible; refusals of grants can never be appealed, even when the grounds for refusal contravene the granting agencies' own published rules for funding.[19] Discussion in committee is confidential; participants are asked to disclose nothing of what is said. In some agencies, committee members vote in secret so that no member will know the final rating of a project.

The veil lifts occasionally to reveal some abusive practices. Recently a medical researcher described the operation of an "old boys' network" at the Medical

Research Council of Canada, suggesting that the network discourages projects of "outsiders."[20] Many granting agencies have peer review committees name their successors, ensuring that the same group stays in control of funds.

One of us recently had an unhappy but enlightening experience when she sat on a provincial government agency peer review committee which judged environmental and occupational health projects. She was the only biologist and the only person with any experience in toxicology or any contact with workers. The ten other participants were all from medical faculties: six members of the same department of epidemiology and biostatistics and four physicians in clinical practice. She was struck by the inability of this group to understand important parameters involved in health research and by her inability to get any of her points across in regard to occupational health.

Occupational health professionals generally come from either engineering or medical backgrounds. Those from medical faculties have been trained in diagnosis of pathologies with little understanding of exposure; no part of their education ensures that they have ever set foot in a factory. The engineers typically have no training in the workings of the human body. The length and cost of training these professionals ensure that they usually do not come from families with any contacts with workers.

That six members of a provincial peer review committee came from a single department in only one of the province's four medical schools may be because this university has a reputation for "excellence."[21] It may also be because this department, like many other high-prestige medical faculties, has a long history of collaboration with industry and government in the field of occupational health. One former chairman, J. Corbett Macdonald, became controversial when it was discovered that funding for his work showing no relation between asbestos exposure and lung cancer had been obtained from the mine owners' association, a source which Macdonald avoided identifying publicly.[22] The current chairman has become a specialist in back pain and has collaborated with the Occupational Health and Safety Commission to show that the best cure for compensable back pain is a swift return to work.[23] The influence of these and similarly trained epidemiologists on North American occupational health cannot be overestimated. They learn and teach a set of rules for properly conducted research which they enforce through their presence on peer review committees.

We present here several of the rules for funding occupational health research, drawn from comments on grant proposals. These rules make it very difficult to fund research of interest to workers.

Relevance to Health

The first formal criterion is that the proposed research be relevant to human health. This is interpreted by the medically trained committees to mean that it must deal with pathologies rather than indicators, signs or symptoms of deterioration in physical or mental states. The pathologies may be studied in humans or animals, but they must be diseases with clear diagnoses. For example, because only symptoms were considered, a recent study of determinants of wrist pain among cashiers was dismissed with the comment: "It is highly unlikely that the superficial survey question truly identifies carpal tunnel syndrome."[24]

The requirement for pathology has two consequences. First, it forces the researcher to consider events that are rare among populations still at work. Ill people often leave the workplace. They can still be studied, but relating their illnesses to their prior working conditions requires a great deal of extrapolation. Second, in view of the requirement for pathology, researchers must study working populations of enormous size in order to find a few cases of illness. Studies become extremely expensive and are restricted to the largest workplaces, the very best-recognized scientists, and the best-defined risks.

Another consequence of waiting for pathology to occur is human suffering. Prevention strategies are most successful before pathology occurs. Symptoms of pain and distress may indicate that musculoskeletal disease is just around the corner. Women with irregular menstrual cycles caused by variable schedules or men with pesticide-associated low sperm counts may some day have fertility problems. Waiting for the disease to happen is a delaying tactic that gets in the way of improving working conditions.

Is This Health Problem Real?

The "reality" of health problems is an especially important issue in occupational health, where both compensation for work-related diseases and improvements in working conditions are often based on a cost analysis. Complaints of workers, especially women or minority workers, about their own health may be ignored, treated as individual problems, or attributed to hysterical overreaction, described in scientific language as "mass psychogenic illness."[25]

The initial manifestations of many work-related health problems are often "nonspecific" symptoms such as headache, fatigue, and pain. As mentioned, these symptoms may be warning signs of an eventual illness or may reflect diminished well-being. However, they are not diagnosed illnesses and are viewed as of little importance. Neurotoxic exposures may lead to such nonspecific symptoms, which are received with much skepticism.[26] Other common problems which scientists have found easy to ascribe to neurosis are cumulative trauma (musculoskeletal) disorder found among assembly-line workers doing repetitive tasks,[27] solvent-related neurotoxic effects,[28] and menstrual disorders (see below).

These judgments by scientists find their way into compensation cases. Compensation can be refused on the grounds that the worker is not "really" ill. We still remember with horror the story told by a woman who worked for twenty years on an assembly line in a cookie factory. She was proud of her job and had never been late or absent. However, when the repetitive motions she made while wrapping small cakes finally caused pain which made her unable to work, the company contested her compensation case with scientific testimony that she was not really ill. Her distress was enormous: how could the company for whom she had done honest work for so long call her a liar? The human consequences of scientific "rigor" in defining occupational illnesses are borne not by the scientists but by the workers.

"Some of These Girls Have Become Rather Anxious"

The type of reasoning that results in restrictive scientific definitions can be demonstrated in relation to one of our research areas which has never been specifically funded and for which grant requests have been refused: the study of workplace effects on menstrual function.[29] Well over half of European and North American women of reproductive age now do paid work, and 30 to 90 percent of menstruating women report lower abdominal and lower back pain around the time of their menstrual periods.[30,31,32] It is therefore surprising that the scientific literature on menstruation has rarely concerned itself with the effects of workplace risk factors on menstrual cycle symptoms and that Western occupational health literature has almost never included menstrual symptoms among outcome variables.

Sexist researchers have been skeptical about the existence of occupational effects. For example, after reporting that beginning airline hostesses underwent unfavorable changes in the menstrual cycle 3.5 times as often as favorable changes, researchers commented, "There is not enough information to explain the pathophysiology of dysmenorrhea. The frequent association of dysmenorrhea with other [sic] neurotic symptoms is indicative of its psychological origin."[33] Earlier researchers were more frank: "In the course of the discussions which form an integral part of the annual medical examination of Hostesses it has become evident that some of these girls have become rather anxious because of [articles in the lay press on the effects of flying on air hostesses]. . . . Let us consider the assertion that flying, particularly in jets, has a definitely adverse effect on the menstrual function of hostesses. . . . When the individual tables are studied, it can be seen that the greatest deterioration occurred in regularity, followed by dysmenorrhea. The former is not at all surprising [in view of jet lag]; the latter, however, is not a measurable entity but is highly subjective and it is just here that the psychological aspect may be of the greatest importance."[34]

When we showed by standard epidemiological techniques that the menstrual cycle could be affected by schedule irregularities,[35] the results were termed "cute" by some occupational health physicians, who suggested that publicizing the results would injure the scientific credibility of the research team. This contempt for women's suffering was reflected in the tone of the comments we received when our requests for funding of studies of the effects of working conditions on dysmenorrhea were refused.

Sufficient Evidence to Justify a Study

The final criterion for funding a study is, paradoxically, demonstration that there is sufficient evidence to link an exposure with an effect. Refusals of our grant applications and those of our colleagues on the effects of manganese and styrene on chromosomes were justified on the grounds that insufficient evidence linked these substances with these effects. Grants were given to the same teams for studying the extremely well-known effects of ionizing radiation on chromosomes. These acceptances and refusals may be motivated by the fact that government and employers do not lose money in compensation by reconfirming a known causal link. As a member of the scientific council of one of these agencies once explained to us, "We don't want to spend our money preventing problems that haven't even happened yet!"

"Rigor": The Rules

In this section, we present the rules for standard conduct of occupational health research, drawn from textbooks and comments on grant requests or submissions to learned journals. Although the discussion is technical, we feel it is important to demonstrate that even the dull, disembodied rules for statistical significance conceal class- or sex-biased assumptions which increase the suffering of workers. We wish to underline the fact that seemingly arcane debates on control groups and dependent variables are important in decisions to deny compensation to victims of occupational hazards. What passes for scientific rigor among researchers may in fact be as much a symptom of racism, sexism, or class bias as a burnt cross or a thrown stone, and may do as much psychological and physical damage.

Appropriate Study Design

Epidemiology textbooks suggest two types of study, both of which divide populations into categories.[36] The case-control study divides people into well and not-well, and looks at the proportions of exposed and not-exposed among them. The cohort study divides workers into exposed and not-exposed, and looks at the proportions of well and not-well among them. The relevant other variables are "adjusted for" (are the workers and controls the same age, do they smoke and drink to equal extents?). If the differences are statistically significant, a relationship is established.

However, categorizing exposures may prevent recognition of complex dose-response relationships. The categories chosen may be inappropriate or out of synchrony with biologically important parameters such as the time of exposure relative to the effect being studied. For example, radiation effects on blood cells will be detectable only during the life span of the cells, so recent exposure is important when considering some effects but not others.[37] The inability to nuance these dose-response relationships may lead to inhumanity. One compensation case turned on the fact that a scientist had said that bladder cancer could be produced after nineteen years' exposure to a specific toxic environment. A worker who had been exposed only 17.5 years was refused compensation, although bladder cancer is very rare and other workers at the jobsite also had bladder cancer.[38]

Dichotomization and categorization are thus good ways to blur occupationally induced health problems. In our work, we have developed alternative methods, whereby exposure levels are related continuously to biological parameters.[39] This work has permitted us to demonstrate that only when exposure has been carefully monitored can it be related to chromosomal effects.[40] Yet these methods have not been accepted as relevant by occupational health scientists and cannot be used in compensation cases.

CONTROL GROUPS

To divide workers into exposed and unexposed, most peer review systems will insist on "reference" populations, comparable groups which differ from workers only by not being exposed to the agent under study. We have often been criticized for studies which compare workers in different parts of the same plant. For example, we presented a study where 720 poultry processing workers reported in detail their work schedules and their menstrual cycles.[41] We found that those who had variable schedules (about half) had more irregular cycles than those who went to work and left at the same time every day. Reviewers criticized the study because we did not compare poultry processing workers to workers in some other industry: "Menstrual cycle abnormalities were studied retrospectively . . . in 726 women working in poultry slaughterhouses or canning factories. The most serious problem with this study is that a control population was not studied."

Such criticisms represent a belief on the part of reviewers that some factory environments involve no risk factors. We think that such comments are evidence that the reviewers (supposedly chosen for their proven expertise in the field) have little notion of what most factory jobs are like. It is hard for us to imagine the appropriate control group these referees ask for.

KEEPING THE SAMPLE UNIFORM

Despite their demand for outside controls and the refusal to use in-factory comparisons, researchers insist that populations examined be "uniform." Uniformity is ensured by eliminating any unusual parts of the population. It is interesting to see which criteria are thought to make populations nonuniform. In cancer research, uniformity might be sought by requiring study subjects to share an urban or rural environment, some nutritional habits, or medical history.

However, other less justifiable criteria are frequently used; for example, women are frequently eliminated. In 1988, Gladys Block and her colleagues published a study of cancer among phosphate-exposed workers in a fertilizer plant.[42] Among 3,400 workers, 173 women were eliminated with the comment that "females accounted for only about 5 percent of the study population, and were not included in these analyses." Yet, the thirty-eight male workers in the drying and shipping department were not considered too small a population for study: a significant rise in their death rate was noted.

Another example was the study paid for with $2 million in public funds, relating cancers to a huge number of occupational exposures. When we asked the researcher why his study excluded women, he replied, "It's a cost-benefit analysis; women don't get many occupational cancers." He did not react when we pointed out that his argument was circular, and that for women taxpayers, the cost-benefit of a study excluding them was infinitely high. The resulting papers, published in peer-reviewed journals, made no attempt to justify the exclusion of women.[43] Much occupational health research is still done on male-only samples without justification.

There are, in fact, some well-identified occupational cancers among women. A study of members of the American Chemical Association shows that women members have significantly increased rates of ovarian and breast cancer.[44] A Canadian study revealed that hairdressers are especially likely to get leukemia and ovarian cancer.[45] But information is still lacking in this area because of the "scientific" elimination of women from studies of occupational cancer.

Thus scientists reinforce the notion that women's jobs are safe, that women's concerns about environmental influences on breast cancer (for example) are unfounded, and that it is justifiable to exclude women's jobs from prevention efforts. Is this because scientists feel far removed in sex and/or social class from women factory workers and hairdressers?

SPECIFICITY: ONE FACTOR, ONE DISEASE

The specificity of a response to a chemical or a working condition is touted as an important criterion for establishing cause and effect in occupational disorders. Although this may appear scientifically valid, exposure to quite high levels of a particular substance or condition is necessary before pathology is observed. This only happens following spills of toxic substances or through long-term exposure to near-pure substances such as asbestos, silicates, or manganese in mines and foundries. By contrast, most work entails many low-level exposures,[46] which produce what are disparagingly called nonspecific symptoms (discomfort, malaise).

A good example is heat exposure, which when present at dramatic levels in foundries can produce serious disorders or even death when the body can no longer maintain its normal temperature. Legal limits for heat exposure are based on studies carried out in very hot industries and take into account the dynamic, muscular, heat-generating work typical of these industries. Few studies have focused on women's work in such places as industrial laundries, kitchens, and the clothing industry, where heat is insufficient to produce specific heat-related illness. However, when coupled to postural and gestural constraints involved in lifting laundry or moving cooking pans (rapidly moving upper limbs with little displacement of the lower body), heat puts stress on the heart, resulting in nonspecific symptoms of discomfort and distress.[47] Although cardiac strain among women laundry workers was shown to exceed recommended levels 40 percent of the time during the warm summer months and was similar to levels recorded for miners, these exposures are allowed to continue because of the nonspecific nature of the health problems.

Among the most recent examples of the way in which scientists have contributed to the hardships of workers is the failure to recognize work-related disorders in the microelectronics industry. This industry employs hundreds of thousands of women worldwide in the manufacture and assembly of printed circuit boards and their microcomponents. The work is visually and physically demanding, sometimes requiring workers to look continuously through a microscope while using their hands and feet in awkward positions. Most often paced by an assembly line, the work is very repetitive, requiring manual dexterity and high levels of concentration. In addition to the ergonomic and organizational constraints, many different chemical substances with known neurotoxic properties are continually used in the work process, exposing workers to what can be described as a toxic cocktail.

In the early 1970s, reports began appearing that women workers in this industry were complaining about losing their memory; they also described mood changes, visual loss, and menstrual difficulties.[48] The scientific community at first ignored these reports. Later, when symptoms appeared frequently in different microelectronics plants, studies were done, but they concluded that symptom reports were due to mass hysteria, since there were no previously identified health risks and the majority of complaints were from women. In a review of the literature on mass psychogenic illness, Brabant et al. noted that the microelectronics industry was overrepresented in reported episodes.[49]

By 1984 there were at least five outbreaks of severe symptoms in the microelectronics industry. Given the multiplicity of risk factors and the wide range of symptoms, the methods developed within the context of the dominant medical model of one causal agent/one disease cannot provide evidence of a link between the women's working conditions and their symptoms. Scientific articles began to appear on the potential health hazards in the microelectronics industry,[50] but no systematic study had been done. Microelectronics workers who were becoming ill could not get work-related compensation since there was no scientific evidence relating their health problems to their work.

In 1988, our research group began to study a group of former microelectronics workers from Albuquerque, New Mexico. Most were women of Mexican-American origin; they had worked in a microelectronics plant for periods ranging from one to sixteen years and were suffering from what they believed to be work-related illnesses. All of these workers had undergone extensive pre-employment medical screening which included an assessment of visual capacity, manual dexterity, and mental stability, important attributes for carrying out the required tasks. During the years they worked in the plant, many of these previously healthy women had experienced skin problems as well as dizziness, fainting, nausea, and excessive menstrual bleeding.

Steve Fox tells the stories of some of these women and about how a local lawyer, struck by the repeated occurrence of certain illnesses, undertook to help them receive workers' compensation.[51] Compensation was granted to 115 women in an out-of-court settlement, but the amounts awarded were insufficient to cover their medical costs or provide a decent living, since many could no longer work. One available recourse in the U.S. legal system was toxic tort litigation (suing the manufacturers of the products which had harmed them rather than their own employer, who was protected by the settlement).

Our study compared these workers to women from the same region who were similar in age, educational level, ethnicity, smoking habits, and number of children. While they worked at the plant, the microelectronics workers had had a four times higher rate of miscarriage compared both to their matched referents and to their own pregnancies prior to working in the plant.[52] At the time of the study, they suffered from personality disturbances and affective disorders which persisted over time, and also presented evidence of severe nervous system dysfunction.[53] Their disorders were consistent with organic solvent-induced damage to the nervous system, although other factors may have played a role in producing them.

Although these workers had been exposed to very high levels of a single chemical with known neurotoxic properties, other chemicals were also present. The workers lost their case, since it was impossible to make the link between one particular chemical and a health effect, despite the association between their working conditions and their illnesses. Thus in this case, and in many others, the only ones left to suffer are the workers. By neglecting the reports of women workers, by not developing methods of analysis that can take into account the reality of this work situation (similar to many others occupied by women workers), and by following the lead of industry-financed research, scientists have thus contributed to the health problems of microelectronics workers.

Adjustment for Relevant Variables

"Adjusting" for a variable while analyzing data means using a mathematical procedure to correct for its effect. It is reasonable, for example, to adjust for smoking when examining the relationship of dust exposure to lung damage, because smoking is an independent determinant of lung damage and might confuse the issue if those exposed to dust smoked more or less than those not exposed. We may need to add a correction factor to the lung function of smokers before testing the relationship between dust exposure and lung damage. This procedure allows us to determine the effect of dust on the lungs while taking into account the deleterious effects of smoking.

"Overadjusting" occurs when the variable adjusted for is a synonym for the exposure. This would happen if night shift workers were both more frequent smokers because of boredom and more exposed (because the chemical in question was used more often on the night shift). Adjusting for the effect of smoking would then diminish the possibility of finding an effect of the chemical exposure on lung cancer.

ADJUSTING FOR SEX

Overadjusting has been applied widely and abusively to sex differences. Studies which examine the health of workers often find that women workers report more symptoms of poor health or psychological distress than their male counterparts.[54] The approach to these differences is to adjust for sex, without any attempt at justification.[55] Adjusting would be appropriate only if sex were an independent determinant of poor health reports—for example, if women were weaker or complained more than men.

But an independent contribution from sex hormones toward occupational health is rarer than effects related to the sexual division of labor. In a study of men and women workers in poultry processing plants,[56] women workers reported a higher prevalence of some symptoms (muscular and articular pain, stress-related disorders) than men, while the prevalence of other symptoms such as hearing difficulties was similar among members of both sexes. When we compared working conditions, we found that noise levels were similar for men and women, but other aspects of their jobs differed. Women mainly worked standing still at the machine-paced assembly line; their work, which was carried out at very rapid speed, required swift, accurate arm and hand movements. Men's jobs allowed them more freedom of movement and were carried out at a less rapid pace. In areas of this workplace where men and women were doing similar jobs, there were no significant differences in symptoms. Adjusting for gender in this study would have obscured the differences in working conditions and would have made it look as if the women reported more symptoms because they were complainers or weaker.[57]

Data obtained from a health survey carried out in Québec provide a good example of the consequences of adjusting for sex.[58] While preparing a book on workplace factors affecting mental health[59] the authors examined data from a questionnaire administered to over 10,000 workers. Results from a psychological distress index showed that 9 percent of working men and 18 percent of working women registered high levels, a very significant difference.[60] Initial analyses, adjusting for sex, indicated that no women's job titles were associated with psychological distress. When data were reanalyzed without adjusting for sex, it turned out that jobs such as nurses' aide, cashier, stenographer, typist, secretary, and data entry clerk were associated with high levels of psychological distress.

Thus the standard techniques can obscure the types of suffering women experience at their jobs and help maintain the illusion that women are physically, mentally, and emotionally "the weaker sex."[61] Such analyses reduce the amount of money allocated to prevent occupational disease in women.

ADJUSTING FOR AGE

Body parts tend to wear out with use, and the probability of having experienced some damaging event such as a disease or an accident increases with time. Therefore almost all health indicators get worse with age.

Age is one of the complicating factors in studying workplace effects on health, because people of different ages are not distributed randomly in the workplace. Older people usually have more seniority and have been more exposed to workplace toxins. However, some may be less exposed, if they have supervisory positions where they are less exposed to chemicals. Their behavior may also be different: they may be better trained and more prudent, and they have often learned to protect themselves better. Adjusting for age may well involve adjusting for exposure and thus rendering invisible a relationship between exposure and effect.

In addition, the "healthy worker effect" is well known.[62] As workers become ill and unable to work, they leave the workplace, leaving only those who are "healthy" (i.e., who can tolerate the working conditions). This phenomenon produces a bell-shaped curve relating age or seniority to health problems: the frequency of problems is initially low due to pre-employment screening, rises as the working conditions produce their effects, and falls as those affected by the working conditions leave.

However, researchers in occupational health often adjust for age without taking these effects into account, thus underevaluating dose-response effects.

ADJUSTING FOR SOCIOECONOMIC STATUS (SES)

In epidemiological studies, reference is often made to subjects' socioeconomic status or social class. This is a loose concept which may mean different things, depending on how it is defined operationally. SES may be determined by reference to individuals' income or educational level, or by their residence in an area of a given average income.

However, SES can also be a surrogate for workplace exposure. "Foundry workers" of different educational levels may be doing jobs with very different exposure levels. Adjusting for education may mean adjusting for exposure, with consequent diminution of significant results.

Despite the vagueness and the difficulties associated with determining SES, income is definitely associated with health status and with the presence or absence of some risk factors in the physical environment.[63,64] Pregnancy outcome is much worse among the poorest Canadian women than among the richest.[65] The poor do not eat as well as the rich, they smoke more, and they are more likely to be exposed to environmental pollutants.

In studying workplace effects, it is hard to know how to take all this into account. People of different social classes also have very different work environments. One only has to think of secretaries in large corporations to realize that the work environments even of those who hold the same job title may vary with social class. Some epidemiologists adjust for social class when studying the effects of work on health. For example, large and often-cited studies on pregnancy and work[66] or fertility and work[67] have used this technique. However, since job status is an important determinant of social class, adjusting for SES may obscure real effects of the work environment.[68,69] In fact, some health effects previously attributed to poverty may in fact be due to unhealthy working conditions.

There are new problems in determining SES now that most women work outside the home. In the past, women's jobs were ignored when determining family SES, and women were assigned the status of their husband or father. There has recently been some discussion of how to take the evolving family into account.[70] However, because women's jobs are very different from men's and because women's income does not reflect education to the same extent as men's does, these problems have not yet been resolved.

In all these ways, adjusting for social class acts to obscure relationships between exposures and effects and leads to underestimates of occupational disease, especially for women workers.

Evaluation of Exposure

We have already mentioned that dividing workers dichotomously into ill and well masks biologically important phenomena which could help identify hazardous situations before pathology occurs. Classifying workers as exposed or unexposed (or even into several exposure categories) is likely to involve errors and thus to conceal important occupational hazards.

Exposure to toxic substances or stressful conditions can be monitored in several ways. Taking measurements in the workplace is recognized as the best method, although it does not always relate well to health effects. Working the Monday night shift may involve different products, concentrations, and methods from work during the day shift on Thursday. Workers even a few meters apart may have quite different exposures. A supervisor may stand back from a tub of solvent while the worker may have her face directly over it. Cold drafts may play on the neck of one worker but not another nearby. Studies which sample at one time or one site per department may in fact produce very inaccurate results.

Another difficulty with establishing a link between exposure and effects is that they occur on different time scales. Health deterioration results from accumulated assaults on the body, while most exposure measures are current. Using current exposures as a surrogate for past exposures weakens the capacity to detect a link between exposure and effects.

In many cases, workers have a fairly good idea of their exposure levels, but their judgments are not considered to be objective. To relate working conditions to illness, researchers refer to "experts," as in the following example, taken from the occupational health literature. In a study of 13,568 workers, experts were asked to class certain jobs according to whether they involved exposure to dust, by using tables relating exposures to job titles. The tables had been derived seven years previously in another country. The researchers offered the following justification: "Although [an expert exposure estimate] cannot be considered an ideal reference, it is not biased by misclassifications of exposure according to personal factors as it is based only on job titles and industry sectors."[71] This study goes on to correlate reported symptoms of dust exposure (such as difficulty breathing and asthma) with ratings by the experts. The study also correlated symptoms of dust exposure with the workers' own reports of dust exposure. The researchers found that workers' reports were much better correlated with symptoms than were the experts'. In fact, for women workers and less-educated workers, the correlations with experts' ratings were quite low. Did the researchers conclude that the experts' estimates were incomplete, class-biased, wrongly applied to women's jobs, or out of date? Not at all: when self-reported exposure was better related to symptoms than expert-reported exposure, the self-reports of exposures and symptoms were taken to be wrong. Because educated men's estimates were closer to the experts', the men were said to have made fewer errors than the women or the less-educated workers of both sexes. But there is no reason why an "expert" who has never been in a workplace is better able than a worker to describe dust exposure; experts consistently underestimate both women's exposures and women's health problems.

Contempt for nonexperts' reports of exposure has a scientific name: "recall bias," a supposed tendency of ill people to overreport exposures. It is frequently invoked (as it was in the above-cited article) to contest workers' reports, but the existence of this bias has never been satisfactorily demonstrated.[72]

Another example of inaccuracy in studies using surrogates for exposure involves using job title as a measure of exposure. It is common to compute standardized death ratios by profession, comparing causes of death with respect to professional categories. This information is usually taken from death certificates, where information is notoriously inaccurate.[73] Due to the common habit of classifying women as housewives unless they are actually at work,[74] retired women's professional categorization at the time of death is even less valid than men's. The fact that farm women are not classed as farmers but as farmers' wives, for example, has kept them out of large studies of pesticide-induced cancer. Also, the fact that women's job titles are less nuanced than men's makes it harder to describe women's exposures. "Clerical workers," for example, can work in many different industries with concomitant exposure to many different circumstances. When attempting to examine the relation between work and health, overgrouping can dilute associations of effects with hazardous conditions.

In summary, imprecision in estimating exposure blurs relationships between exposures and effects[75] and is a cause of underestimation of toxic effects of working conditions. These imprecisions are caused in part by a class-biased reluctance to believe worker reports.

Statistical Significance

Before a drug, cosmetic, or food product can be marketed in North America, it must undergo extensive testing on animals. Although these tests do not guarantee that these substances are safe for humans (or even for the same animal species in their natural environment), they do show a concern for human consumers. Contrary to the situation with regard to drugs and cosmetics, no law in Canada requires employers to be sure that an exposure is safe before imposing it on workers. For example, tens of thousands of women worked with video display terminals (VDTs) before the first study of VDT effects on pregnancy. Even now, no one is yet absolutely sure that VDTs do not pose a danger for pregnant women. But pregnant women have not stopped working with VDTs while waiting for the evidence to come in. In this and many other cases, the burden of proof is on the worker rather than on the employer: an agent must be proved dangerous before it will be removed.

This antiworker bias has been heartily endorsed by scientists in their requirements for statistical tests. The necessity for choice in statistical methods arises from the fact that there is usually a rather long interval between the first suspicions about particular working conditions and the time when the final word is in on the exact level of risk. Initial studies may show a weak relationship between an exposure and an effect, so that scientists have had to set a standard for when they will believe a relationship exists. Scientific practice is to accept that there is a risk if there is less than one chance in twenty that the observed association was due to happenstance. In other words, in order for scientists to accept that Agent X causes problems for pregnancy, a study must establish the toxic effects with 95 percent certainty. A study which shows that the researcher would have one chance in ten of being wrong in concluding that there is a risk is considered to be "negative," that is, the study has failed to show a risk. This is true even if the group being studied is so small that there is virtually no chance of demonstrating anything.[76,77]

For scientists to be really sure of their results, more than one study must show the same relationship. Given the small number of workers in most women's workplaces and the large number of potential hazards, it is no wonder that very few dangers for pregnancy or fertility have been established.[78] The burden of proof is on the side of minimizing the costs of improvements in hygiene rather than on minimizing questionable exposures. This is a political decision, but in the scientific literature it is presented as a scientific decision about "the standard level of statistical significance" and is never justified or explained.

Lack of Bias

Probably the most frequent question our research group hears about our work is about our relationship with unions.[79] We have often heard rumors that discussion of our grant applications is replete with references to our close relationship with trade unions. That our studies are often initiated by workers is thought to lead to bias. In three cases we know about, this was a reason for refusing our grant requests.

Whenever workers are involved in efforts to improve their working conditions, this is thought to be evidence that they will fake symptoms to gain their point. Our work has been criticized on these grounds, even when we use elaborate study designs to take these factors into account and even when the patterns of physiological change are so specific to the toxic effect that the worker would have to be a trained biologist to produce them.

Many studies are published which have been initiated or inspired by employers, and the major granting agencies have put forward programs for university-industry collaborative research. Clearly, bias is only perceived when it acts opposite to that of industry and established scientists.

A recent example from a scientific paper illustrates some of the bias in interpretation which passes peer review when the employers' perspectives are supported.[80] In a study of 2,342 Québec workers who had been compensated in 1981 for a spinal injury, records for the following three years were examined to see whether the worker was compensated subsequently for any spinal injury. This paper shares many of the flaws mentioned above: it adjusts for sex and age without any information on working conditions, and without justifying these choices. No exposure information at all is presented: we have no idea which workers continued during follow-up at the same job which injured their spine the first time around.

The authors found that the shorter the initial absence, the less likely was a recurrence. They mention that 70 percent of those returning to work still feel pain. Our interpretation would be that workers whose injuries are more severe and who return to work before being fully recovered have a high risk of recurrence and/or that workers with poorer working conditions may have worse and more frequent injuries. They may return due to problems with compensation (no information was sought on the reason for return to work) or to machismo (sex differences were not presented; they were "adjusted for," so we cannot consider this hypothesis). Their interpretation: "A longer duration of the initial episode *resulted in* [our italics] more absence from work in the three year study period, both in terms of total cumulative duration of absence and risk of recurrence." The authors suggest that physicians should return injured workers rapidly to the workplace in order to lower the likelihood of recurrence.

This image of workers as lazy malingerers, coddled by their colluding physicians, was published in the clinical literature, where it can be cited to block compensation to injured workers who stay out "too long." Thus human suffering can be justified by a class-biased view of workers presented in the guise of "objective" scientific research.

The Other Myths: Responsibility and Excellence

We have explained how the myths of rigor and objectivity justify concealment of suffering and of life-threatening risks to exposed workers, especially women and minority workers. We will note two other myths which act to maintain this system in place: "responsibility" and "excellence."

Responsibility is a name often given to concealment of risks. A scientist is supposed to confine publication to refereed journals, in order to keep unvetted information out of the public domain. We have been criticized and threatened with legal action for giving information to workers about our findings regarding damage to their own cells. A grant was refused on the grounds that examining sperm counts of metal-exposed workers would be irresponsible, since it would give them information they "didn't need" (since a low sperm count does not necessarily prevent conception and does not affect well-being of workers who are not trying to conceive) and which would needlessly upset them. Such "responsibility" keeps workers from being able to organize to prevent biological damage.

Excellence is a name given to exclusionary practices that keep women, working-class people, and minorities out of university departments and thus maintain social distance between researchers and the workers they study. Excellence of candidates for academic positions is judged by proven grant-getting ability and the ability to publish in peer-reviewed journals with high impact factors.[81] Thus ability to accept the biases of the dominant culture in occupational health is a bona-fide occupational qualification for a career in research.

Even before they enter research, academics in occupational health have been selected for membership in a privileged social class, quite distant from that of the workers they will be studying. Much has been written about the class, sex, and racial composition of the medical profession.[82,83] Similarly, sexism, class bias, and racism can directly block access to university positions.[84] In North America, the hope that women and ethnic minorities will be increasingly included as members of the occupational health establishment comes from affirmative action; no ways have yet been developed to favor access of working-class people to university positions. But even these minor steps toward the democratization of scientific institutions have recently been opposed by a discourse on "excellence" which can conceal the fact that people perceive some scientists as nonexcellent for reasons related to their sex, class, or color.[85]

Working-class people or people of color may express themselves in somewhat different language, which is often misperceived as inaccurate or sloppy. Women who combine child rearing with scientific training or people who need to earn money while studying can be misperceived as not serious about science. Bias which makes it difficult to perceive excellence in such cases can only be compensated by efforts to recruit and maintain a broader spectrum of scientists.

The Rat Talks Back: Forces for Change

How is it that we at the Centre pour l'Etude des Interactions Biologiques entre la Santé et l'Environnement (CINBIOSE) have come to favor techniques which differ from those of other scientists, and yet we survive? Is our $ 1,000,000 in current grants proof that we have either sold out or that we are exaggerating the forces acting against worker-initiated research?

We think that our relative success is due to the existence of some important structures which have been put in place in Québec after struggles by working people and progressive scientists: (1) agreements between our university and the three major Québec labor unions; (2) granting agencies or programs which include representation from the groups being studied; (3) feminist and worker solidarity.

The Université du Québec à Montréal was founded in 1969 with a mandate to serve the Québec community, including "those sectors of the community not usually served by universities." It has signed agreements with the three major Québec unions, the CSN,[86] the FTQ,[87] and the CEQ,[88] providing resources for responding to union requests, such as released time for professors who participate in educational activities and university seed money for research.[89] These agreements have given us the opportunity to participate in union-organized workshops on work and health; to produce information on noise, radiation, and solvents; to work with the unions' women's committees to write brochures on protection of pregnant women, women's occupational health, and health risks for women in nontraditional jobs; to carry out research in response to needs expressed by workers; and to provide expert testimony in litigation involving occupational health.[90]

During the 1970s, professors at the Université du Québec à Montréal joined the CSN union. We negotiated clauses in our collective agreement recognizing work done in the context of the university-union agreements. Because our work in occupational health is carried on as part of our regular workload (for forty-five hours of teaching in unions we can be released from forty-five hours of university teaching), many professors have been available to work with unions. The unions benefit from the prestige and perquisites of professors' status, such as access to university services and grants. Union-initiated research, which is often received with hostility by the scientific community, must be recognized by our employer, at the risk of grievance procedures in the context of our collective agreement.

Features of the university-union agreement have been important in making it productive: explicit recognition of the power imbalance between career researchers and community groups, with structures in place to guarantee that the needs of both are recognized throughout the entire project, and guarantees of scientific credibility through peer review. Although peer review committees need to be sensitized to the specific difficulties of community-based research, they provide an important quality guarantee to the community group and help maintain the scientific credibility of the researchers, title to the research results held by the researcher and by the initiating organization, and seed money available for feasibility studies, often necessary with the radically new questions posed by community groups.

Agreements between the University and the Community Organizations

Two Québec granting agencies incorporate representatives from labor or community groups in the determination of funding. The Québec Institute for Research in Occupational Health and Safety incorporates representatives from labor and management into allocation of grants; the two sides explicitly and openly negotiate to fund research that they consider important. This process paradoxically appears to allow for less research bias in funding, since the practical consequences of the projects are put on the table from the outset.

During one period, one of the labor representatives was in charge of women's affairs for her union and we were able to get funding for some feminist studies. However, since her departure we (and the union women's committees) have turned to the Québec Council for Social Research, which incorporates community representatives in its decision-making processes. It is currently funding a study of the impacts and consequences of methods used to reconcile family and professional responsibility, initiated by the FTQ union, and a study of ways to make women's work more visible, in partnership with women's committees from all three unions.

Most important, because of the relationships built up over the years, we have been able to count on helpful hints and research suggestions from union members, from health and safety committees, and from the very active women's committees of the three major trade unions in Québec. With them, we have been able to come to a better understanding of the needs of working women and provide information that has been helpful to them. Their solidarity has been a source of strength to us over many years, and their insights have been invaluable. From the cleaner who explained to us why the job called "light work" was heavier than that called "heavy work," to the technicians who told us why time-and-motion studies were not a good way to organize radiotherapy treatments, to the factory workers who made us understand why being able to go to the bathroom when you need to is an important health practice,[91] workers have constantly put us in touch with real occupational health problems. If our colleagues had similar opportunities and information, perhaps they would change the rules for occupational health research. They might be more able to believe workers' accounts of risks and symptoms and become convinced that it is important to protect workers' health.

Most of our colleagues in other universities and other countries do not have access to such advantageous conditions for worker-initiated research. They are forced by financial and time constraints to work on projects initiated or inspired by industry and government priorities. In consequence, their results are unlikely to lead to improvements in working conditions. Luckily, workers are less helpless than laboratory rats when faced with such scientists. We cannot describe in detail all the struggles waged by these workers, but we would like to close with an extract from a letter from a microelectronics worker, which describes her efforts to join with other workers in fighting back.

> ... When I heard you and Dr. Bowler on the stands [giving expert testimony on the effects of solvent exposure] everything fell into place in my life, why my ovaries were petrified, why I was tender to the touch and every muscle on my body hurt. Why I lost my strength and had chronic fatigue. And why there were pieces of my life missing.
>
> I'm sure that you are aware by now we lost the case for the 13 people. But not all is lost because we have formed a non-profit organization called Toxic Victims' Assistance Corp. for us that were exposed to chemicals in the working environment. We have decide to go public to talk about what happened to us.
> ...
>
> I do not want any woman to go through what I have gone through ... We have found out that other people are getting sick the way we were. The Sandia labo-

ratories ladies. Some ladies in Phoenix Arizona and some Asian women in some valley in Calif. And some ladies in the Honeywell plant. . . .

And I remember the Dow lawyer saying that the rat when exposed to V.G. [trichloroethylene] only lost its grip. And he made a fist with his hand. And I keep seeing that hand opening and closing on my mind. And I remember I couldn't brush my teeth standing up. For three years I brush my teeth laying down cause I was so tired. The rat couldn't speak but we can.

Isabel S Aragon

Notes

We thank Dr. Nancy Waxler-Morrison, Dr. Lucie Dumais, and Dr. Serge Daneault for helpful comments. KM thanks the Social Sciences and Humanities Research Council of Canada and the Conseil Québécois de Recherche Sociale for research support.

1. A search of the Medline data base for the past ten years revealed no publication under the name of this professor, so it appears the study yielded no results.

2. Gina Kolata, "Testing for Cancer Risk," *Science*, 207 (1980), pp. 967–969.

3. The journal did, however, mention our work in the same space because we were responsible for a technical advancement; see Jean L. Marx, "Detecting Mutations in Human Genes," *Science*, 243 (1989), pp. 737–739.

4. Kolata, "Testing," p. 969.

5. This is one reason why employers in the United States have been anxious to restrict women's access to nontraditional jobs which they suspect of causing reproductive damage. A fetus cannot be covered by workers' compensation and a malformed child could potentially sue an employer for millions of dollars.

6. A historical and political treatment of workers' compensation can be found in Katherine Lippel, *Le Droit des accidentés du travail à une indemnité: Analyze historique et critique* (Montréal: Thémis, 1986).

7. In countries such as Canada which have paragovernment compensation agencies and state health insurance, workers have better access to health and social benefits. The fact that workers have difficulty in obtaining compensation from employer-funded insurance results in a transfer of the costs of occupational illness from employers to the public purse, but at least the workers' health costs are covered.

8. Karen Messing, "Union-Initiated Research in Genetic Effects of Workplace Agents: A Case Study," *Genewatch*, 6 (1990), pp. 8–12.

9. In fact, women's domestic work overload is often used as a reason for refusing compensation. Recently the employer presented evidence in a musculoskeletal injury case purporting to show that the woman had in fact injured herself by carrying heavy loads in her kitchen.

10. Carole Brabant, "Heat Exposure Standards and Women's Work: Equitable or Debatable?" *Women and Health*, 18 (1992), pp. 119–130.

11. Karen Messing, Lucie Dumais, and Patrizia Romito, "Prostitutes and Chimney Sweeps Both Have Problems: Toward Full Integration of the Two Sexes in the Study of Occupational Health," *Social Science and Medicine*, 36 (1992), pp. 47–55.

12. Céline Tremblay, *Les Particularités et les difficultés de l'intervention préventive dans le domaine de la santé et de la sécurité des femmes en milieu de travail*, communication présented at the 58th Annual Meeting of the Association Canadienne-Française pour l'Avancement des Sciences, Université Laval, Québec, May 14, 1990.

13. In 1988–1990, only eleven of eighty-six claims for psychological and psychosomatic pathologies were accepted; Michel Vézina et al., *Pour donner un sens au travail: Bilan*

et orientations du Québec en santé mentale au travail (Québec: Editions Gaetan Morin, 1992), p. 109.

14. Katherine Lippel, *Le Stress au travail* (Cowansville, Québec: Les Editions Yvon Blais, 1993), p. 228.

15. There were eight claims per year in 1986–1988 and twenty-nine in 1989–1990; Vézina et al., *Pour donner un sens au travail*, p. 109.

16. Karen Messing and Donna Mergler, "Determinants of Success in Obtaining Grants for Action-Oriented Research in Occupational Health," Proceedings of the American Public Health Association, Las Vegas, 1986.

17. Karen Messing, *Occupational Health and Safety Concerns of Canadian Women: A Review* (Ottawa: Labour Canada, 1991), and "Doing Something about It: Priorities in Women's Occupational Health," Proceedings of the Round Table on Gender and Occupational Health, Health Canada, Ottawa 1993, pp. 155–161.

18. National Health Research and Development Program (1986–1987), "Research Support to Women's Health Issues," supplemented with information supplied by Mme. Raymonde Desjardins.

19. We were once refused a grant on the grounds that our work was too applied, insufficiently related to basic science objectives. We replied by quoting long sections of the granting agency's own appeal for projects, which clearly and emphatically stated that applied work should be funded. We were told to submit our grant proposal again the following year. Our new proposal was rejected on the same grounds. We then submitted the same proposal without the applied part: the request was enthusiastically accepted and funded (among the top five proposals).

20. Gilles Ste-Marie, "La Recherche scientifique au Canada: Un trafic déloyal," *Le Devoir* (newspaper), May 23, 1992.

21. The university senate recently came out explicitly in favor of "excellence," stating that it considered quality of academic staff to be weakened by the various equal-opportunity programs indulged in by other universities.

22. Paul Brodeur, *The Expendable Americans* (New York: Viking, 1974), part 3.

23. Walter O. Spitzer et al., "Scientific Approach to the Assessment and Management of Activity-Related Spinal Disorders," *Spine* (suppl.), 12 (1987), pp. 1–59.

24. Philip Harber et al., "The Ergonomic Challenge of Repetitive Motion with Varying Ergonomic Stresses," *Journal of Occupational Medicine*, 34 (1992), pp. 518–528.

25. Carole Brabant, Donna Mergler, and Karen Messing, "Va te faire soigner, ton usine est malade: La place de l'hystérie de masse dans le problématique de la santé des travailleuses," *Santé mentale au Québec*, 15 (1990), pp. 181–204.

26. Patricia Boxer, M. Singall, and R. Harble, "An Epidemic of Psychogenic Illness in an Electronics Plant," *Journal of Occupational Medicine*, 26 (1984), pp. 381–385.

27. Yolande Lucire, "Neurosis in the Workplace," *Medical Journal of Australia*, 145 (1986), pp. 323–327.

28. Rosemarie Bowler et al., "Stability of Psychological Impairment: Two Year Follow-up of Former Microelectronics Workers' Affective and Personality Disturbance," *Women and Health*, 18 (1992), pp. 27–48.

29. The quotation in the heading comes from R. G. Cameron, "Effect of Flying on the Menstrual Function of Air Hostesses," *Aerospace Medicine*, September 1969, pp. 1020–1023.

30. N. F. Woods, A. Most, and G. K. Dery, "Prevalence of Premenstrual Symptoms," *American Journal of Public Health*, 72 (1982), pp. 1257–1264.

31. G. Sundell et al., "Factors Influencing the Prevalence of Dysmenorrhea in Young Women," *British Journal of Obstetrics and Gynaecology*, 97 (1990), pp. 588–594.

32. S. Pullon, J. Reinken, and M. Sparrow, "Prevalence of Dysmenorrhea in Wellington Women," *New Zealand Medical Journal*, February 10, 1988, pp. 52–54.

33. R. E. Iglesias, A. Terrés, and A. Chavarria (1980), "Disorders of the Menstrual

Cycle in Airline Stewardesses," *Aviation, Space and Environmental Medicine*, May 1980, pp. 518–520.

34. Cameron, "Effect of Flying."

35. Karen Messing et al., "Menstrual Cycle Characteristics and Working Conditions in Poultry Slaughterhouses and Canneries," *Scandinavian Journal of Work, Environment and Health*, 18 (1992), pp. 302–309.

36. This subsection draws heavily for its examples of orthodoxy on Richard R. Monson, *Occupational Epidemiology* (Boca Raton, Fla. CRC Press, 1980).

37. Karen Messing et al., "Mutant Frequency of Radiotherapy Technicians Appears to Reflect Recent Dose of Ionizing Radiation," *Health Physics*, 57 (1989), pp. 537–544.

38. Katherine Lippel, "L'Incertitude des probabilités en droit et médecine," *Revue de Droit, Université de Sherbrooke*, 22 (1992), pp. 445–472.

39. For example, Messing et al., "Mutant Frequency of Radiotherapy Technicians," and Guy Huel et al., "Méthodologie épidémiologique et statistique en pathologie industrielle," report submitted to the Institut de Recherche en Santé et en Sécurité du Travail.

40. Ana Maria Seifert et al., "HPRT-Mutant Frequency and Lymphocyte Characteristics of Workers Exposed to Ionizing Radiation on a Sporadic Basis: A Comparison of Two Exposure Indicators, Job Title and Dose, *Mutation Research*, 319 (1993), pp. 61–70.

41. Messing et al., "Menstrual Cycle Characteristics."

42. Gladys Block et al., "Cancer Morbidity and Mortality in Phosphate Workers," *Cancer Research*, 48 (1988), pp. 7298–7303. This is only one of many examples.

43. Jack Siemiatycki et al., "Cancer Risks Associated with Ten Organic Dusts: Results from a Case-Control Study in Montreal," *American Journal of Industrial Medicine*, 16 (1989), pp. 547–567.

44. J. Walrath, F. P. Li, and S. K. Hoar, "Causes of Death Among Female Chemists," *American Journal of Public Health*, 75 (1985), pp. 883–885.

45. J. J. Spinelli et al., "Multiple Myeloma Leukemia and Cancer of the Ovary in Cosmetologists and Hairdressers," *American Journal of Industrial Medicine*, 6 (1984), pp. 97–102.

46. M. G. Ott, M. J. Teta, and H. L. Greenberg, "Assessment of Exposure to Chemicals in a Complex Work Environment," *American Journal of Industrial Medicine*, 16 (1989), pp. 617–630.

47. Brabant et al., (1992).

48. R. Baker and S. Woodrow, "The Clean, Light Image of the Electronics Industry: Miracle or Mirage?" in Wendy Chavkin (ed.), *Double Exposure* (New York: Monthly Review Press, 1985).

49. Brabant, Mergler, and Messing, "Va te faire soigner."

50. J. Ladou, "Health Issues in the Microelectronics Industry," *State of the Art Reviews in Occupational Medicine*," 1 (1986), pp. 1–11.

51. Steve Fox, "Toxic Chemicals and Stress: Anatomy of an Out-of-Court Settlement for Women Workers at GTE Lenkurt Electronics Plant," Ph.D. dissertation, University of New Mexico, 1988.

52. Guy Huel, Donna Mergler, and Rosemarie Bowler, "Evidence of Adverse Reproductive Outcomes among Former Microelectronics Workers," *British Journal of Industrial Medicine*, 47 (1990), pp. 400–404.

53. Bowler et al., "Stability of Psychological Impairment," and Rosemarie Bowler et al., "Affective and Personality Disturbance among Women Former Microelectronic Workers," *Journal of Clinical Psychology*, 47 (1991), pp. 41–52. See also Donna Mergler et al., "Visual Dysfunction among Former Microelectronics Assembly Workers," *Archives of Environmental Health*, 46 (1991), pp. 326–334.

54. Brabant et al., "Va te faire soigner."

55. For example, all the studies of carpal tunnel syndrome cited by Mats Hagberg et al. control for sex; see "Impact of Occupations and Job Tasks on the Prevalence of Carpal Tunnel Syndrome," *Scandinavian Journal of Work Environment and Health*, 18 (1992), pp. 337–345, esp. table 2.

56. Donna Mergler et al., "The Weaker Sex? Men in Women's Working Conditions Report Similar Health Symptoms," *Journal of Occupational Medicine*, 29 (1987), pp. 417–421.

57. Similar errors are often made when people of different races are studied. See, for example, Nancy Krieger, "Social Class and the Black/White Crossover in the Age-Specific Incidence of Breast Cancer," *American Journal of Epidemiology*, 131 (1990) pp. 804–814, and Nancy Krieger and Diane L. Rowley, "Re: 'Race, Family Income and Low Birth Weight,' " *American Journal of Epidemiology*, 135 (1992), p. 501.

58. Enquête Santé Québec, *Et la santé: Ça va?* (Québec: Editeur Officiel, 1988).

59. Vézina et al., *Pour donner un sens au travail*.

60. F. W. Ilfeld, "Methodological Issues in Relating Psychiatric Symptoms to Social Stressors," *Psychology Reports*, 39 (1975), pp. 1252–1258.

61. The appropriate procedure is to analyze the data separately for both sexes, considering them together only if the same relationships appear to be operating in both sexes. See Margrit Eichler, "Nonsexist Research: A Metatheoretical Approach," *Indian Journal of Social Work*, 53 (1992), pp. 329–341.

62. Mel Bartley, "Unemployment and Health: Selection or Causation—a False Antithesis?" *Sociology of Health and Illness*, 10 (1988), pp. 41–67.

63. Russell Wilkins, "Health Expectancy by Local Area in Montréal: A Summary of Findings," *Canadian Journal of Public Health*, 77 (1986), pp. 216–220.

64. D. Blane, G. D. Smith, and Mel Bartley, "Social Class Differences in Years of Potential Life Lost: Size, Trends and Principal Causes," *British Medical Journal*, 301 (1990), pp. 429–432.

65. Russell Wilkins, Greg J. Sherman, and P. A. F. Best, "Findings of a New Study Relating Unfavorable Pregnancy Outcomes and Infant Mortality to Income in Canadian Urban Regions in 1986," *Health Reports*, 3 (1991), pp. 7–31 (Statistics Canada Catalogue 82–003). This study classes SES according to residence.

66. Alison D. Macdonald et al., "Occupation and Pregnancy Outcome," *British Journal of Industrial Medicine*, 44 (1987), pp. 521–526.

67. P. Rachootin and J. Olsen, "The Risk of Infertility and Delayed Conception Associated with Exposures in the Danish Workplace," *Journal of Occupational Medicine*, 25 (1983), pp. 394–402.

68. Chantal Brisson, D. Loomis, and N. Pearce, "Is Social Class Standardization Appropriate in Occupational Studies?" *Journal of Epidemiology and Community Health*, 41 (1987), pp. 290–294.

69. Jack Siemiatycki et al., "Degree of Confounding Bias Related to Smoking, Ethnic Group and Socioeconomic Status in Estimates of the Associations between Occupation and Cancer," *Journal of Occupational Medicine*, 30 (1988), pp. 617–625.

70. B. R. Blishen, W. K. Carroll, and C. Moore, "The 1981 Socioeconomic Index for Occupations in Canada," *Revue Canadienne de Sociologie et Anthropologie*, 24 (1987), pp. 465–488.

71. Mohammed Hsairi et al., "Personal Factors Related to the Perception of Occupational Exposure: An Application of a Job Exposure Matrix," *International Epidemiology Association Journal*, 21 (1992), pp. 972–980.

72. Michael Joffe, "Male- and Female-Mediated Reproductive Effects on Occupation: The Use of Questionnaire Methods," *Journal of Occupational Medicine*, 31 (1989), pp. 974–979; Michael Joffe, "Validity of Exposure Data Derived from Interviews with Workers," *Proceedings of the 23d International Congress on Occupational Health* Montréal, 1990, p. 61;

and Susan MacKenzie and Abby Lippman, "An Investigation of Report Bias in a Case-Control Study of Pregnancy Outcome," *American Journal of Epidemiology* 129 (1989), pp. 65–75.

73. Irving J. Selikoff, "Death Certificates in Epidemiological Studies Including Occupational Hazards: Inaccuracies in Occupational Categories," *American Journal of Industrial Medicine,* 22 (1992), pp. 493–504. Unfortunately this study appears to have included only white males; although this is not clearly stated, it is implied by the fact that death rates are compared with those of white males.

74. Elizabeth G. Marxhall et al., "Comparison of Mother's Occupation and Industry from the Birth Certificate and a Self-Administered Questionnaire," *Journal of Occupational Medicine,* 34 (1992), pp. 1090–1096.

75. N.J. Birkett, "Effect of Nondifferential Misclassification on Estimates of Odds Ratios with Multiple Levels of Exposure," *American Journal of Epidemiology,* 136 (1992), pp. 356–362.

76. H. L. Needleman, "What Can the Study of Lead Teach Us about Other Toxicants?" *Environmental Health Perspectives,* 86 (1990), pp. 183–189.

77. Karen Messing, "Environment et santé: La santé au travail et le choix des scientifiques," in *L'Avenir d'un monde fini: Jalons pour une éthique du développement durable,* Cahiers de Recherche Ethique, no. 15 (Montréal: Editions Fides, 1991), pp. 107–110.

78. Not only pregnant women are affected; any powerless group may have the burden of proof placed on it without explicit justification.

79. The second-most frequent refers to our feminism. A critique of one of our articles on feminist perspectives in occupational health began, "This emotional article . . . " We think the emotion was the reviewer's!

80. Michel Rossignol, Samy Suissa, and Lucien Abenahaim, "The Evolution of Compensated Occupational Spinal Injuries," *Spine,* 17 (1992), pp. 1043–1047.

81. The guesswork has now been taken out of the task of evaluating candidates' curricula vitae by the use of impact factors for the journals in which a candidate has published; the impact factor is calculated as the average number of times an article in the given journal is cited.

82. J. L. Weaver and Sharon D. Garrett, "Sexism and Racism in the American Health Care Industry: A Comparative Analysis," in Elizabeth Fee, *Women and Health* (Farmingdale, N.Y: Baywood, 1983), pp. 79–104.

83. Claudia Sanmartin and Lisa Snidel, "Profile of Canadian Physicians: Results of the 1990 Physician Resource Questionnaire," *Canadian Medical Association Journal,* 149 (1993), pp. 977–984.

84. Anne Innis Dagg, "Women in Science—Are Conditions Improving?" in Marianne G. Ainley, *Despite the Odds: Essays on Canadian Women and Science* (Montréal: Vehicule Press, 1990), pp. 337–347.

85. Karen Messing, "Sois mâle et tais-toi: L'Excellence et les chercheures universitaires," *Women's/Education/des femmes,* 9 (Autumn 1991), pp. 49–51.

86. Confédération des Syndicats Nationaux, a Québec union with 200,000 members, about half of whom are women.

87. Fédération des Travailleuses et Travailleurs du Québec, a Québec union with 350,000 members, about 30 percent of whom are women.

88. The Centrale de l'Enseignement du Québec (CEQ) includes all of Québec's primary and secondary school teachers, as well as some junior college and university lecturers and professors and some support staff. It has a large majority of women members.

89. Comité Conjoint UQAM-CSN-FTQ, *Le Protocole d'entente UQAM-CSN-FTQ: Sur la formation syndicale,* 1977 Services à la Collectivité, Université du Québec à Montréal, CP 8888, Succ. A, Montréal, Québec H3C 3P8. Comité Conjoint UQAM-CSN-FTQ, *Le Protocole UQAM-CSN-FTQ: 1976–1986, Bilan et perspectives,* 1988, Services à la Collectivité, Univer-

sité du Québec à Montréal, CP 8888, Succ. A, Montréal, Québec H3C 3P8. 1988. Karen Messing, "Putting our Two Heads Together: A Mainly Women's Research Group Looks at Women's Occupational Health," in J. Wine and J. Ristock, *Feminist Activism in Canada: Bridging Academe and the Community* (Toronto: James Lorrimer Press, 1991); reprinted in *National Women's Studies Association Journal*, 3 (1991), pp. 355–367.

90. The university has an analogous agreement with women's groups, called Relais-femmes, about which information can be found in two reports: Université du Québec à Montréal, *Le Protocole UQAM-Relais-femmes*, 1982, Services à la Collectivité, Université du Québec à Montréal, CP 8888, Succ. A, Montréal, Québec H3C 3P8, and Marie-Hélène Côté, *Bilan des activités 1987–88 et perspectives pour la prochaine année*, 1988, Services à la Collectivité, Université du Québec à Montréal, CP 8888, Succ. A, Montréal, Québec H3C 3P8. In this context, we have furnished expertise to the local women's health center (Centre de Santé des Femmes) and to groups involved with employment access, such as Action-Travail des Femmes. We also work with predominantly male unions. These experiences are outside the scope of this chapter, and are described in two papers: Donna Mergler, "Worker Participation in Occupational Health Research: Theory and Practice," *International Journal of Health Services*, 17 (1987), pp. 151–167, and Karen Messing, "Union-Initiated Research on Genetic Effects of Workplace Agents," *Alternatives: Perspectives on Technology, Environment and Society*, 15 (1987), pp. 15–18.

91. Doctors testified otherwise, but the workers won their case.

3 | Democratizing Biology
Reinventing Biology from a Feminist, Ecological, and Third World Perspective

Vandana Shiva

Why Biology Needs to Be Democratized

THE DOMINANT PARADIGM of biology is in urgent need of reinvention and democratization because it is inherently undemocratic. There are three aspects to this. In the first place, it is socially undemocratic. The dominant paradigm casts patterns of human social behavior as biologically determined, thus ignoring the ways in which they are in reality outcomes of social prejudice and bias on the basis of race, class, and gender. Notions that human behavior is the product of biological determinism make such prejudice immune to democratic questioning and transformation, thus perpetuating social and economic inequalities.

Second, biology is the basis of all food production systems that are intimately linked to survival and to women's work and knowledge in the Third World. All of these are threatened by reductionist biology, particularly in the form of genetic engineering. The reinvention of biology is therefore also an economic imperative so that alternative production systems which are more socially just and ecologically sustainable have a chance to flourish.

Third, the dominant approach to studying biology is undemocratic with respect to nonhuman species. It is based on the metaphor of "man's empire over inferior creatures" rather than the metaphor of "the democracy of all life." It therefore contributes to a "war against species," manipulating them without limits if they are economically useful and pushing them to extinction if they are not. It also erodes biodiversity, leaving us materially impoverished. Last but not least, intellectual property rights in the area of life forms that are constructed on the basis of a reductionist biology deepen the exclusion of other knowledge systems.

The democratization of biology requires that culturally and socially determined behavior and characteristics are removed from the domain of biological determinism. This in turn makes it an imperative that excluded groups such as women and Third World communities have a role in reinventing biology on democratic principles. Democratizing biology from the perspective of the democracy of all life requires that we reinvent biology to account better for the intrinsic worth and self-organizing capacities of all living organisms.

The democratization of biology involves the recovery of pluralism of knowledge traditions, both within the modern Western traditions and within ancient, time-tested, non-Western traditions of agriculture and health care. Democratizing biology requires that the latest application of biology, genetic engineering, be evaluated in the context of alternatives; it also needs to be evaluated in the context of empirical evidence that is undermining the basic tenets of genetic determinism and the assumption of immutable, unchanging genes. Democratizing biology, in addition, involves coherence and honesty both in the "owning" of benefits and the "owning" of risks of genetic engineering. There is therefore a need to have a coherent theory of "novelty" in the areas of intellectual property rights (which deal with ownership of benefits) and biosafety (which deals with the "owning" of risks and hazards).

These issues of the democratization of biology go to the very heart of democracy in the twenty-first century.

The Undemocratic Paradigm of Biological Determinism

The dominant paradigm of biology has been an imperialist one. Biological difference between human and nonhuman species, between white and colored peoples, and between men and women has been seen as reason and justification for the rule of the white man over nature, women, and all nonwhite races. Not only is the dominant biology based on these three exclusions; the exclusions themselves are interwoven and interlinked. And the exclusions are shaped by and in turn shape the mode of knowing and thinking about the world. It is these multiple and complex relationships between science, gender, and ecological survival that I want to explore.

Feminist critiques of dominant science emerged from analysis by feminist scientists, which showed how biology as a science was constructed on the basis of the gender biases of patriarchy, and these biased social constructions were then used to justify women's continued oppression.[1]

Central to the patriarchal construction of biology was the association of activity and creativity with the male and passivity with the female. Furthermore, this construction was based on an artificial mind–body dichotomy, a dichotomy which was also seen as gendered. These distortions have led to the equation of "biology" with lack of mind and intelligence, and intelligence as something outside biological organisms. The world was thus split into "thinking humans" and "vegetating other species." The relationship of domination of humans over other species and of white men over other humans is based on and justified through the myth of the disembodied mind.

The Enlightenment was also a period in which human beings became seen as separate from other species. The ecological separation from the earth body went hand in hand with the epistemological separation from the human body. The human body was treated as "nature"; the disembodied mind alone was a truly human faculty. The mind–body split mirrored the culture–nature dichotomy in the Cartesian project of creating an "objective" science. As Susan Bordo has explained in "The Cartesian Masculinization of Thought," Descartes split the world into two ontological orders based on the mutual exclusion of *res cogitans* and *res extensa,* and it made possible the conceptualization of complete intellectual transcendence of the body, which was viewed as a source of deceptive senses and distracting commotion. This also dislocated and excluded the natural world from the realm of the human. Nature for Descartes is pure *res extensa,* totally devoid of mind and thought.[2]

This is in total contrast to other cultures of knowledge which do not split the mind from the body. A view which is simultaneously an ecological, feminist, and Third World perspective is beautifully captured in a poem by José María Marguedas titled "A Call to Certain Academics:"

> *They say that some learned men*
>   *are saying this about us*
> *These academics who reproduce*
>   *themselves*
> *In our own lives*
> *What is there on the banks of*
>   *these rivers, Doctor?*
> *Take out your binoculars*
> *And your spectacles*
> *Look if you can.*
> *Five hundred flowers*
> *From five hundred different types*
>   *of potato*
> *Grow on the terraces*
> *Above abysses*
> *That your eyes don't reach*
> *Those five hundred flowers*
> *Are my brain*
> *My flesh.*

The Western patriarchal conceptualization of biology was based on robbing intelligence from organisms if they were female or nonhuman. Intelligence and "higher" faculties, for the production of "culture" and "science," were kept as a monopoly of men, particularly European men. By being robbed of their intelligence and minds, women, like Third World peoples, have been treated as not fully human. They have in fact been seen as closer to nature while godlike men have the exclusive capacity for creation of culture.

The treatment of other species mirrors the treatment meted out to the "others" of the human species. Furthermore, in artificially constructing "nature" out of these excluded others, the privileged knower is also artificially constructed as the creator of "culture and science," produced by a disembodied mind. Out of these multiple exclusions mutually exclusive dualities are carved:

Nature is opposed to culture
Human is opposed to nature
Human is opposed to animal
Man is opposed to woman
Mind is opposed to Body
Science is opposed to superstitions.

In most non-Western cultures, all species have been seen as part of an earth family. We have called it Vasudhaiva Kutumbkam in India. For native Americans, "kinship with all creatures of the earth, sky and water was a real and active principle."[3] As Carrie Dann stated at the Right Livelihood award ceremony, Earth has many children and humans are only one of them.[4] These non-Western perceptions have been viewed as a block in establishing "man's empire" over other species in the interactions of European men with other cultures in the process of colonization.

Robert Boyle, the famous scientist who was also governor of the New England Company, saw the rise of mechanical philosophy as an instrument of power, not just over nature but also over the original inhabitants of America. He explicitly declared his intention of ridding the New England Indians of their ridiculous notions about the workings of nature. He argued that "the veneration, wherewith men are imbued for what they call nature, has been a discouraging impediment to the empire of man over the inferior creatures of God."[5]

Modern Western science and technology have in fact been the continuation, not a break, with the Judeo-Christian myth of creation, according to which all species were made for man's use. The world view within which Western science is practiced is based on imperialism, not democracy. As Lyn White, Jr., has stated, "by gradual stages a loving and all-powerful God had created light and darkness, the heavenly bodies, the earth and all its plants, animals, birds, and fishes. Finally, God had created Adam and, as an after-thought, Eve to keep man from being lonely. Man named all the animals, thus establishing his dominance over them. God planned all of this explicitly for man's benefit and rule: no item in the physical creation had any purpose save to serve man's purposes."[6]

In Genesis, Eve (woman) created sin by forcing Adam (man) to eat the forbidden fruit. This story of "original sin" still underlies the dominant paradigm of biology. It is, for example, the metaphor in such popularizations of modern biology as Matt Ridley's *Red Queen*[7] and Richard Dawkins's *Selfish Gene*[8] and *The Blind Watchmaker.*[9]

The exclusion of all women and of men from non-Western cultures from the category "human" continues as biological determinism is articulated at newer and deeper levels. Non-Western men and all women are the "others" among the human species who have been excluded from the human family, just as the non-human species have been excluded from the earth family. "Western man's" relationship with nonhuman species mirrors his relationships with those who are cast as "others" of his own species.

Furthermore, the assumption of superiority of white men over other humans and other species has justified violence to and extinction of nonhuman species and non-Western cultures. Violence in human society has often been justified on the grounds that some humans are closer to "nature" and other species, and hence not fully human.

Biology and the Third World

The Western world view has sanctified European man as being made in the image and likeness of God. The Christian theology of man as the master of all species justified the domination over, and even the decimation of, all life that happened to be nonwhite and nonmale. St. Augustine said that "man but *not* woman was made in the image and likeness of God." And all non-Europeans were, of course, not in God's likeness because they were colored and in every way so different from the colonizers. Europeans had either to conceive of the naturalness of cultural diversity and invent cultural tolerance to go with it or to assume, given Christian dualism, that non-Europeans, being different, were not in God's image and therefore were "in league with Hell." Most Europeans made the latter choice. As Crosby observes, "Again and again, during the centuries of European imperialism, the Christian view that all men are brothers was to lead to persecution of non-Europeans—he who is my brother sins to the extent that he is unlike me."[10]

Whenever Europeans "discovered" the native people of America, Africa, or Asia, they projected upon them the identity of savages who needed redemption by a superior race. Even slavery was justified on these cosmological grounds. It was considered wise that Africans should be carried into slavery, since they were carried, at the same time, out of an "endless night of savage barbarism" into the embrace of a "superior civilization."[11] All brutality was moralized on the basis of this assumed superiority and the exclusive status of European men as fully human.

The decimation of original peoples everywhere was justified morally on the grounds that indigenous people were not really human. They were part of the fauna. As John Pilger has observed, the *Encyclopaedia Britannica* appeared to be in no doubt about this in the context of Australia, stating that "man in Australia is an animal of prey. More ferocious than the lynx, the leopard, or the hyena, he devours his own people." In another Australian textbook, *Triumph in the Tropics*, Australian aborigines were equated with their half-wild dogs. Being animals, the original Australians and Americans, the Africans and Asians, possessed no rights as humans. They could therefore be ignored as people and exterminated. Their lands could be usurped as "Terra Nullius"—lands empty of people, "vacant," "waste," "unused."[12] European men were thus able to see their invasions as "discovery," piracy and theft as "trade," and extermination and enslavement as their "civilizing mission."

Since Columbus arrived in North America, the indigenous populations continue to be decimated, largely on the grounds of their not being treated as fully human. As Carrie Dann observes, "Since that time, the indigenous people of the Western Hemisphere have been described as savages, heathens, infidels and basically labelled as one of our relatives, the four legged, the wolf or coyote."[13]

The original inhabitants, who had populated the land for thousands of years, were decimated after the arrival of Europeans in America. From a total population of seventy-two million in 1492, their population declined to about four million in a few centuries. Their land was conquered and colonized, their resources raped and destroyed. The world view of man's empire over lesser creatures rendered each act of invasion into other people's land as "discovery": in America in 1492, in Australia in 1788, in Africa and Asia through the past five hundred years of colonization.

In Australia, the aboriginal population numbered 750,000 in 1788. In the years following the European invasion, 600,000 of them had died. In Africa, an official of the Belgian Commission, reporting in 1919, reached the conclusion that the population of the Belgian Congo has been reduced by half since the beginning of the European occupation in the 1880s.

In Africa, depopulation was an obvious effect of the slave trade, which involved the deportation and death of several tens of millions of Africans over three or four centuries. The rate of mortality among slaves taken from Africa to the Americas was so high that whole "slave populations" had to be replaced every few years. The long-sustained destruction of African people is set forth in the German imperial records. "I know these African tribes," wrote Von Trotha, the general entrusted with the task of putting down the Herero and Nama in South Africa. "They are all the same. They respect nothing but force. To exercise this force with brute terror and even with ferocity was and is my policy. I wipe out rebellious tribes with streams of blood and streams of money. Only by sowing in this way can anything new be grown, anything that is stable."[14]

In 1904 it was estimated that there were 80,000 Herero and 20,000 Nama. In 1911, the estimates showed 15,130 Herero and 9,871 Nama remaining alive. Nearly 75,000 of them had paid the price for a "new" order based on the domination of the Europeans, who regarded themselves as naturally superior to the Africans and could use the most brutal forms of oppression and exploitation for their civilizing mission.

Western science continues to turn non-Western peoples into less than human objects of scientific enquiry. For example, U.S. scientists have started a Human Genome Diversity Project to collect human DNA samples from indigenous communities around the world. The economic opportunity to collect and the push to preserve human genetic diversity has been fired by the development of new biotechnologies and the formation of the Human Genome Organization (abbreviated HUGO). Medical science has long been aware that there is not just one human genetic map. Each ethnic community may have a slightly different genetic composition. Some of the differences and mutations could someday prove to be invaluable to medicine.

Officials of the Human Genome Diversity Project estimate that an initial five-year sweep of relatively accessible populations will cost $35 million and will allow sampling from 10,000 to 15,000 human specimens. At an average total cost of $2,300 per sample, the project will spend more money gathering the blood of indigenous peoples than the per capita GNP of any of the world's poorest 110 countries.

White blood cells from each person will be preserved in vitro at the American Type Culture Collections in Rockville, Maryland. Human tissue (scraped from the cheek) and hair root sampling will be used in shorter-term studies. The project's leaders, concerned that human blood can survive only forty-eight hours outside of storage, are planning their collections carefully. "One person can bleed fifty people and get to the airport in one day," they calculate.

In the draft report of the Human Genome Diversity Project, preservation is the dominant theme, and there is an assumption that many or most of the human populations are inevitably going to disappear. The project's emphasis on preservation and its insensitivity to indigenous peoples is best exhibited by the term used to describe indigenous communities that have been targeted for human DNA sampling: "isolates of historic interest" (IHIs).

Sometimes the interest is more than historical. It is also directly economic, as illustrated in the case of the patent claim WO 9208784 A1 lodged by the U.S. secretary of commerce for the human T-lymphotropic virus type 2, drawn from the "immortalized" DNA of a twenty-six-year-old Guami Indian woman from Panama. The original blood sample is cryogenically preserved at the American

Type Culture Collection in Rockville, Maryland. Under citizen pressure, the secretary of commerce was forced to withdraw the patent claim. However, no ethical or legal framework exists to prevent such patenting in the future. Meanwhile, similar patents have been claimed for indigenous communities from Papua New Guinea and the Solomon Islands.

The use of other peoples as raw material is one of the aspects of an imperialistic science of biology. Other aspects include the targeting of Third World women for population control and the treatment of Third World biodiversity and Third World biological knowledge as raw material for the economic empires of northern corporations in the age of biology. These issues, related to "intellectual property rights" and in the domain of biology, are dealt with later in further detail.

Women's Indigenous Knowledge and the Conservation of Biodiversity

The links between gender and diversity are many. The construction of women as the "second sex" is linked to the same inability to cope with difference as the development paradigm that leads to the displacement and extinction of diversity in the biological world. In the patriarchal world view, man is the measure of all value and there is no room for diversity—only hierarchy. Woman, being different, is treated as unequal and inferior. Nature's diversity is not seen as intrinsically valuable in itself. It gets value only through economic exploitation for commercial gain. Within a commercial value framework, diversity thus is seen as a problem, a deficiency. Destruction of diversity and the creation of monocultures becomes an imperative for capitalist patriarchy.

The marginalization of women and the destruction of biodiversity go hand in hand. Diversity is the price paid in the patriarchal model of progress which pushes inexorably toward monocultures, uniformity, and homogeneity. In the perverted logic of progress, even conservation suffers. Agricultural "development" continues to work toward the erasure of diversity, while the same global interests that destroy biodiversity urge the Third World to conserve it. This separation of production and consumption, with "production" being based on uniformity and "conservation" desperately attempting to preserve diversity, guarantees that biodiversity will not be protected. It can only be protected by making diversity the basis, the foundation, the logic, of the technology and economics of production.

The logic of diversity is best derived from biodiversity and from women's links to it. It helps look at dominant structures from below, from the ground of diversity. From this ground, monocultures are not productive, they are unproductive; and the knowledge that produces monocultures is not sophisticated, it is primitive.

Diversity is, in many ways, the basis of women's politics and the politics of ecology. Gender politics is, to a large extent, a politics of difference. Ecopolitics too arises from the fact that nature is varied and different, while industrial commodities and processes are uniform and homogeneous.

The two politics of diversity converge in a significant way when women and biodiversity meet in fields and forests, in arid regions and wetlands. Diversity is the principle of women's work and knowledge. It is the reason that women's knowledge and work have been discounted in the patriarchal calculus. Yet it is also the matrix from which an alternative calculus of "productivity" and "skills" can be built, one that respects diversity instead of destroying it.

In Third World economies, many communities depend on biological resources for their sustenance and well-being. In these societies, biodiversity is simultaneously a means of production and an object of consumption. It is the survival base that has to be conserved. Sustainability of livelihoods is ultimately connected to the conservation and sustainable use of biological resources in all their diversity.

However, biodiversity-based technologies of tribal and peasant societies have been viewed as backward and primitive and have been displaced by technologies which use biological resources in such a way that they destroy diversity and people's livelihoods. There is a general misconception that diversity-based production systems are low-productivity systems. However, the high productivity of uniform and homogeneous systems is a contextual and theoretically constructed category, based on taking only one-dimensional yield and output into account. The low productivity of diverse, multiply dimensional systems and the high productivity of uniform, one-dimensional systems of agriculture, forestry, and livestock are therefore not based on a neutral, scientific measure, but are biased toward the commercial interests for whom maximizing of one-dimensional output is an economic imperative.

This push toward uniformity, however, undermines the diversity of biological systems which form the production system. It also undermines the livelihoods of the people whose work is associated with diverse and multiple-use systems of forestry, agriculture, and animal husbandry.

As an example, in the state of Kerala, which derives its name from the coconut palm, coconut is cultivated in a multistoried, high-intensity cropping system along with betel and pepper vines, bananas, tapioca, drukstick, papaya, jackfruit, mango, and vegetables. Compared to an annual labor requirement of 157 person days per year in a monoculture of coconut palm, the mixed cropping system increases employment to 960 person days per year. In the dry-land farming systems of the Deccan, the shift from mixed cropping of millets with pulses and oilseeds to eucalyptus monocultures has led to a loss of employment of 250 person days per year.

When labor is scarce and costly, labor-displacing technologies are productive and efficient. When labor is abundant, labor displacement is unproductive because it leads to poverty, dispossession, and destruction of livelihoods. In Third World situations, sustainability has therefore to be achieved at two levels simultaneously—natural resources and livelihoods. Biodiversity conservation has to be linked to the conservation of livelihoods derived from biodiversity.

In India, agriculture employs 70 percent of the working population and about 84 percent of all economically active women.[15] For example, in the tribal economy of Orissa—shifting cultivation (*bogodo*)—women spend 105.4 days per year on agricultural operations compared to men's 59.11 days.[16] According to Vir Singh's assessment, in the Indian Himalaya a pair of bullocks work for 1,064 hours, a man for 1,212 hours, and a woman for 3,418 hours a year on a one-hectare farm. Thus a woman works longer than men and farm animals combined![17]

K. Saradamoni's study of women agricultural laborers and cultivators in three rice growing states—Kerala, Tamilnadu, and West Bengal—shows that both groups of women make crucial contributions to production and processing.[18] Joan Mencher's studies in the Palghat region of Kerala reveal that apart from ploughing, which is exclusively men's work, women have a predominant role in all other processes. On the basis of this study, it is estimated that more than two-thirds of the labor input is female.[19] Bhati and Singh, in a study of the gender division of labor in hill agriculture in Himachal Pradesh, show that, overall, women contribute 61 percent of the total labor on farms.[20] A detailed study by Jain and Chand in three villages each in Rajasthan and West Bengal, covering 127 households over twelve months, highlights the fact that women in the age group nineteen to seventy spend longer hours than do men in a variety of activities.[21]

Women's work and livelihoods in subsistence agriculture are based on multiple use and management of biomass for fodder, fertilizer, food, and fuel. The collection of fodder from the forest is part of the process of transferring fertility for crop production and managing soil and water stability. The work of the women engaged in such activity tends to be discounted and made invisible for all sectors.[22] When these allied activities which are ecologically and economically critical are taken into account, agriculture is revealed as the major occupation of "working" women in rural India. The majority of women in India are not simply "housewives" but farmers.[23]

Women's work and knowledge is central to biodiversity conservation and utilization because women work between "sectors" and perform multiple tasks. Women have remained invisible as farmers in spite of their contribution to farming, as people fail to see their work in agriculture. Their production tends not to be recorded by economists as "work" or as "production" because it falls outside the so-called production boundary. These problems of data collection on agricultural work arise not because too few women work but because too many women have to do too much work. There is a conceptual inability of statisticians and researchers to define women's work inside and outside the house (and farming is usually part of both). This recognition of what is and is not labor is exacerbated by the great volume of work that women do. It is also related to the fact that although women work to sustain their families and communities, most of their work is not measured in wages.

Women's work is also invisible because women are concentrated outside market-related or remunerated work and are normally engaged in multiple tasks. Time allocation studies, which do not depend on an a priori definition of work, reflect more closely the multiplicity of tasks undertaken and the seasonal, even daily, movement in and out of the conventional labor force which characterize the livelihood strategy for most rural women. Studies with a gender perspective which are now being published prove that women in India are major producers of food in terms of value, volume, and hours worked.

In the production and preparation of plant foods, women need skills and knowledge. To prepare seeds they need to know about germination requirements, seed preparation, and soil choice. Seed preparation requires visual discrimination, fine motor coordination, and sensitivity to humidity levels and weather conditions. To sow and strike seeds one needs to know about seasons, climate, plant requirements, weather conditions, microclimatic factors, soil enrichment; sowing seeds requires physical dexterity and strength. To care for plants properly, one needs information about the nature of plant diseases, pruning, staking, water supplies, companion planting, predators, sequences, growing seasons, and soil maintenance. Plant propagation also requires persistence and patience, physical strength, and attention to plant needs. Harvesting a crop requires judgments in relation to weather, labor, and grading and knowledge about preserving, immediate use, and propagation.

Women's knowledge has been the mainstay of the indigenous dairy industry. Dairying, as managed by women in rural India, embodies practices and logic rather different from those contained in the dairy science imparted at institutions of formal education in India, since the latter is essentially an import from Europe and North America. Women have been experts in the breeding and feeding of farm animals, which include not just cows and buffaloes but also pigs, chickens, ducks, and goats.

In forestry too, women's knowledge is crucial to the use of biomass for feed and fertilizer. Knowledge of the feed value of different fodder species, of the fuel value of firewood types, food products, and species, is essential to agriculture-related forestry in which women are predominantly active. In low-input agriculture, fertility is transferred from the forest and farm trees to the field by women's work, either directly or via animals.

It is in the "in between" spaces, the interstices of "sectors," the invisible ecological flows between sectors, that women's work and knowledge in agriculture are uniquely found, and it is through these linkages that ecological stability and sustainability and productivity under resource-scarce conditions are maintained. The invisibility of women's work and knowledge arises from the gender bias which has a blind spot for realistic assessment of women's contributions. It is also rooted in the sectoral, fragmented, and reductionist approach to development, which treats forests, livestock, and crops as independent of each other.

The focus of the "green revolution" has been to increase grain yields of rice and wheat by techniques such as dwarfing, monocultures, and multicropping. For an Indian woman farmer, rice is not only food; it is also a source of fodder for cattle and straw for thatch. High-yield varieties (HYVs) can increase women's work. The destruction of biological diversity undermines women's diverse contributions to agriculture by eroding biological sources of food, fodder, fertilizer, fuel, and fiber. The shift from local varieties and local indigenous crop-improvement strategies can also take away women's control over seeds and genetic resources. Women have been the seed custodians since time immemorial, and it is their knowledge and skills which should be the basis of all crop improvement strategies.

Women have been the custodians of biodiversity in most cultures. They have been selectors and preservers of seed. However, like all other aspects of women's work and knowledge, their role in development and conservation of biodiversity has been represented as nonwork and nonknowledge. Their labor and expertise has been defined into nature, even though it is based on sophisticated cultural and scientific practices. Women produce, reproduce, consume, and conserve biodiversity in agriculture. Women's role in the conservation of biodiversity, however, differs from the dominant patriarchal notions of biodiversity conservation in a number of ways.

The recent concern with biodiversity at the global level has grown as a result of the erosion of diversity due to the expansion of large-scale monoculture-based production in agriculture and the vulnerability associated with it. However, the fragmentation of farming systems which was linked to the spread of monocultures continues to be the guiding paradigm for biodiversity conservation. Each element of the farm ecosystem is viewed in isolation, and conservation of diversity is seen as an arithmetic exercise of collecting variety.

In contrast, biodiversity in the traditional Indian setting is a relational category in which each element gets its characteristics and value through its relationships with other elements. Biodiversity is ecologically and culturally embedded. Diversity is reproduced and conserved through the reproduction and conservation of culture, in festivals and rituals. Besides being a celebration of the renewal of life, these festivals are the platform for carrying out subtle tests for seed selection and propagation. These tests are not treated as scientific by the dominant world view because they are not embedded in the culture of the lab and the experimental plot; they are carried out not by men in lab coats but by village women. Yet they are reliable and systematic, and they help to preserve a rich biological diversity in agriculture.

Women have been the selectors and custodians of seed. When they conserve seed, they conserve diversity; and when they conserve diversity, they conserve a balance and harmony. *Navdanya*, or nine seeds, are the symbol of this renewal of diversity and balance, not just of the plant world but also of the social world. It is this complex, relational web which gives meaning to biodiversity in Indian culture and has been the basis of the conservation of that diversity over millennia.

Women's work in organic agriculture also supports the work of decomposers and soil builders which inhabit the soil. It is based on partnership with other species. Organic manure is food for the community of living beings which depend on the soil. Soils treated with farmyard manure have from two to two-and-a-half times as many earthworms as untreated soils. Farmyard manure encourages the buildup of earthworms through increasing their food supply, whether they feed directly on it or on the microorganisms it supports. Earthworms contribute to soil fertility by maintaining soil structure, aeration, and drainage and by breaking down organic matter and incorporating it into the soil. The work of earthworms in soil formation was Darwin's major concern in later years. When finishing his book on earthworms he wrote: "It may be doubted whether there are many other animals which played so important a part in the history of creatures."[24]

The little earthworm working invisibly in the soil is actually the tractor and fertilizer factory and dam combined. Worm-worked soils are more water stable than unworked soils, and worm-inhabited soils have considerably more organic carbon and nitrogen than parent soils. By their continuous movement through soils, earthworms make channels which help in soil aeration. It is estimated that they increase soil-air volume by up to 30 percent. Soils with earthworms drain four to ten times faster than soils without earthworms, and their water-holding capacity is higher by 20 percent. Earthworm casts, which can be 4.36 tons dry weight per acre per year, contain more nutritive materials containing carbon, nitrogen, calcium, magnesium, potassium, sodium, and phosphorus than the parent soil. Earthworms' work on the soil promotes microbial activity which is essential to the fertility of most soils. Yet the earthworm was never seen as a worker in "scientific" agriculture.[25] The woman peasant who works invisibly with the earthworm in building soil fertility has also not been seen as doing "productive" work or providing an "input" to the food economy. We need to look beyond the mentality that tells us that fertility is "bought" from fertilizer companies; we need to look beyond the fertilizer factory for maintaining soil fertility; and we need to recover the work of women and peasants who work with nature, not against it. In regions of India which have not yet been colonized by the green revolution, women peasants continue to work as soil builders rather than soil predators, and it is from these remaining pockets of natural farming that the ecological struggles to protect nature are emerging.

However, these sophisticated systems of agriculture which used biological diversity to provide internal inputs for pest control, fertility renewal, and soil and water conservation were displaced by the chemical-industrial model from the West, financed by aid and pushed by planning from international agencies like the World Bank under the label of the green revolution.

At the technological level, the instrumental and functionalist approach to nonhuman species tends to lead to the extinction of those species which capitalist patriarchy does not value. Distortion and mutilation in the name of "improvement" is likely to be the fate of those species that are found useful. In either case, entire communities of species become victims of imperialism. In this way, the so-called green revolution led to the displacement of thousands of varieties of crops and seeds. Wheat, maize, and rice were treated as the only crops of "value." To increase their commodity value, these crops were engineered to become dwarf varieties so that they could take up more chemical fertilizers. The engineered crops were vulnerable to pests and disease, so they needed more pesticides and fungicides.

The new plant biotechnologies will follow the path of the earlier HYVs of the green revolution in pushing farmers onto a technological treadmill. Biotechnology can be expected to increase the reliance of farmers on purchased inputs even as it accelerated the process of polarization. It will even increase the use of chemicals instead of decreasing it. The dominant focus of research in genetic engineering is not on fertilizer-free and pest-free crops but on pesticide- and herbicide-resistant varieties. For the seed and chemical multinational companies, this might make commercial sense, since it is cheaper to adapt the plant to the chemical than to adapt the chemical to the plant. The cost of developing a new crop variety rarely reaches $2 million, whereas the cost of a new herbicide exceeds $40 million.[26]

Like green revolution technologies, biotechnology in agriculture can become an instrument for dispossessing the farmer of seed as a means of production. The relocation of seed production from the farm to the corporate laboratory relocates power and value between the North and South and between corporations and farmers. It is estimated that the elimination of homegrown seed would dramatically increase the farmers' dependence on biotechnology industries by about $6 billion annually.[27]

It can also become an instrument of dispossession by selectively removing those plants or parts of plants that do not serve commercial interests but are essential for the survival of nature and people. "Improvement" of a selected characteristic in a plant also constitutes a selection against other characteristics which are useful to nature, or for local consumption. "Improvement" is not a class- or gender-neutral concept. Improvement of partitioning efficiency is based on the enhancement of the yield of the desired product at the expense of unwanted plant parts. The desired product is, however, not the same for rich people and poor people, or rich countries and poor countries; nor is efficiency. On the input side, richer people and richer countries are short of labor and poorer people and poorer countries are short of capital and land. Most agricultural development, however, increases capital input while displacing labor, thus destroying livelihoods. On the output side, which parts of a farming system or a plant will be treated as "unwanted" depends on what class and gender one is. What is unwanted for the better off may be the wanted part for the poor. The plants or plant parts which serve the poor are usually the ones whose supply is squeezed by the normal priorities of improvement in response to commercial forces.

The destruction of people's livelihood and sustenance goes hand in hand with the erosion of biological resources and their capacity to fulfill diverse human needs while regenerating and renewing themselves. Attempts to increase commodity flows in one direction generate multiple levels of scarcities in related outputs. Increase of grain leads to decrease of fodder and fertilizer. Increase of cereals leads to decrease of pulses and oilseeds. The increase is measured. The decrease goes unnoticed, except by those who are deprived by virtue of the creation of new scarcity. Both people and nature are impoverished; their needs are no longer met by the one-dimensional production systems, which replace biologically rich and diverse ecosystems and put added burdens on remaining pockets of biodiversity that could satisfy these needs.

The extinction of people's livelihoods and sustenance is closely connected with the erosion of biodiversity. Protection of biodiversity can only be ensured by regenerating diversity as a basis of production in agriculture, forestry, and animal husbandry. The practice of diversity is the key to its conservation.

There are two conflicting paradigms on biodiversity, and from them emerge different paradigms of biology. The first paradigm is held by communities whose survival and sustenance is linked to local biodiversity utilization and conservation. The second is held by commercial interests whose profits are linked to utilization of global biodiversity for production of inputs into large-scale homogeneous, uniform, centralized, and global production systems. For local indigenous communities, conserving biodiversity means to conserve the integrity of ecosystems and species, the rights to their resources and knowledge, and their production systems based on biodiversity. For commercial interests, such as pharmaceutical and agricultural biotechnology companies, biodiversity in itself has no value. It is merely "raw material" to provide "components" for the genetic engineering industry. This leads to a reductionist paradigm of biology and production based on biodiversity destruction, since local production systems based on diversity are displaced by production based on uniformity. The reductionist paradigm of biology leads to the paradigm of genetic engineering for production and reproduction. It is also closely associated with the treatment of living organisms as manufactured commodities, as "products of the mind," needing "intellectual property protection."

Intellectual Property Rights and Intellectual Imperialism

Even as feminists, environmentalists, and Third World scientists reshape our ideas of knowledge, the Cartesian and Baconian project of the disembodied mind as the model knower and control and domination as the goal of knowledge continues. Intellectual property rights (IPR) as related to biological organisms are the ultimate expressions of the Cartesian mind–body split and of knowledge as invasion to establish "man's empire over lesser creatures." In intellectual property rights, the legacies of Descartes, Locke, and Hobbes meet to create an antinature view of "creation." *Creation* here does not refer to the rich diversity of life but to the products of "godlike" acts of one group of humans. And through this distorted definition of *creation*, this group claims ownership of life in all its diversity.

The freedom that transnational corporations are claiming through intellectual property rights protection in the GATT agreement is the freedom that European colonizers have claimed since 1492, when Columbus set the precedent of treating the license to conquer non-European peoples as a natural right of European men. The land titles issued by the pope through European kings and queens were the first patents. Charters and patents issued to merchant adventurers were authorizations to "discover, find, search out and view such remote heathen and barbarous lands, countries and territories not actually possessed of any Christian prince or people."[28] The colonizers' freedom was built on the enslavement and subjugation of the people with original rights to the land. This violent takeover was rendered "natural" by defining the colonized people into nature, thus denying them their humanity and freedom.

Locke's treatise on property[29] effectively legitimized this same process of theft and robbery during the enclosure movement in Europe. Locke clearly articulates capitalism's freedom to build on the freedom to steal; he states that property is created by removing resources from nature through mixing them with labor. But this labor is not physical labor, but labor in its "spiritual" form as manifested in the control of capital. According to Locke, only capital can add value to appropriated nature, and hence only those who own capital have the natural right to own natural resources, a right that supersedes the common rights of others with prior claims. Capital is thus defined as a source of freedom, but this freedom is based on the denial of freedom to the land, forests, rivers, and biodiversity that capital claims as its own. Because property obtained through privatization of commons is equated with freedom, those commoners laying claim to it are perceived to be depriving the owner of the capital of freedom. Thus peasants and tribals who demand the return of their rights and access to resources are regarded as thieves.

Within the ambit of IPRs, the Lockean concept of property merges with the Cartesian concept of knowledge to give shape to a perverted world which appears "natural" in the eyes of capitalist patriarchy. During the scientific revolution, Descartes fashioned a new intellectual world order in which mind and body were deemed to be totally separate, and only the male European mind was considered capable of complete intellectual transcendence of the body. Intellectual and manual labor were thus pronounced to be "unrelated," even though all human labor, however simple, requires a degree of unity of "head and hand." But capitalist patriarchy denies the "head," the mind, to women and Third World peoples. The application of IPRs to agriculture is the ultimate denial of the intellectual creativity and contribution of Third World peasants, women and men who have saved and used seed over millennia.

The implication of a world view that assumes the possession of an intellect to be limited to only one class of human beings is that they are entitled to claim all products of intellectual labor as their private property, even when they have appropriated it from others—the Third World. Intellectual property rights and patents on life are the ultimate expression of capitalist patriarchy's impulse to control all that is living and free.

Corporations have patented naturally occurring microorganisms. Merck has a patent on soil samples from Mount Kilimanjaro in Kenya for production of an antihypertensive and a patent for soil from Mexico for production of testosterone. In Pimpri, India, Merck found a soil bacterium that led to patents for use in treatment of gastrointestinal and appetite disorders.[30] Corporations have patented the biopesticide and medicinal products from neem (*Azadirichta indica*), though Indian women in every village and every household have been processing and using neem products for crop and grain protection and as medicine for centuries.[31]

In cases where patent claims are not based on natural products or prior knowledge of non-Western cultures but on genetically modified organisms, they are still false. In the case of plants that are not genetically engineered, patents given for medical and agricultural uses are often based on a theft of knowledge from non-Western cultures that use nonreductionist modes of knowing. Sir Walter Bodmer, director of the Imperial Cancer Research Fund and a major actor in the Human Genome Project, told the *Wall Street Journal* that "the issue [of ownership] is at the heart of everything we do." IPRs determine the issue of ownership.[32]

A shift to a postreductionist paradigm of biology that recognizes that biological organisms are complex and that ways of knowing their properties can be plural would undermine the epistemological basis of IPRs for life forms. In the area of IPRs and life forms, the issue is not merely *who* will own life but *whether* life can be owned. IPRs are therefore an issue not just of ownership but also of ethics.

Reinventing Biology, Reinventing Creativity

Reinventing biology to include concern and respect for all species and all humans requires other reinventions. Knowledge systems which view humans as members of an earth family locate creativity in understanding the relationships between different organisms. This generates different ways of knowing and different claims to knowledge. Third World, feminist, and environmental approaches to science are converging in a reinvention of biology based on the recognition of creativity across cultures.

Third World, non-Western scientific traditions and feminist perspectives have sought to evolve noninvasive modes to know other organisms which are seen as live, not dead, matter. This tradition of seeing trees and plants as alive has been continued into modern times by eminent Indian scientists such as J. C. Bose, who did detailed experiments to show

> that the pretension of man and animals for undisputed superiority over their hitherto "vegetative brethren" does not bear the test of close inspection. These experiments bring the plant much nearer than we ever thought. We find that it is not a mere mass of vegetative growth, but that its every fibre is instinct with sensibility. We are able to record the throbbings of its pulsating life, and find these wax and wane according to the life conditions of the plant, and cease in the death of the organism. In these and many other ways the life reactions in plant and man are alike.[33]

Western scientific method to study biological organisms acts as if they were dead matter and confirms that assumption with its invasive and destructive methods of experimentation.

Mae Wan Ho has called this the "cataclysmic violence of homogenisation."[34] She, like Bose, is working toward a biology that allows organisms to inform. If organisms have intelligence, they can inform us: "This is the reason why sensitive, non-invasive techniques of investigation are essential for really getting to know the living system." The reinvention of biology is inspired by a convergence of non-Western scientific traditions and feminist approaches to science, exemplified so powerfully in Barbara McClintock's "feeling for the organism" and Rachel Carson's "listening to nature."[35]

The dominant paradigm of biology excludes knowledge in which organisms are treated as subjects, not mere objects of manipulation. In the genetic engineering revolution phase, species are being even further manipulated to serve the distorted and narrow ends of a small class of humans. Plants are being engineered to become poison factories, cows are being engineered to produce human protein in their milk, pigs are being engineered with human genes governing growth. Carp, catfish, and trout have also been engineered with a number of genes from humans, cattle, and rats to increase their growth. Mammals are being genetically engineered to secrete valuable pharmaceuticals in their milk. As Robert Bermol of the University of Wisconsin has stated, "the mammary gland can be used as an impressive bioreactor."[36] Living organisms are being reduced to mechanical systems to be manipulated at will. A biotechnologist has said that "a cow is nothing but cells on the hoof."[37]

And while scientists play god with living organisms, they want no questions asked about the ethical and ecological implications of their "god tricks." Experts at fragmenting life take on the arrogant stance of being experts at everything. As James Watson has said, "Although some fringe groups thought this was a matter to be debated by all and sundry, it was never the intention of those who might be called the molecular biology establishment to take the issue to the general public to decide."[38] Besides excluding people from decisions of public concern, the new technologies have also worked out the exclusion of ethical concern which reinforces "man's empire over lesser creatures."

All life is precious. It is equally precious to the rich and the poor, to white and black, to men and women. Universalization of the protection of life is an ethical imperative. On the other hand, private property and private profits are culturally and socioeconomically legitimized constructs holding only for some groups. They do not hold for all societies and all cultures. Laws for the protection of private property rights, especially as related to life forms, cannot and should not be imposed globally. They need to be restrained.

Double standards also exist in the shift from private gain to social responsibility for environmental costs. When the patenting of life is at issue, arguments from "novelty" are used. Novelty requires that the subject matter of a patent be new, that it be the result of an inventive step, and not something existing in nature. On the other hand, when it comes to legislative safeguards, the argument shifts to "similarity," to establishing that biotechnology products and genetically engineered organisms differ little from parent organisms.

What counts as "nature" is constructed differently in patriarchal systems, depending on whether it is rights or responsibilities which have to be owned. When property rights to life forms are claimed, it is on the basis of them being new, novel, not occurring in nature. However, when environmentalists state that being "not natural," genetically modified organisms will have special ecological impacts, which need to be known and assessed, and for which the "owners" need to take responsibility, the argument is that they are not new or unnatural. These organisms are "natural," and hence safe. The issue of biosafety is therefore treated as unnecessary.[39] Thus when biological organisms are to be owned, they are treated as not natural; when the responsibility for consequences of releasing genetically modified organisms is to be owned, they are treated as natural. These shifting constructions of "natural" show that the science that claims the highest levels of objectivity is actually very subjective and opportunistic in its approach to nature.

The inconsistency in the construction of the natural is well illustrated in the case of the manufacture of genetically engineered human proteins for infant formula. Gen Pharm, a biotechnology company, is the owner of the world's first transgenic dairy bull, called Herman. Herman was bioengineered by company scientists while an embryo to carry a human gene for producing milk with a human protein. This milk is now to be used for making infant formula.

The engineered gene and the organism of which it is a part are treated as nonnatural when it comes to ownership of Herman and his offspring. However, when the issue is safety of the infant formula containing this bioengineered ingredient extracted from the udders of Herman's offspring, the same company says, "We're making these proteins exactly the way they're made in nature." Gen Pharm's chief executive officer, Jonathan MacQuitty, would have us believe that infant formula made from human protein bioengineered in the milk of transgenic dairy cattle is human milk: "Human milk is the gold standard, and formula companies have added more and more [human elements] over the past twenty years." Cows, women and children are merely instruments for commodity production and profit maximization in this perspective.[40]

As if all this were not enough, Gen Pharm has now got ethical clearance for using Herman for breeding on grounds that the modified version of the human gene for lactoferrin might be of benefit to patients with cancer or AIDS. This change of direction has brought heavy criticism of both the company and the committee.[41]

However, this kind of opportunistic biology, in which species are manipulated arbitrarily for profits, is not inevitable. In its place we could have democratized biology in which diversity is recognized as the very basis of life and is treated as a reason for celebration rather than a reason for exploitation and in which ordinary citizens have a say in biotechnology policy.

Democratizing biology involves recognition of the intrinsic value of all life forms and their inherent ability and right to survival, independent of gender, race, and species differences. It also involves the recognition of the rights of all citizens in determining how we relate to diverse species. Through such democratization we could create sciences that respect all "others," and include all "others." In a democratized biology, knowledge of different cultures and groups has equal standing, and no arbitrary assumptions are made about the creative and self-organizational capacities of nonhuman species or about the expertise and ignorance of technocrats and citizens. The colonization of other species, other cultures, and all societies has threatened both biological and cultural diversity. The democratization of biology offers an opportunity to undo these colonizations and to create possibilities for the flourishing of diversity in nature and in our minds.

Notes

1. Lynda Birke, *Women, Feminism and Biology* (Brighton: Wheatsheaf Books, 1986); Evelyn Fox Keller, *Reflections on Gender and Science* (New Haven, Conn.: Yale University Press, 1985); Sandra Harding and Jean F. O'Barr, *Sex and Scientific Inquiry* (Chicago: University of Chicago Press, 1987); Ruth Bleier, *Science and Gender* (New York: Pergamon Press, 1984); Ruth Hubbard, *The Politics of Women's Biology* (New Brunswick, N.J.: Rutgers University Press, 1990).

2. Susan Bordo, "The Cartesian Masculinisation of Thought," in Harding and O'Barr.

3. Chief Luther Standing Bear, quoted in *Touch the Earth*, compiled by T. C. McLuhan (London: Abacus, 1982), p. 6.

4. Speech by Carrie Dann at the Right Livelihood award ceremony, Stockholm, December 9, 1993.

5. Quoted in Brian Easlea, *Science and Sexual Oppression: Patriarchy's Confrontation with Woman and Nature* (London: Weidenfeld and Nicolson, 1981), p. 64.

6. Lyn White, Jr., "The Historical Roots of Our Ecologic Crisis," in Ian Barbour, ed., *Western Man and Environmental Ethics* (Reading, Mass.: Addison Wesley, 1973), p. 25.

7. Matt Ridley, *The Red Queen* (Harmondsworth: Penguin Books, 1993).

8. Richard Dawkins, *The Selfish Gene* (Oxford: Oxford University Press, 1976).

9. Richard Dawkins, *The Blind Watchmaker* (Harmondsworth: Penguin Books, 1988).

10. Alfred Crosby, *The Columbian Exchange* (Westport, Conn.: Greenwood Press, 1972), p. 12.

11. Basil Davidson, *Africa in History* (New York: Collier Books, 1974).

12. John Pilger, *A Secret Country* (London: Vintage, 1989), p. 26.

13. Carrie Dann, acceptance speech, Right Livelihood award ceremony, Stockholm, December 9, 1993.

14. Davidson, *Africa in History*, pp. 178–179.

15. Vandana Shiva, "Most Farmers in India Are Women," FAO, Delhi, 1991, p. 1.

16. Ibid., p. 2.

17. Vir Singh, "Hills of Hardship," *Hindustan Times Weekly*, Delhi, January 18, 1987.

18. K. Saradamoni, "Labour, Land and Rice Production: Women's Involvement in Three States," *Economic and Political Weekly*, vol.22, no.17, April 25, 1987.

19. Joan Mencher, "Women's Work and Poverty: Women's Contribution to Household Maintenance in Two Regions of South India," in *A Home Divided: Women and Income Control in the Third World*, ed. D. H. Dwyer and J. Bruce (Stanford University Press, 1987).

20. J. B. Bhati and D. V. Singh, "Women's Contribution to Agricultural Economy in Hill Regions of North West India," *Economic and Political Weekly*, vol.22, no.17, 1987.

21. Devaki Jain and Malini Chand Seth, "Domestic Work: Its Implications for Enumeration of Workers," in Saradamoni (ed.), *Women, Work and Society*, Indian Statistical Institute, Delhi, 1985.

22. Vandana Shiva, *Staying Alive: Women, Ecology and Development* (London: Zed Books, 1988).

23. Vandana Shiva, "Women's Knowledge and Work in Mountain Agriculture," paper presented at Conference on Women in Mountain Development, ICIMOD, Kathmandu, 1988.

24. Charles Darwin, *The Formation of Vegetable Mould through the Action of Worms with Observation on Their Habits*, (London: Faber and Faber, 1927).

25. J. E. Satchel, *Earthworm Ecology* (London: Chapman and Hall, 1983).

26. C. E. Fowler et al., "The Laws of Life", *Development Dialogue* nos. 1–2 (1988): 1–350.

27. J. Kloppenburg, *First the Seed* (Cambridge: Cambridge University Press, 1988).

28. Djelal Kadir, *Columbus and the Ends of the Earth* (Berkeley: University of California Press, 1993), p. 90.

29. John Locke, *Two Treatises of Government*, Peter Caslett (ed.) (Cambridge: Cambridge University Press, 1967).

30. RAFI, "Conserving Indigenous Knowledge," UNDP, New York, 1994.

31. Vandana Shiva and Radha Holla Bhar, "Intellectual Piracy and the Neem Patents," Research Foundation for Science, Technology and Natural Resource Policy, Dehra Dun, 1993.

32. See R. C. Lewontin, *The Doctrine of DNA: Biology as Ideology* (London: Penguin, 1991), p. 75.

33. J. C. Bose, quoted in M. S. Randhawa, *A History of Agriculture in India* (New Delhi: Indian Council for Agricultural Research, 1980), p. 97.

34. Mae Wan Ho, "The Physics of Biology," manuscript, 1992.

35. See Evelyn Fox Keller, *A Feeling for the Organism: The Life and Work of Barbara McClintock* (New York: Freeman, 1983).

36. Quoted in Andrew Kimbrell, *The Human Body Shop* (New York: HarperCollins, 1993), p. 185.

37. Quoted in Kimbrell, *Human Body Shop*.

38. Quoted in Michael W. Fox, "Super Pigs and Wonder Corn" (New York: Lyons & Burford, 1992).

39. UNEP, Report of Panel IV, Expert Group on Biosafety, Nairobi, 1993.

40. Rural Advancement Fund International Communique, June 1993, Ontario.

41. *New Scientist*, January 9, 1993.

PART II

Personal Accounts

To do science at all, science students have to be socialized into the ways of behaving considered appropriate to this thing called science. We learn to generate and test hypotheses, to present and write papers, and to persuade others of the veracity of our claims. We must also learn to distance ourselves, often to lose any overt sense of empathy with suffering we may inflict on the organisms we study. Such desensitization is a major theme in this part, in which three biologists give accounts of the intellectual and emotional changes they underwent as they moved through scientific training and into research and teaching in biology.

Each of us brings to this training a personal background conditioned by our ethnic and family histories, our different experiences with nature, and the fact of having grown up female. The three of us went through our scientific training at times when there was undoubted discrimination against women, but the scarcity of women in science was not thought of as an issue. Today it is widely discussed, but little attention is paid to possible reasons. Much of the rhetoric about women in science focuses on finding ways to encourage them to enter scientific training, without looking too deeply at reasons why girls may feel alienated from science. In relation to biology, could this be, in part, because girls tend to be put off by the way life is treated in the laboratory? If so, Ruth Hubbard argues, this is not because of intrinsic differences between girls and boys but may be due to the fact that empathy is permissible or even encouraged in girls more than in boys.

The solution to this is not to toughen up girls but to alter the training as well as the practice in biology so that domination of nature is not paramount. Though undergraduate training now may involve less actual dissection than it did in our time, the focus has shifted, if anything, even further away from living organisms by virtue of the vast amounts of attention being paid to genes and other molecules. While this may, at first glance, accommodate "feminine squeamishness," in practice it moves biology further and further from the study of life. Any students who now come to university biology thinking that they will be able to work among, and learn about, living organisms are bound to be disappointed.

Given the continued tendency to think of science as more suitable for males than females, it becomes easier for women students to translate their unease into a shift toward sociology or literature. Trying to persuade more girls or women into biology probably is not the answer, unless the direction and practices of biological research change substantially.

Though Lynda Birke was not particularly aware of these complexities at the start of her studies, her lifelong love of animals and plants presented her with many profound dilemmas in deciding to become a biologist. Her turn to research in animal behavior seemed a partial way out of such conflicts. Yet her involvement with feminist and environmental politics has also sharpened those conflicts.

For Ruth Hubbard, too, engagement with feminism and other liberation politics changed the course of her professional life. Having started as a devoutly reductionist biologist, interested primarily in the relationships among molecules, she realized in midlife that she needed to explore how both politics and personal history shape the practice of science.

Betty J. Wall's history as a Chinese-American girl growing up in a small town in Louisiana and the first in her family to go to a university brought many special sensibilities to her training. Her journey into and out of research on the nerves and muscles of insects forms the backdrop to her present life as a massage therapist and healer.

4 | On Keeping a Respectful Distance
Lynda Birke

In the lore of the dolphins it is recorded that at some moment in time a few individual human beings will break through to a new, transhuman level of consciousness, [and] will become true philosophers comprehending the whole in all its parts . . .

—J. Rodman[1]

WE BEGAN THIS project by asking what science would have looked like if it had approached the study of nature with respect for the lives of the organisms biologists study. In some ways this question is unanswerable simply because that is not how the science of biology has developed. It is also unanswerable because science is not separate from the wider society; science has both justified and drawn upon a deeply held cultural preoccupation in Western culture with keeping a distance from the rest of the living world and seeing some humans as superior to it. Keeping a distance has entailed the use and abuse of other creatures rather than respect for their integrity. Science is no exception.

In this chapter I want to address the issue of respect for animals in two ways. The first is to outline my own personal history and the dilemmas I felt about becoming a biologist. Why is it that I came to consider respect for animals (or nature more generally) in and through science to be an important issue? Partly it comes from a long history of concern about the issues and a long history of concern about animals; partly too it comes from my involvement with feminist critiques of science in which feminists have necessarily engaged with all kinds of systems of domination, including the ways in which humans dominate animals. That science fails to respect the animals studied goes hand in hand with a failure to respect other people; and both are served by a reductionism in biological thinking that fails to see the "whole in all its parts." What students of biology learn today is not respect for the wonderful lives of other organisms but a respect for the wizardry of DNA technologies.

Second, I want to address the study of animal behavior (ethology), the area in which I was formally trained and in which I did research for many years. As in other areas of science, there are ways in which the study of behavior—especially in the laboratory—typically entails doing things to animals that we might call disrespectful of their integrity. Yet ethology is, I suggest, deeply ambivalent toward animals, and many of its practitioners are motivated by a strong respect for the living world.

I want to end, however, by asking what we mean by respect. As the various chapters of this book indicate, having respect for other organisms can mean many different things. It might mean respect for others as individuals; it might mean respecting the species. We might also differentiate between respecting the integrity of another creature in its own right and respecting it for what it represents for us. In science it is the second of each of these pairs that has more salience.

Toeing the Line, or Becoming a Scientist

A respect for nature in general and individual animals in particular was critical for my wanting to do biology; it also posed many dilemmas. In this sense, what I am outlining here is not so much a change in direction as a progressive working out of these dilemmas.

Despite growing up in London (or perhaps because of it), I developed an interest in natural history. Perhaps it was the effect of realizing how easily nature could recolonize the postwar devastation; perhaps it was the joy of discovering that even in urban parks I could find abundant wildlife. But another important influence was my own close association with animals; even in London it was possible to learn to ride horses, and horses have had a profound influence on my life. The budding eight-year-old scientist could recite the names of every single bone in the vertebrate skeleton—provided it had a horse's form around it.

Alongside that interest in horses and natural history, I had a growing interest in science. No one in my family thought that girls could not do science (even when I set fire to the curtains); on the contrary, I was given chemistry sets, and together with my father, built radio sets. So that taught me, if nothing else, how to use an inquiring mind.

I had long been drawn to natural history; yet I was ambivalent about doing biology precisely because of the need to do things to animals. Twice over, in my last years at school and then subsequently when I began my undergraduate career, I tried to concentrate on the physical sciences; but always, something drew me back to biology. The fascination with the living world won out, even though I had to steel myself against the need to do dissections—or worse. Nearly thirty years later, I still can feel vividly that sense of horror and disgust when we were confronted in school with a white rabbit with pink ears and expected to dissect it. And of course I said nothing. You were simply expected to get on with it, not to show emotion; I did, after all, want to do science. Alongside the sense of revulsion, however, was another emotion: a sense of fascination at the beauty of once-living tissues, at how they are put together.

An undergraduate career in science does involve the use of animals, and it does involve a subtle process of desensitization. Slowly you need to learn to suppress the emotional reactions to the use of animals, living or dead. Insofar as aesthetic or emotional reactions are encouraged in scientific training, they are likely to be responses to what nature has become after the processes of science. We might, for instance, express pleasure at the colors or the orderliness of cells in a photograph taken with an electron microscope; we might be excited by the way a new hypothesis appears to explain what we see or infer.

The other kind of emotional reactions, expressing revulsion at having to kill an animal, for example, or distress at being expected to do something to an animal that would be labeled as cruel by the world outside science—those emotions are ones the budding scientist must learn to suppress. They are considered "unmanly," as the entomologist Miriam Rothschild once noted in a lecture.[2] Whatever else it involves, becoming a scientist entails learning to acquire, or fit into, the macho culture of the laboratory[3] and forswearing such "feminine" responses as empathy with the animals.[4]

Biological education is, I think, deeply ambivalent in its approaches to nature. Much of it is in accord with the long traditions of experimental, laboratory-based practice. Indeed, with the growth of techniques in recombinant DNA and other gene technologies, genetics and molecular biology now dominate biological research. Like much of science, these approaches are profoundly reductionist, seeking causes deep within the organism or its constituent molecules. Another strand of biology, however, has come from natural history traditions, which have emphasized observation of nature "as we find it" rather than in the highly artifactual form we construct in laboratories. This less reductionist tradition appears in ecology, in evolution and systematics, in the study of animal behavior (though it may often coexist with an increasingly experimental approach).

That natural history strand seems (in principle at least) to start from a more respectful stance. At the very least, it is in closer accord with the respect and awe we might feel in response to (say) beautiful mountains, or a glimpse of some shy, wild creature. And perhaps it was that element of natural history that drew me away from physiology (which I found fascinating in theory) toward studying animal behavior. Yet even if some areas of biological training afford greater respect for living organisms, the ways in which biology is taught and practiced limit how respectful we can be.

In the first place, the reductionism of science, combined with its authority within the wider society, ensures that students rarely learn to question what they are taught. To pass exams, we learn to regurgitate the facts. How those "facts" arose and what creatures were dismembered to provide them are rarely part of the pedagogy. Second, the wider context in which scientific knowledge is constructed is largely absent. Students might learn how recombinant DNA technology can be used in industry, but they are less likely to learn much about its history, about the social dynamics of constructing that knowledge, or the ethical problems posed by it.

With all these contradictions and dilemmas in the background, I began a research career with some unease. Somehow I ended up doing animal behavior research but doing it in the laboratory, with all the ambivalences that brought. Despite all my turbulent feelings about animals and nature, I had been sufficiently desensitized to toe the line: ambivalence notwithstanding, I did do laboratory-based research.

Moreover, whatever else was involved in deciding the course of research, my feminism had an influence in choosing problems to pursue. My interest in women's health, for example, prompted me to do research on effects of hormones then being used in contraceptives. At the time, the feminist interest in women and their health took priority; only later did I ask other questions.

The impact of feminism on my laboratory work raises an important question for me: what kinds of issues might influence decisions to follow a particular line of research? Funding obviously is centrally important and is firmly located in a wider politics: biomedical scientists can usually obtain funding only if they are willing to justify their work in terms of potential clinical use. At that time, I used somewhat similar arguments to myself; I put my anxieties about animals on hold and justified the research to myself in relation to feminist politics. It seemed somehow all right, as long as I did not do anything too nasty to the animals and made sure they were well cared for. Only later did I explicitly question the use of animals altogether and the fact that keeping them in laboratories must inevitably mean their subsequent deaths.

It seems to me that it is an abuse of animals, not respect (let alone the economic considerations), that allows large numbers of animals to be bred up only to be wasted. Animals are killed routinely in laboratories. Some are "sacrificed" in the course of an experiment;[5] many more are killed simply because no one uses them on time or because scientists from one laboratory in the building don't particularly talk to those in another, so that in different laboratories, animals are killed for different parts of their bodies, when laboratories could cooperate and thus save lives. I find it odd that the number of animals killed because they are not "needed" for experiments seems to merit far less attention from animal rights activists than the animals killed during particular experimental procedures.

Still, those concerns about killing did not for a long while actually stop me from doing science. I knew that animals were going to have to be killed. I knew that some of the procedures I might have to use were somewhat invasive. Yet I swallowed my feelings about those for many years—such is the power of the desensitization that comes through scientific training. And, of course, individual scientists, myself included, want to conform. You learn to toe the line because you want to believe that you are being "a real scientist." Empathy is not part of the game.

For all that, doing science in the labs is also fun; it is about solving puzzles. I think that point needs emphasizing. In the zeal to stress the multifarious ways in which science is problematic, the feminist critiques of science have tended to ignore that (and I own up to having done that, too). But doing science has also meant for me a sense of alienation, sometimes as a woman in a still largely male world, and more often as someone who cares deeply about animals. It is distressing to be in a lab around people who are being cavalier with animals. There is a disrespect in the way some people—but not all—handle the animals they use. To them, the animals are tools, means to an end. Perhaps such people don't mean to be cruel—they don't, I suppose, think that they are, but stunning a rat by swinging it round by the tail while cracking jokes is hardly a sign of respect. On the other hand, I don't know that anyone who wants to stun a rat would be able to be anything other than cavalier. I don't know how your emotional reactions would be if you tried to have respect for the animal you are about to stun and decapitate. Cracking jokes and the macho stance may be ways of coping with doing something that, in other contexts, would be considered quite horrible.

Alongside the laboratory work, I was living and working with animals at home—my precious horses. To be a scientist in the lab thus meant having two quite different relationships to animals. My experience of those animals with whom I lived was so much at odds with my experience of animals in the laboratory. In lab work, you end up treating animals in groups. Animal 39/2/F is just a number in a cage. She represents a group or a treatment or a species, but you know nothing about her own history, about her life with her companions.

Where individuals in the laboratory start to be respected as individuals by humans is where they pass over the boundaries from the "analytic" animal that is destined to become data to one who perhaps becomes a pet.[6] Researchers working with animals sometimes designate particular animals as pets, removing them from the realms of potential experimental animals.[7] I can well recall the occasional animal that passed through our hands that would become special "like a pet"—whose death we would mourn in a way that we did not mourn for all of the other animals, who remained numbers in cages.

By contrast to the numbered lots of rats in the lab, I knew the animals at home, my horses and dogs, as individuals; I worked with them and knew their idiosyncrasies. I trained the horses daily. That gave me a very different perspective, based in their individuality. Scientific accounts based on such individual narratives would be considered insufficient for any generalizations about the species horse. Yet, after many years of working with horses, I have a strong suspicion that I know that species far, far better (and thus in a way that is more predictive of its behavior) than I know any of the species that I worked with in the laboratory.[8]

That said, the laboratory work had its own value in the development of my thinking about our relationship to animals and what that means. First, it was through working with rats that I came to appreciate better what fine animals they are. I know full well the cultural loathing of these animals, which is played on by organizations defending animal use in science as they point to the fact that most experiments are done on rats and mice.[9] I accept, too, that people who have not had such privileged lives as I have had may have good reason to hate them, as Ruth Hubbard points out in her chapter. Rats there are aplenty in the stables, but I have none infesting my house, and I cannot imagine what it must be like to have them nibble my toes. So I grew to like them, to appreciate their curiosity and watchful eyes, their playfulness, and their obvious intelligence in spite of their impoverished lives in laboratory cages.

Second, it was research with animals that enabled me to focus on the multiplicity of factors involved in the development of behavior. That might seem a truism if we were talking about child development. But all too often the development of, for example, gender differences in behavior in animals is attributed simply to hormones. Experiments are then done that support the hypothesis. It is one thing, however, to discover that hormones are involved in behavioral development; it is quite another to assume, as much of the available literature tends to do, that they are the only or the primary factor. Differences emerged in our rats out of a complex set of social interactions—mother and siblings as well as hormones.[10] Apart from my belief that such accounts are likely to yield a better description of what happens as infant animals develop, they also allow us to see animals as not being determined simply by their internal biology—a point to which I will return below.

Ultimately, it was a fascination with questions like that—trying to find out how animals get to be the way they are, even in the laboratory—that kept me doing science. I was interested in the animals themselves rather than in any putative use to which the knowledge might be put (and concepts of use of scientific knowledge about animals usually mean usefulness to humans). But there was always a tension, the fascination versus my gut feelings that gains in knowledge about animal behavior could not be justified, for me, if that meant using animals.

I no longer work in laboratories. Sometimes I miss the fun of doing science; at other times I recognize that what drove me into it in the first place was a love of nature and respect for animals—and laboratory work seemed to fly in the face of that love.

Yet however personally I have experienced the ambivalences, I think they are also embedded in the area of biology I chose to study: ethology. There are several ways in which the practice of ethology differs little from the aspects of science that give me cause for concern; it can, for example, be reductionist, and it has lent itself to mechanistic explanations of behavior. But it has also, in recent years, generated a renewed interest in cognitive ethology—studying animal thought processes and minds. While this may still be problematic, in that it remains embedded in the rhetoric and practice of science, thinking about animal minds does begin to accord them some respect. Perhaps from there, we can begin to think about them differently.

Behaving like Clockwork

I want to turn now to the subject matter of ethology, the study of animal behavior itself, the subject in which I was trained. How respectful has that been? I want first to explore some of the ways in which ethology might be criticized for its apparent lack of respect for the organisms it studies. But from there I want to move on to ways in which I think ethology might also offer some room for improvement and scope for change.

Ethology has had somewhat different histories in North America and Europe (though very much blurred by now). In Europe, its roots were largely in natural history, while in North America, the roots have tended to be more strongly experimentalist, coming from a history of behaviorism. With the natural history tradition comes some respect for the animal in its wild state; the early ethological studies of Niko Tinbergen were largely based, for instance, on his observations of the behavior of wild animals such as herring gulls. The animals were free to do what they willed. Behaviorism, by contrast, seems on the face of it to have been much less respectful of animal lives; it purported to study the laws of learning but did so under highly constrained experimental conditions in the laboratory.

I think that the practice of ethology has retained some respect for animals in their natural habitats. Most people in my experience go into it out of an interest in what wild animals do, what makes them tick, even if they subsequently work in laboratories (as I did). Nevertheless, there has been, in the last twenty years or so, a move toward more mechanistic and reductionist explanations. Even though animals may be studied in their wild habitat, the dominant narratives now emerging in behavior and ecology (sometimes called sociobiology) are ones that seem to construct animals as though they were economic calculators, optimally efficient and strategizing. They must, furthermore, do so mindlessly: it is evolution through natural selection that has ensured that they act "as if" they were making appropriate decisions. However well this jargon may help scientists to describe their findings, such a narrative constructs stories of animals as automata; we may marvel at the clockwork of such wonderful machines, but I'm not sure that we can accord them respect.[11]

Scientific training in ethology also fosters a belief that we can study mechanisms underlying behavior. Such claims are at the heart of attempts to locate genes "for" particular characteristics in humans (such as homosexuality), as well as underlying physiological studies of the mechanical "bases" of behavior. Neurophysiology, for instance, has given us stories about how nervous impulses underlie the withdrawal behavior of the seaslug, *Aplysia*, as well as stories about the activities of our own brains.

There are certainly issues of respect here. It does seem profoundly disrespectful to the integrity of an animal to study how its brain works by sticking electrodes into it, as the frequent use of such imagery by antivivisectionist organizations would testify. And tinkering with nervous systems undoubtedly runs the risk of causing pain and suffering. I share that disquiet, and could not personally do such experiments. But having said that, I do wonder whether at least some of the emotional reaction to those images is precisely because it is the brain that is involved.

We associate the brain with the mind (at least in ourselves); we thus identify the brain with the essence of us. That, of course, is one reason why people are typically much more anxious at the thought of brain transplants or disembodied brains than they are at the prospect of having a heart transplant. I do not know how physiology in general (and neurophysiology in particular) could have developed if the prevailing ethos had been one of respecting the organisms it studied. If knowing something about possible underlying mechanisms of behavior can contribute to respect on our part, then it is likely to be a respect for the intricacies of the mechanism involved rather than for the behaving organism itself. I have often had the "isn't nature marvelous" reaction, a sense of awe and respect for those intricacies of mechanism as I learned about them. But I doubt very much that such knowledge contributes much to our sense of respect for the whole behaving animal, any more than understanding the physics of sound waves produced by a violin helps us to appreciate a string quartet.

I have the same double-edged reaction to learning about experiments on songbirds. In some experiments, songbirds, such as white-crowned sparrows or chaffinches, were isolated; in others, they were surgically deafened. The scientists' purpose was to investigate how song developed. What they found was, in some ways, interesting and important, because it challenged simplistic views that bird song was just something built into the animal. For most songsters, what was needed was opportunity to practice and learn from others, in relation to a kind of rough guide that the animal seemed to have been born with. The bird had species-specific expectations, in other words, against which it could match its own efforts, even if it was in solitary confinement. Listening to others helped to refine the song and to add embellishments and local dialects. Birds who were deaf could not match up their own songs; they produced sounds that were little like those of their wild counterparts.

That is a useful story with which to challenge any simplistic notion of nature versus nurture; the birds need subtle interactions between both inbuilt "templates" and listening experience. Other studies tell us about how particular parts of the songbird brain are involved in singing. No doubt that knowledge, too, is useful in some way. I can't help feeling, though, that my respect for the song of the nightingale far outstrips anything I might think about the intellectual claims of science. All that science has done is to silence the singer.

The search for mechanisms underlying behavior may well tell us something about how that behavior comes about. But it does two other things; apart from obviously inflicting permanent injury on these creatures, it helps to perpetuate a narrative that their behavior is "like clockwork," mere mechanism.

Endangered Minds?

What does exist in ethology, I would argue, is an undoubted respect for nature and for the species that comprise it. For that reason, the study of ethology goes hand in hand with wider environmental considerations that focus on conservation and the need to preserve endangered species. What is less clear is that the study and practice of ethology necessarily foster respect for individual animals, except as exemplars of a species. This is in contrast to the position taken by those who campaign for animal rights, for whom it is the rights of the individual that matter. Tom Regan, for example, bases his advocacy of animal rights on the notion that every animal (or at least those who are sentient) has, in principle, a right to life.[12] His position is clearly at odds with the kind of environmental reasoning which prioritizes the species or the group, for which individuals matter little.

That tension, between individuals and groups, is played out in ethology. Until recently there was much more interest among ethologists about species than there was about the welfare of individuals. But that has begun to change. In recent years there has been a growing concern for the welfare of animals in, for example, zoos, circuses, and the laboratory itself.[13] One facet of this has been an interest in providing animals held in captivity with enriched environments—an idea that would have been unthinkable a few years ago, especially for laboratory animals. Primates particularly seem to benefit; if you give them something to do with their time, such as searching for food among wood chips, they are much less likely to develop abnormal, stereotyped behavior.[14]

There is, then, some concern among those who study animal behavior to develop applied research for the benefit of animals who are living under captive conditions. That, at least, seems to be congruent with greater respect for individuals. The animal behavior world has, moreover, been at the forefront of developing ethical guidelines; the journal *Animal Behaviour* was probably the first in the world to publish guidelines (in 1981) on the ethical use of animals. Animals can still be used in research, of course, but a greater concern for ethics might well herald a more respectful stance.

Another important development has been the rise of cognitive ethology, which emphasizes the notion that animals might think, in ways that we must take seriously (this is further discussed by Lesley Rogers in this volume). Donald Griffin thus writes about "animal minds"; others point out how some animals might have consciousness, or even self-consciousness.[15] Perhaps unsurprisingly, many writers have resisted such claims, attempting to refute any evidence that shows animals to be clever.[16] There is, it would seem, too much invested, both scientifically and culturally, in the notion of animal irrationality and inability.

Apart from obvious hostility to the notion that at least some animals may have minds is the problem of how to interpret research findings. Humans are rather too good at disparaging what an animal does, especially if it fails to perform a task in the way that we would do it and on our terms. If the horse Clever Hans was responding to his trainer's cues rather than counting, then I think that is pretty clever; I would not use the story to dismiss his abilities, merely because he did not seem to "count" the way we do.[17]

Deception, too, seems to be widespread in nature. (Perhaps Clever Hans was cleverly deceiving the crowds who came to look.) At its simplest, this involves mimicry, perhaps unconsciously; we cannot know, for example, how self-aware the plover is when she feigns a broken wing to distract predators from her nest. A possibly clearer example is provided by vervet monkeys, who give "fake" alarm calls when another monkey troupe is near, so that when this causes them to flee, the call-giver can wander, apparently unconcerned, into the open space vacated by the other troupe. The heritage of behaviorism would have us deny consciousness on the part of these animals; on the other hand, as Griffin notes, "perhaps it is best to keep an open mind and not dismiss such possibilities out of hand."[18]

Indeed, those who train animals might wonder why it has taken science so long to catch up with what they have long known about animal thinking. They might sometimes adopt the languages of science—talking behavioristically of conditioning, for example—while simultaneously believing in the animal's abilities to form complex concepts. Admittedly, the kinds of animals that we train in depth are nearly always mammals or birds; hence we know relatively little about the concept formation of other kinds of animals.

There is a strong belief that animals are simply not as smart as we are. Yet interpreting "stupidity" is not easy, even among ourselves. In looking at "animal consciousness," Radner and Radner note the interpretations given in scientific writing to examples of alleged animal automatons.[19] They refer, for example, to the case of a species of bee that was fooled by experimenters into repeating a particular behavior pattern over and over again. (The bees respond to the odor of oleic acid, which indicates to them that there is a dead bee in the hive that needs to be removed. The experimenters daubed a live bee with oleic acid and found that the bees repeatedly tried to remove it). Now, the behavior can be thought of as illustrative of bees failing to recognize a problem, a logical mismatch. But as the Radners point out, why are we so sure that they are simply being stupid?

The Radners note, for example, our own uncertainties about how to recognize death, even with the aid of high-tech medical apparatus. And, more important, we make allowances for humans to be credulous or gullible even when they persist in irrational beliefs, while "animals . . . are expected to be perfect little scientists. In order to earn the epithet 'conscious' they must be proficient in logic, ever ready to change their beliefs in the face of available evidence, careful to take all considerations into account. When people fail to live up to this idea, we say they are all too human. When animals fail, they are said to be machine-like."[20]

Whose Knowledge Counts?

Casting nonhuman animals as lacking some of our mental abilities allows us to accord them less respect than we might give to at least some humans. Our knowledge—and especially scientific knowledge—counts for everything, theirs for nothing unless it is used by us. (Thus we value the skill of a dog to locate odors when it helps people to track a missing person.) But seeing nonhumans as somehow lacking also serves indirectly to reinforce disrespect toward those people who are themselves cast as closer to nature, to animals, as other chapters of this book note. Feminists, among others, have noted, for instance, how often these biological claims of proximity to "nature" serve to disempower women.[21]

Yet part of the problems posed by biological determinism has been the tendency for animals to be seen as much more on the nature side of the nature/nurture dichotomy. In a sense, that is how we define nature; animals we define as being in it in a way that we are not. So some critics think it illegitimate to move from talking about the biological bases of behavior in animals to talking about humans, because to do so seems both to imply a determinism and to reduce us "to the level of animals." But I would argue that the move seems inappropriate because of the way biology is taken to include everything about animals, and only some things about us. We can thus talk about "the biology of the rat," to include its physiology and its behavior. Yet if we talk about the biology of human beings, we largely mean physiology and biochemistry. Behavior, intellectual capabilities, we see as being somehow separated from the rest of that biology. That, indeed, is the basis of critiques of biological determinism: it is our minds that we seek to separate from biology.

It is certainly harder to have respect for a creature that you have defined a priori as being the product of its genes or hormones, as being *in* nature—at least not in the same way you might have respect for the free will of another human being. It is harder still if, by using an experimental method that ignores the wider context in which the animal lives, scientists seek simple causes within the individual. In her chapter, Anne Fausto-Sterling outlines some of the problems with using such an approach in studies of primate sexual behavior. By definition, by method, nonhuman animals come out looking as though they are driven like machines.[22] And how can we respect these clockwork creatures?

One answer is, of course, that many of us don't, simply because we cannot believe in these tales. The animals whose lives I know well are not the clockwork machines of scientific experiments. Moreover, those people who are outside science may well be skeptical of the more mechanistic accounts of what animals are, which does not always accord with people's perception of at least some animals around them.

Skepticism might also be fueled by growing discontent with science. Several writers, including many scientists, have commented on public unease about science and its claims. Writing about the rise of the animal rights movement, for example, James Jasper and Dorothy Nelkin have suggested that there is a growing feeling of antiinstrumentalism, particularly in relation to new social movements such as feminism and the environmental movement.[23] This in turn leads, they suggest, to revolt against what is perceived as the instrumentality of science; among other things, that includes the way in which laboratory animals are seen as means to an end.

Making science more respectful might mean taking seriously these wider public concerns. Part of the relative success of the animal rights movement in mobilizing public opinion is, I suspect, due to its ability to tap into widespread anxieties about where science is going, about its lack of accountability to the people who pay for it, and about people's powerlessness to challenge its authority.

Vicki Hearne, in *Adam's Task*, has emphasized the contrast between the claims of science, with its authoritative voice and generalizations from groups, and the way in which knowledge of animals is gained by animal trainers working with individual animals. Science should, she stresses, pay more heed to such collective experience. Yet only one source of knowledge is granted authority in our culture, only one is thought of as approximating "truth." Thus what I know from my experience of horses, and my involvement in the world of people who live and work with horses, cannot enter the hallowed portals of scientific knowledge; at some level, it does not "count."

The development of science has been based on ignoring the knowledge claims of people outside the institutions of science (such as animal trainers) or on appropriating and renaming their knowledge. In the heyday of imperialist expansion in the nineteenth century, for instance, British scientists ignored the extensive knowledge of indigenous people of the animals and plants around them, while seeking to give species names that honored the scientist and his culture.[24] Science remains ignorant of many of the pharmacological properties of plants well known to the indigenous peoples of the world and continues to ignore most of the accumulated knowledge of people who live daily with animals.

In that sense, science is doubly disrespectful. For it fails to pay heed to the wisdom of those people who best know animal lives,[25] as well as showing little respect for the creatures whose lives and bodies provide it with data. The word *science* comes from Latin roots meaning "knowledge" or "wisdom." Sadly, modern science seems rather short of both wisdom and respect. We can only wonder at what it might have looked like if science had been built on wiser, more respectful foundations.

Notes

1. J. Rodman, "The Dolphin Papers," *Antaeus on Nature* (London: Collins Harvill, 1986).

2. M. Rothschild, *Animals and Man: The Romanes Lectures 1984–1985* (Oxford: Clarendon Press, 1986).

3. Arnold Arluke has noted the "macho" culture of some primate labs, for instance, and the "cowboy" ways that animals are treated in such labs in "The Ethical Culture of Primate Labs," a paper given at a conference on Science and the Human-Animal Relationship, Amsterdam, March 1992.

4. See Z. T. Halpin, "Scientific Objectivity and the Concept of 'the Other,' " *Women's Studies International Forum,* 12 (1989), pp. 285–294, and L. Birke, "Science, Feminism, and Animal Natures," *Women's Studies Internation Forum,* 14 (1991), pp. 443–449.

5. See M. Lynch, "Sacrifice and the Transformation of the Animal Body into a Scientific Object: Laboratory Culture and Ritual Practice in the Neurosciences," *Social Studies of Science,* 18 (1988), pp. 265–289.

6. Lynch.

7. See A. Arluke, "Sacrificial Symbolism in Animal Experimentation: Object or Pet?" *Anthrozoos,* 2 (1988), pp. 97–116, and L. Birke and M. Michael, "The Researchers' Dilemma," *New Scientist,* April 4, 1992, pp. 25–28.

8. Also see V. Hearne, *Adam's Task: Calling Animals by Name* (London: Heinemann, 1987).

9. This is the point emphasized, for instance, in publicity material produced by the Research Defence Society in Britain. The main thrust of the argument is to focus on the medical benefits brought by animal-based research. Nevertheless, the society does play into cultural antipathy toward rats and mice by noting that more than 85 percent of experiments in Britain use these species.

10. See L. Birke, "How Do Gender Differences in Behaviour Develop? A Reanalysis of the Role of Early Experience," in P. Bateson and P. Klopfer (eds.), *Perspectives in Ethology 8: Whither Ethology?* (London: Plenum, 1989).

11. The narratives, moreover, are constructed in ways that obscure what Donna Haraway has noted as the intersections of class, race and gender in her study of the development of primatology. The science, she insists, is hardly innocent of a colonial his-

tory in which white Western scientists study nature in other parts of the world. See her *Primate Visions* (London: Routledge, 1989).

12. Regan's position does force him to think in terms of those animals that have sentience and therefore potential rights, and to contrast them with those that he believes to lack sentience. Such drawing of lines is inevitably problematic, as is the use of the concept of "rights." For further discussion, see L. Birke, *Feminism, Animals and Science: The Naming of the Shrew* (Buckingham: Open University Press, 1994).

13. For example, see M. S. Dawkins, *Animal Suffering: The Science of Animal Welfare* (London: Chapman and Hall, 1980).

14. A. Chamove et al., "Deep Woodchip Litter: Hygiene, Feeding and Behavioral Enhancement in Eight Primate Species," *International Journal for the Study of Animal Problems*, 31 (1982), pp. 308–318.

15. E.g., D. Griffin, *Animal Minds* (Chicago: University of Chicago Press, 1992), and S. Walker, *Animal Thought* (London: Routledge, 1983).

16. See, for example, J. S. Kennedy, *The New Anthropomorphism* (Cambridge: Cambridge University Press, 1992), and M. P. T. Leahy, *Against Liberation: Putting Animals into Perspective* (London: Routledge, 1991).

17. For further discussion, see L. Birke, *Feminism, Science and Animals: the Naming of the Shrew* (Buckingham: Open University Press, 1994).

18. Griffin, p. 57; see also D. Cheney and R. Seyfarth, *How Monkeys See the World* (Chicago: University of Chicago Press, 1990).

19. D. Radner and M. Radner, *Animal Consciousness* (New York: Prometheus Books, 1989).

20. Radner and Radner, pp. 180–181.

21. See, for example, R. Hubbard, *The Politics of Women's Biology* (New Brunswick, N.J.: Rutgers, 1990); L. Birke, *Women, Feminism and Biology* (Brighton: Wheatsheaf, 1986); and A. Fausto-Sterling, *Myths of Gender* (New York: Basic Books, 1992).

22. Ironically, alternative endings to these stories can be found in scientific accounts themselves. See, for example, Anne Fausto-Sterling, *Myths of Gender* (New York: Basic Books, 1992), especially chapter eight.

23. J. Jasper and D. Nelkin, *The Animal Rights Crusade* (New York: Free Press, 1992). Also see Ulrich Beck, *Risk Society* (London: Sage, 1992).

24. See H. Ritvo, "The Power of the Word: Scientific Nomenclature and the Spread of Empire," *Victorian Newsletter*, Spring 1990, pp. 5–8, and *The Animal Estate: The English and Other Creatures in the Victorian Age* (Harmondsworth: Penguin, 1987).

25. In Western culture, that would include all kinds of people, including farmers, pet-keepers, amateur naturalists; it also includes the ecological understandings of, for example, various native, or non-Western, peoples. See D. Suzuki and P. Knudtson, *Wisdom of the Elders: Sacred Native Stories of Nature* (New York: Bantam, 1993), who note, for example, how the !ko Bushmen of southern Africa were well aware of the predatory behavior of the spotted hyena—long before science gave up its stories of the hyena as feeding entirely on carrion (p. 113).

5 | The Logos of Life

Ruth Hubbard

THE *American Heritage Dictionary* defines *biology* as "the science of life and life processes." Since we began to plan this collection, I have thought a good deal about how I experienced this science while learning it as a young student, practicing it as a researcher, and teaching it.

City born and bred, I came to biology not out of love for living organisms but almost by accident. My father and I often went walking in the woods and meadows surrounding the Vienna of my childhood. Our Sunday adventures were special. We sang as we went along, and my joy at being with my father was part of enjoying the beautiful countryside. But for both of us this enjoyment was more of an aesthetic appreciation of sights, smells, and sounds than an immersion, or even an interest, in the life of the plants and animals around us. The spring flowers and cuckoo's calls were a backdrop for my pleasure in our Sunday excursions.

In the high school I attended after we emigrated to the United States, I studied physics, not chemistry or biology, and I entered Radcliffe College (the institution in which women students had the questionable privilege of taking courses from the all-male Harvard faculty) thinking I might major in philosophy and physics. But this was the fall of 1941: the Germans were marching toward Stalingrad, the United States was entering the Second World War, and I wanted to get ready to do something useful and practical. With little imagination, given that both my parents were physicians, I decided to head for medical school and therefore became a premedical student. That meant taking a sprinkling of chemistry, physics, and math, and lots of biology.

So, like many biology undergraduates, I studied biology not to learn about animals, plants, or "life," but to get into medical school. The organisms I observed in the laboratory were means to an end. I would not be surprised if the (in)famous goal-directedness of premedical students is responsible for the fact that generations of biology undergraduates put up with being made to dissect pickled dogfish, cats, or piglets that reek of formaldehyde (which we now know to be a carcinogen). After all, premeds know that this insult to their sensibilities is nothing compared with what they will experience in medical school anatomy, when they meet a human corpse—their "cadaver," in medspeak.

I do not remember encountering recognizable animals in any undergraduate biology course, except comparative anatomy, where I met the proverbial smelly dogfish and cat. The rest all happened with prepared slides, which we viewed under the microscope. In botany, we dissected onions and the stems, roots, and blossoms of flowering plants, but to a city child like myself, who had never grown plants and thought of them mainly as flowers to buy for special occasions and vegetables to cook for dinner, these laboratory experiences raised no questions. And none were raised by my instructors about either the plants or the animals we "used" in our studies.

The fact that I never had to seek out the organisms in their native habitats and collect them myself no doubt had a lot to do with my obliviousness. Like the meat and vegetables we buy, usually prepackaged, in the supermarket, the organisms we study in the laboratory come from quasi-industrial supply houses. And like animals grown to be eaten, laboratory animals are grown for teaching and research.

Our instructors seemed as distanced from the organisms as we were. Just as we were shown how to measure electric currents in physics lab or to collect and weigh precipitates in chemistry, so we were taught what to look for when we cut up our pickled frog or dogfish. The fact that the animals were once alive and someone had to kill them in order for us to draw pictures of their insides was never brought up—not by our instructors and not by us students.

Becoming a Research Biologist

During my last year at Radcliffe, I enrolled in a course that allowed me to work in a research laboratory at Harvard. Initially this simply seemed more fun than taking yet another biology course with lectures and preset labs. But gradually this experience made me decide to become a research biologist instead of a physician.

Though the work was not particularly interesting, it gave me my first chance to handle live animals on a regular basis. (What a commentary on city living that it took a laboratory experience to put me in daily touch with animals!) I was working with male white laboratory rats—animals bred to be sufficiently passive and stupid that novices like me can pick them up without getting hurt; but I liked handling the furry little creatures. Most of all, though, I enjoyed the ambience of that laboratory. My immediate supervisor was a returning graduate student who had taught many years in a large state university. She was getting her Ph.D. at Radcliffe/Harvard in order to improve her chances of a faculty appointment at a research-oriented school. The professor in charge of the lab was George Wald—young and lively, and always ready to engage in conversations not only about our experiments but also about politics, books we were reading, and whatever else interested one of us at the moment.

The research itself consisted in feeding groups of rats diets we prepared so that they lacked one or another specific vitamin or mineral. The question we were trying to answer was whether, and to what extent, the nutritional deficiencies the animals developed affected their physical activity. To make such measurements, we kept the rats in individual cages, connected to a running wheel, which was equipped with a counter. Since the rats were free to go back and forth between the stationary part of the cage and the running wheel, this arrangement enabled us to determine the "distance" each animal decided to run. Once a day, at roughly the same time, we would go to the animal room, measure how much food and water each rat had consumed, weigh it, and note how much it had run.

On most of the diets, the rats eventually developed nutritional deficiencies, which usually meant that they ate less and lost weight, their fur became matted, and they began to look sick. What interested us was how their nutritional state affected their running. Preliminary experiments with rats deprived of thiamin (a member of the vitamin B complex) had shown that, contrary to expectation, as the animals became thiamin deficient, the distance they ran increased, even though they were clearly having trouble getting about. When the animals had become quite sick and we had observed what this did to their running, we would return them to a complete diet or let them get worse and die. In either case, we would clock the effects on their activity.

Each time we considered an experiment finished, that group of rats would be gassed by us or by the men who took care of the animal rooms. Then we would design another experiment, which we would start with a new batch of rats. In all this, we looked on the rats much as we did on the diets—means to getting answers to our questions.

We had lively discussions about what factors to omit from the diet and what the animals' reactions might mean. But we never talked about what we were doing to the rats, much less about whether the information we were likely to get from the experiments justified torturing or killing them. The fact is, we would never have admitted to ourselves or anyone else that by depriving the rats of essential nutrients, we were torturing them. Torture to us somehow implied an intent to hurt, and the hurts we were inflicting were incidental accompaniments of what we considered to be a much worthier goal.

We bought our rats from a supply house that bred animals to sell to research labs. Their destiny was to be used in more or less informative experiments, during or at the end of which they would be killed. How they lived or whether they died mattered only insofar as it affected the cost of the experiments.

I was mildly interested in the experiments we were doing. But their chief importance to me, as well as, I believe, to the graduate student with whom I worked, was that we would be able to translate them into my bachelor's and her Ph.D. theses.

I am sure I would not have stated it that way at the time. I was too engrossed in the glamor of doing research to think critically about it. "Where are you going?" "To the lab," as I happily trotted off—weekends as well as during the week. My parents, my brother, my friends all respected my odd schedule, convinced that, because it was research, what I was doing must be both important and interesting.

Part-way through the semester I had a strange experience. A researcher from another laboratory, who was removing the ovaries from anesthetized baby rats in order to test the effects of various hormones after the animals recovered, invited me to learn this sugical procedure. A budding Dr. Arrowsmith, I eagerly agreed. I ovariectomized my first baby rat and everything went fine. I did another and that worked, too. But with the third, I overdid the anesthesia and my patient died on the operating table.

I was devastated. I had killed a baby rat; just gone and killed it by being careless. All the while I agonized over this mishap, I sensed the contradiction. By then I had killed many rats; yet here I was all upset at having killed this one. It was as though I, personally, had not killed the others; they had been fated to die on the altar of science. This one, I had killed.

I thought about it; talked about it. Yet the experience did not lead me to raise questions about the intended suffering and death of the rats we were "using" in our nutrition experiments.

Dissecting Animals in Teaching Labs

In 1946, as a beginning graduate student, I took the physiology course Harvard Medical School then required of its first-year medical students. There we dissected anesthetized animals—usually cats—several times a week for a whole semester. Six students worked in a group and rotated tasks, so that everyone got a turn at the various procedures. One of us always had the job of maintaining a proper level of anesthesia and giving "our" cat a lethal dose of anesthetic at the end of the procedure.

I suspect the ratio of students to cats was dictated by economics more than concern for the animals, but in terms of the learning experience, it seemed about right. I cannot imagine learning an equivalent amount from demonstrations or models, or from watching videos. After all, finding out how tissues feel and smell is an integral part of the experience. Both then, and thinking back on it now, I do not think we mistreated the animals—certainly not by comparison with what we did to the poor rats in our nutrition experiments.

For one thing, the animals were always thoroughly anesthetized and never allowed to regain consciousness. For another, if we accept the idea that physicians need to know the way various organ systems in the human body function—and I do—it makes sense to me that a good way to begin is to have students observe these functions in live mammals. That is where they learn the effects different drugs have on the heart, on nerves, and on muscles and begin to see concretely how the internal organs function under various conditions. Working with these animals also offers practice with surgical skills. We learned to insert a canula into a blood vessel and to isolate nerves and muscles, all the while trying not to injure adjoining tissues or provoke excessive bleeding. Provided the animals are properly anesthetized and as long as not too many of them are used, this still strikes me as appropriate training.

In her chapter in this collection, Hilary Rose describes the nineteenth-century controversies about these practices and speculates how medicine would have developed if the "vivisectors" had not won out. Clearly, I have not overcome my own vivisectionist socialization. So far, I would simply insist that, since this pedagogy implies that it is all right to kill animals so as to learn skills that are expected eventually to benefit humans, we owe it to our students to make them aware of the fact that different people draw the lines differently and encourage them to think critically about the ways they and their teachers justify the use of animals in teaching and research.

This kind of sensitivity certainly was not encouraged when I took medical school physiology. On the contrary, all the medical students were urged to attend the hearing of the Massachusetts legislature at which antivivisection groups each year proposed legislation to regulate experiments with live animals. The role of the students was to applaud the scientists' testimony about the need for animal experimentation and to giggle or hiss when people testified about animal suffering in the laboratory.

As a graduate student and after I had become a practicing scientist, I occasionally taught undergraduate biology labs in which the students were expected to dissect frogs whose brain and spinal cord one of us instructors had destroyed in a procedure called "pithing" or to dissect anesthetized rats which were killed before they could wake up. We explained to the students why it is safe to assume that these animals cannot feel pain.

In the 1950s and '60s, students hardly ever raised questions about such labs and I, too, did not question the learning potential of these laboratory exercises. For me, still, a special thrill of understanding goes with seeing a pulsing heart or blood flowing through the capillaries of a frog's skin, provided the necessary precautions have been taken so that the animal cannot feel pain. However, during the early 1970s, when the antiwar movement was at its peak and many students and teachers were questioning everything about our society, it was not unusual for a student to object when asked to dissect an animal. Though we always excused students who did not want to participate in dissections, other issues came up as well. For example, some of my African-American students, who had grown up in rat-infested inner-city housing thought it absurd to cringe at dissecting an anesthetized rat. Some had known children who had been bitten by rats and looked on rats as vermin that should be killed. They considered it an expression of class privilege to have finer feelings about these animals. Perhaps they were being macho, but I tend to take their feelings at face value.

A question I have often wondered about is whether most students who feel uneasy about dissecting anesthetized animals actually do so out of sympathy with the animals. Would they do whatever is necessary to help an injured animal, or are they just uneasy about handling live animals? I wonder about this especially when I am told that women students are more reluctant to do lab dissections than the men are. Are we really dealing with differences in empathy with other living creatures or with differences in permitted squeamishness? After all, girls are allowed, if not outright expected, to squeal and jump on the table when a mouse runs across the floor or to refuse to touch worms, snakes, frogs, or spiders, whereas boys are supposed to be macho about it.

Clearly, laboratory teaching should not harden students to animal suffering. Therefore it is important to talk about these issues in biology classes. But I am not sure that the kinds of alternatives I have heard proposed are adequate substitutes for animal experimentation. Demonstrations, models, or videos may do for some things, but I do not believe they can always substitute for dissection. Least of all do I think that using tissues obtained from slaughterhouses is an adequate substitute, though I have heard this alternative mentioned. The food industry abuses animals to a much greater extent than most scientific and teaching laboratories do. So, to hold up the use of remnants from slaughterhouses as a humane way to deal with the problem of laboratory dissection seems hypocritical. Here again, it is important to question whether we dissect animals in order to learn about them or about ourselves, and when, if ever, we are justified in using (or rather, exploiting) other animals as surrogates for us humans.

Working on Animal Eyes

I cannot write about these issues without wondering to what extent my own research experience has desensitized me to the realities of animal suffering and has made me too ready to accept the use of animals in research. After all, from 1950 until the mid-1970s, when I stopped doing laboratory experiments, I used animals or their tissues on a daily basis.

During more than twenty-five years, I tried to understand the way the light we see interacts with the colored pigments by which it is absorbed in our eyes. These so-called visual pigments are located inside the rod and cone cells of the retina, which lines the inside of our eyeballs. Earlier experiments had shown that these visual pigments are bleached by the light they absorb. We and other researchers before and since have believed (on fairly solid experimental evidence) that the changes the pigments undergo as they absorb light are somehow transformed into nerve impulses that get transmitted from the retina to the brain. And the brain is where light elicits the sensations of seeing—vision. We therefore described our work by saying that we were working on the chemistry or, more specifically, the photochemistry of vision ("photo" is Greek for light). (I say "we" because I was not working alone but as part of a research group, which included George Wald, myself, and several other researchers and students.)

To do chemical experiments on visual pigments, we had to extract these pigments from animal eyes. So, we had to start by getting our hands on eyes. Most eyes contain very little visual pigment. To do the different kinds of chemical experiments we wanted to do therefore required dissecting the retinas out of several eyes, at times even hundreds. Depending on which animals' eyes we worked with, we either killed the animals ourselves or arranged with a slaughterhouse to get eyes of cattle or other animals that were being butchered for meat.

Initially, I worked mostly with frog or cattle eyes; later, with the eyes of squid. Of all the organs in the body, there is something particularly uncanny about working with eyes, because they seem to be looking at you. But after I got over the initial unease at killing an animal and having disembodied eyeballs stare at me, a considerable aesthetic element entered into dissecting retinas.

To obtain the maximum amount of visual pigment, we had to dissect out the retinas and do many of our experiments under as dim red light as possible to keep the pigment from bleaching. So, imagine sitting in a darkroom illuminated by eerie red light and, with fine scissors, cutting all the way around an eyeball just behind the iris. With delicate forceps you lift off the cornea and lens, and place the rest of the eye in a dilute salt solution. Then you gently insert a specially ground, thin spatula to loosen the retina from the back of the eye to which it is not really attached. And now, if you have done everything just right, as you watch, a beautiful, transparent veil—the retina—floats into the solution. What look like minute bubbles spread all around it: those are rod cells, becoming detached and floating away from the retina. And now you are ready to start the experiment you have decided to do that day.

Occasionally, someone in our laboratory worked on human eyes. To make that possible, we contacted ophthalmologists at the Massachusetts Eye and Ear Infirmary, who would call when they were removing an otherwise healthy eye because of a severe injury or a cancerous growth on the outside of the eyeball. One of us would rush to the hospital, collect the eye, put it in a dark container, and hurry back to the lab to dissect out the retina. We could then freeze the retina and store it until enough human retinas had been accumulated to do experiments.

Early on in my research career, someone called from the Eye Infirmary to alert us that an eye was being removed that afternoon. Though I was not working with human retinas, I volunteered to fetch it. (Perhaps I still had hidden thoughts about becoming a doctor.) As the operation was about to begin, the physician in charge asked whether I wanted to watch. I surely did. The patient was an old man, and the procedure was being done under local anesthesia by an inexperienced resident. The resident was visibly jittery and addressed the patient much too loudly in a strong foreign accent which the man could not understand. As soon as the resident began to retract the man's eyelids in order to inject an anesthetic into the orbit surrounding the eye, I realized I would keel over unless I stepped outside immediately. After a momentary attempt to go back into the operating room, I went out for good and waited to receive the eye after the procedure was finished.

Shortly after the end of World War II, George Wald received reprints of two scientific papers from a German scientist, Gotthilft von Studnitz, in which von Studnitz and a colleague described extracting visual pigments from human eyes. In the "Methods" section of one, they wrote that the eyes had been removed (impersonal passive mode) "immediately following execution";[1] in the other, "immediately post mortem."[2] There was no question but that they could only have obtained the numbers of eyes they needed for their experiments from concentration camps. We were horrified, but the human/animal barrier was too deeply ingrained in our consciousness for us to draw the slightest connection to what we ourselves were doing.

In fact, once I had dissected the retinas and was doing chemical experiments, I was no longer conscious of dealing with animals, or their tissues, at all. From then on, it was a matter of test tubes, chemicals, pipettes, spectrophotometers, and other arcane measuring devices. Like these, the animals were means to my exploration of "the photochemistry of vision."

This is not to say that we did not talk about how to minimize the fear and pain of the animals we killed in order to dissect their retinas. And we developed routines designed to assure as quick and painless a death as possible. With cold-blooded vertebrates, such as frogs, we used sharp scissors or knives to cut off their heads. In addition, we took care to pith both the severed head and the body on the assumption that we didn't really know at what level of their nervous system these animals experience pain. In fact, crueler than the actual killing was the way we kept the frogs. We used to buy several dozens of them at a time and store them, covered in peat moss, in a cold room, never feeding or tending them in any way. We really looked on them as reagents more than living animals.

So, it's not that we did not think about how to minimize pain, but we were easily reassured that we were avoiding needless suffering. And we never discussed whether the knowledge we gained from a specific experiment was worth killing animals for. That our experiments were sufficiently important to justify the killing was a foregone conclusion.

Living with Pets

For several years during this period we had an iguana, who freely roamed throughout our laboratory. It spent a good deal of its days in a large rubber plant we borrowed for it from the biology department's greenhouse. Then, in the late afternoon, it would climb up a pair of skis leaning against a bookcase, to spend the night lying on top. In the morning, it would come back down and wander some more. Gradually, I became its special friend, and it would come and eat out of my hand and ride on my shoulder. When my children were young, we also had assorted small animals at home. There were rabbits and gerbils and guinea pigs, and also frogs and snakes and turtles.

I can still remember my daughter's baby rabbit, which had been born into a friend's household and not learned to be shy. It would come bounding upstairs and jump into our laps in a most unrabbitty way, and left tooth marks on many chairs and on its favorite windowsill, from which it looked out at the passing scene. There was the excitement one morning, when we discovered that our recently acquired garter snake had produced a dozen threadlike babies. I cannot remember what we fed the babies—the mother ate hamburger, occasionally laced with ground eggshells—but they all grew to a respectable size before we released the lot into the marsh behind our house on Cape Cod, hoping they were knowledgeable enough to survive. And there was the occasional baby bird that lost its mom before it was ready to live on its own. I especially remember a little grackle whom my daughter taught to fly by waving her arms up and down as he sat on her hand. We called him Moishe, and when he flew off for the last time, we could only hope he was ready to live a bird's life.

The point is that, as an adult, I lived surrounded by animals who were not that different from the ones I "used" in the lab; but I experienced no contradiction between nurturing our pets and killing my experimental subjects. What was going on? Is it similar to what Robert J. Lifton has called "splitting": the fact that the Nazi doctors could torture and kill men, women, and children "at work," while at home they were loving fathers, husbands, and neighbors?[3] Yes, I know, iguanas and rabbits and grackles aren't people. But isn't the way we define and experience the distinction, and our easy acceptance, or rather invention, of it part of what we are trying to explore in this book?

My Sea Change

Gradually, various lines of thinking led me to begin to question my laboratory activities. For one thing, the U.S. war in Vietnam and the role scientists were playing in making high-technology warfare possible led many American scientists to question earlier, comfortable assumptions about our work. Until that point, I had uncritically accepted the positivist thinking of most practicing scientists. I believed that science is shaped by questions nature throws up for us and proceeds from these questions to answers, which bring up new questions, which we answer in turn, and so onward and upward toward a truer and better understanding of nature. Accepting the line scientists like to draw between science and technology, I also believed that science itself was altogether good and that only its applications needed to be examined for their effects on society.

It took the reevaluations initiated by my participation in the anti-imperialist, civil rights, and women's liberation movements for me to recognize that not only our scientific interpretations but also what we conceptualize as "nature" is embedded in European-American politics and culture.[4] Gradually, I came to understand that our science bears the marks of the men who have produced it, and that it is an essential part of the interlocking systems of domination and exploitation that have resulted in the differences in wealth and power between women and men, poor people and rich people, people "of color" and "white" people, the poor "underdeveloped" countries and the rich industrialized ones.

I began to understand that since the inception of our discipline, biologists have helped to provide scientific backing for the ideologies that justify systems of exploitation by attributing differences in power to inborn differences between the sexes, classes, and so-called races. This quick summary, of course, does not do justice to the analyses many of us have produced in the subsequent decades,[5] but it encapsulates the wake-up call that made me begin to rethink the assumptions I had internalized as part of growing up as a middle-class white person in Europe and the United States and as part of my education as a scientist.

Growing up as a woman involved further layers of denial, since I had not even acknowledged that my gender was an issue. Though I was educated, and later worked, at Harvard—one of the most male-dominated institutions in the United States—I had remained oblivious to the fact that not one of the portraits on its hallowed walls, not one of my professors, and only a small minority of science students were of my kind. Or rather, of course I noticed, but felt flattered to be singled out. It took the political insights of the 1960s for me to understand that Harvard's acceptance of a few tokens like me—hesitant and at arm's length as that acceptance was, since none of us were offered secure jobs with retirement benefits—was a way of showing that the university was accessible to anyone who was good enough, which clearly most women (or African-American, Latino, or poor men) were not. Once many of the beliefs I had accepted uncritically came crashing down, I was forced to take a fresh look at my relationships to the animals I had been using—for food and in my work.

At the same time, I ran into a concrete issue in my laboratory work. Increasingly in the 1960s and '70s, I was exploring the visual pigments of squid. These pigments react to light similarly to the ways vertebrate visual pigments do, yet with sufficient differences to open up new ways to investigate the interactions of light with visual pigments. Working with squid raised two problems. One was how to stun or kill a squid. How do I make sure it doesn't experience pain? Unlike a frog, a squid does not have a central nervous system I can destroy to reassure myself that I am not torturing it when I remove its eyes. I can cut off its head, but what does that do to the way a squid feels? These questions, which no one can answer, began to haunt me. The second problem was that squid are utterly beautiful. I can sit for hours watching them swim. After a while, I could no longer stand killing another squid. I decided that nothing, but nothing, I could find out was worth doing that ever again. And this feeling quickly spread to other animals.

If I was not going to kill any more animals, that left cattle eyes. But by then I had stopped eating meat. To refuse to eat animals but use the by-products of their slaughter in my work seemed hypocritical. And gradually, the realization began to dawn on me that other ways of looking on nature and interacting with it might lead to quite different scientific questions and practices. I could not specify what they would be, but I knew I had to stop doing the old, familiar things and try to understand how scientists create science rather than to continue with my own efforts to try to understand what we define as nature.[6]

In looking back at poems I wrote twenty years ago, it is clear that I am not projecting my present sentiments back to that time. I will quote three fragments from poems I wrote in 1973 to recapture the feeling:

I will not build my castle
on the wasteland
of neatly ordered facts,

answers to rigorously worded
pointless questions.

I will wait for the
questions that come one by one
and won't try to answer them
till I'm ready.

I have learned to recite in my sleep:

Life is the play of
DNA.   DNA makes
RNA, and that
makes proteins.

This, Ladies and
Gentlemen, is the secret of
    Life.   You,
understand, don't you?

In the graveyard of dead parts
we hope to learn how to
escape from death,

while our life's
time ticks by.

Trying to learn about
    life, we
kill the living, yet
    by this act we
    kill ourselves.

How to Resolve Such Problems?

George J. Annas and Michael A. Grodin, two American legal scholars and bioethicists, recently edited a collection of articles about the Nazi concentration camp experiments.[7] In the concluding chapter, they write:

> In the medical arena, some modeling [of actual clinical situations] can be done by using animals for experiments. However, once the leap is made to human experimentation, subject and object merge. It is this merger of the subject and object of human experimentation that makes it problematic; the researcher uses the human "object" as her model for nature. Nonetheless the human subject retains humanity, and the experimenter is also obligated to respect the rights and welfare of this subject-object.[8]

Clearly, these authors do not consider the possibility that nonhuman animals warrant similar respect. Yet, how big a step is it from routine and unacknowledged exploitation of animals to the exploitation of helpless human beings, especially when it is done for the greater good of science or humanity?

At the beginning of 1994, the new U.S. energy secretary Hazel R. O'Leary refused to continue to cover up a report prepared eight years earlier by the U.S. House of Representatives' Subcommittee on Energy Conservation and Power. The report, entitled "American Nuclear Guinea Pigs: Three Decades of Radiation Experiments on US Citizens," described experiments, conducted from the late 1940s into the 1970s, in which members of the military, prison inmates, retarded teen-agers, and pregnant women, among others, were exposed to radiation and a variety of chemicals in order to study their effects on human metabolism and biochemistry.[9] These experiments were sometimes described as "treatments," but usually there was not even a pretense that they would benefit the persons involved. And since little was known about the effects of radiation on humans (which is one reason the experiments were done), no one could predict what harm might result. Such experiments were conducted by established scientists at Harvard, MIT, the University of California, the Atomic Energy Commission's Oak Ridge National Laboratory, and other prestigious institutions.

Eva Mozes-Kor, who was herself used as a subject in Nazi concentration camp experiments, has written that "every time scientists are involved in human experimentation, they should try to put themselves in place of the subject and see how they would feel."[10] The question I want to ask here is whether it is too far-fetched to ask biologists and medical scientists to do the same when working with nonhuman organisms? For until they do, animal and plant life will inevitably be abused. The potential death of these organisms and the possibility of their suffering must be factored into any calculus about what questions scientists can, or should, ask and how to answer them.

Biologists must assume a level of responsibility such that not just "life" in the abstract but real, living organisms command their awe and respect to the point where they "put themselves in the place of their subject and see how they would feel." And where there is doubt how an animal would feel, it would be safer to overestimate its potential suffering.

Making organisms suffer or taking their lives is an act of violence, whether they are animals or people. There may be times when we feel we can justify such violence. But we should never unthinkingly accept suffering or death as necessary for the optimistic, but unproven, goal of furthering human health or improving our understanding of life.

What would my life in biology have been like if I had been encouraged to think this way at the beginning rather than coming to it after decades of unthinking abuse and killing? What if, from the start, I had been made aware of the "splitting" biologists practice? I was neutral and rather oblivious about animals as a young student, but I came to love living with them even as my work involved killing them. By the time I became aware of the dissonances, they overwhelmed me. In addition, I had become too interested in exploring questions I and others had begun to raise in the social studies of science to continue with my laboratory work.

It would be a mistake, however, to conclude that one cannot try to move beyond "the graveyard of dead parts" while working as a biologist. In fact, most of the contributors to this collection are biologists engaged in biological research and teaching while they explore the issues raised here. These essays and this book represent our collective attempt to further discussions and active explorations of biological practices which a group of aware researchers, teachers, and students might devise. By insisting on empathy and avoidance of suffering, and by erecting very high barriers to killing, I would hope that we could invent a logos of the living that would represent biology in its literal meaning.

Notes

1. H. K. Loevenich, "Sehphysiologische Untersuchungen an menschlichen Netzhäuten. 2. Die Vorstufen der Farbsubstanzen," *Klinische Monatsblätter für Augenheilkunde,* 110 (1944), pp. 620–622.

2. H. K. Loevenich and G. v. Studnitz, "Sehphysiologische Untersuchungen an menschlichen Netzhäuten. 3. Zur Spektralabsorption des Sehpurpurs," *Klinische Monatsblätter für Augenheilkunde,* 110 (1944), pp. 622–624.

3. Robert J. Lifton, *The Nazi Doctors* (New York: Basic Books, 1986).

4. See, for example, Ruth Hubbard, *The Politics of Women's Biology* (New Brunswick, N.J.: Rutgers University Press, 1990).

5. Some of the other chapters in this book, such as those by Birke, Rose, and Shiva, elaborate this point.

6. Hubbard, *The Politics of Women's Biology,* esp. pp. 1–5.

7. George J. Annas and Michael A. Grodin (eds.), *The Nazi Doctors and the Nuremberg Code: Human Right in Human Experimentation* (New York: Oxford University Press, 1992).

8. Annas and Grodin, "Where Do We Go from Here?" in *The Nazi Doctors and the Nuremberg Code,* p. 307.

9. Details of these experiments have been described in the *New York Times* and the *Boston Globe,* beginning December 31, 1993, and throughout January 1994 and beyond. See, for example, Keith Schneider, "U.S. Expands Inquiry into Its Human Radiation Tests," *New York Times,* December 31, 1993, p. A18; John W. Mashek, "CIA Launches Radiation Probe," *Boston Globe,* January 5, 1994, p. 14; Edward J. Markey, "Compensating America's Nuclear Guinea Pigs," *Boston Globe,* January 15, 1994, p. 15.

10. Eva Mozes-Kor, "The Mengele Twins and Human Experimentation: A Personal Account," in *The Nazi Doctors and the Nuremberg Code,* p. 58.

6 | More Than the Sum of Our Parts

Betty J. Wall

THIS IS A personal account of my experience with animals and plants as well as with other human beings. I want to describe many transitions in my work as well as in myself. From the time I started graduate school, most of my time was spent studying biology. As I became more proficient as a biologist, I began to learn more about myself. I began to be in touch with how I felt. I began to learn more about my interconnection with other people and with all other life.

Through experimenting in the lab with other organisms, I learned about their lives and how their well-being related to my own life. The experiment became much larger and more significant. The real experiment is with my life, their lives, our lives, all interrelated and influencing each other, each making a meaningful contribution to each other.

My parents are Chinese immigrants. My father's mother died of tuberculosis when he was four years old, and in 1911 at age seven he and his father left China and came to the United States. They went to a small town in Arkansas where they opened a grocery store. My father often said he felt "like an orphan" growing up in the United States because he had no mother and was separated from his father starting at age twelve, which is when he decided to go to St. Louis to attend school because the principal of the school in his town would not allow him to continue school there, even though my father had made straight As, was a bright student, and wanted to go to medical school. In St. Louis my father worked in a bakery at night to support himself. However, soon he was sick with tuberculosis and the doctor recommended he go to a dry climate. So my father dropped out of school and went to California, where he worked on farms and regained his health. During this time my father's father had returned to China, remarried, and returned to Arkansas. My father also returned to Arkansas, but then also opened up other grocery stores, one in Lake Providence, Louisiana, where he worked until he was twenty, at which time he decided to return to China to find a bride. He went to the "marriage broker" in the area where he was from, and was shown a few eligible young women. He said he chose my mother because she was strong.

My mother's family stayed in China. Her father was a scholar, lawyer, and banker. My mother's mother had nine children to care for. As was the custom in China during that time, my mother's father married a second wife when his first wife stopped having children, and had three more children with her. My mother said everyone got along and she grew up in a harmonious home. When my mother was a teenager, she moved to a separate house in the village where young girls received training in homemaking skills. By the time she was twenty-one, when she married my father, she was considered well accomplished.

After marriage my mother moved in with my father's stepmother, and my father returned to the United States with the idea of "making a fortune and going back to China to live like a king." After a few years my father went back to stay with my mother, during which time she became pregnant. Before their child was born, my father returned to the United States.

Their first child was a son, whom my mother adored. When he died suddenly at age seven of appendicitis, my father had never seen him. Their son's death touched both of them deeply. My mother said her mind was never as sharp afterward. My father said it was then that he decided to return to China to bring my mother to the United States so they could be together. He said he realized that they could have lived their whole lives separated, with her in China and him in the U.S.

They went to Lake Providence, where my older brother, Skippy, was born. The next year they moved to the neighboring town of Tallulah, where my father opened another grocery store, and that is where my own journey began.

I was born at home, as were my three brothers. There were always two or three other people in addition to our immediate family, also Chinese immigrants, who lived with us. We spoke primarily Chinese at home, since my mother did not speak English.

One of my earliest memories, when I was two or three years old, was of my mother telling me of my father's gambling and her unhappiness with the situation, that she would take us and leave him if she could speak English. I loved both my mother and father, and was dependent on them. I also felt they both loved my brothers and me. My mother expressing to me how she felt about my father's behavior was the beginning of my mixed feelings toward my father, which took years to understand.

As a child, I was always interested in biological phenomena. My first experiment was conducted as an assistant to my older brother when I was about three years old. I gathered the neighbors' baby chicks together while my brother placed them, one at a time, between two boards, pressing the boards together and flattening the chicks until they were limp. Little did I know I was witnessing changes which brought on death. When my father arrived home, he patiently explained to us that the changes that occurred in the chicks were irreversible and that life should be respected and not needlessly destroyed. This was a positive learning experience, and the neighbors, my father, my brother, and I all felt we kids had learned a valuable lesson.

After this experience I was more aware of the animal and plant life around me. I had a pet duck which taught me many things. I learned I could talk with it, that the duck was a wonderful and sensitive creature capable of expressing its feelings. Then one day my parents decided it was time to eat the duck. I was upset and refused to eat any. When the duck was gone, I transferred my affection to other animals. I learned to treat the vegetables we grew with respect and admiration for their beauty and abundance. I learned how to feed the chickens and gather their eggs, but also how to kill them and clean them for cooking.

Soon after my brothers and I started school, we also started working in the store, helping the customers, making change. My father encouraged us to be friendly and communicative, to look people in the eye when we spoke to them. He taught us about forgiveness by his own example, not to hold grudges. When someone poured paint over the big windows at the store, my father said we were to just forget about it. He trained us in the grocery business, how to work until the job was done, how to be on time. I have always appreciated what my father taught us about working in our family store.

Tallulah was a small town where everybody knew everybody else, and the grocery store was the hub of activity. During the time of hoeing the fields and cotton picking, we would open the store for about an hour at 5:00 A.M. for the laborers to get their day's treats. The trucks would then pick them up in front of the store to take them to the farms and at the end of the day drop them off at the store. In addition to the early morning hour, the store was open 8:30 A.M.–9:30 P.M. weekdays, until midnight on Saturdays and 8 A.M.–noon on Sundays. We children worked after school and all day Saturday. My mother had a kitchen in the back of the store and we ate supper there in shifts. The store sold all kinds of things: screens, plumbing fixtures, firecrackers, salt pork, kerosene, cooking utensils, harnesses. It was truly a general store. My mother and I had dresses made out of flour sacks, pretty prints, and always in style. From the time I first started working in the store, I knew I would always be able to take care of myself because I could be a cashier.

When my brothers misbehaved, my father hit them with a belt. I felt awful during those beatings and we were not allowed to cry. As children we are much like the young spring flower, growing taller to bloom, but easily squashed or broken. Respect and appreciation for life can be destroyed at a young age if we are hurt too much either physically, emotionally, or mentally. This kind of damage can be perpetuated for generations, as often the abuser was once the abused child. I also felt badly about my father's smoking and gambling but had no way to influence him to stop doing either. I often got headaches from the cigarette smoke. The worst thing about the gambling was that my mother got so upset.

Acknowledging feelings, especially unpleasant ones such as sadness, was not practiced in our family. My father had instructed us not to cry when normally a kid would cry, such as when my youngest brother broke his arm. He was teaching us to be strong in all situations. I learned not to feel or at least not to acknowledge what I was feeling. I learned to pretend that everything was fine, and smile no matter what.

In elementary school my grades were not particularly good; sitting still in a classroom for six hours was hard, and the people in school were much more interesting than studying from books was. My brothers all played football in school, but being a girl I played basketball and marched in the band. When I reached seventh grade I was inspired by my teacher, Miss Kitchens. She was an experienced nurturer of young minds who was able to draw out the best in any student, and through her I became awakened to the world of education and aspired to learn. From then on my grades were good, and I was head of my class.

From age thirteen until I graduated from high school, I went on Saturdays to Lake Providence to help my aunt and uncle, who had just come from China, to run their grocery store. I returned to Tallulah by train or bus early Sunday, my day off. My father encouraged us to experience religion, so I went to Sunday school. He went only to Easter sunrise service because it did not interfere with his work schedule and he thought it was a fine time for a service.

When I went away to college, my headaches disappeared. Breathing air without cigarette smoke was what made the difference. Because my father had wanted to be a doctor, he tried to persuade all of his children, all of our relatives' children, and anyone else who would ask him to become doctors. I majored in premed for one semester and realized right away it was not for me because I could not see myself being in the hospital handing out pills to patients. After I studied anatomy, my interest sparked a career in biology. From the first time I started experimenting with animals in the lab, my teachers taught what they thought was respect for the organisms: to anesthetize them and kill them quickly. I was so enthusiastic about dissecting and learning the structure of the smelly stiff cat which had been preserved in formaldehyde that I gave little thought that this had been a living organism before it was killed for science. As an undergraduate, I also had a lab project with chickens, the first animals I used in experiments in the lab.

As a beginning graduate student, I enjoyed teaching undergraduate labs. I also started dissecting cockroaches, searching for their nervous systems in the sea of tissue I saw under the microscope. Although I had never cut open an insect before and had no idea of what I was looking at, I quickly learned to identify the various parts. My interest in anatomical structure continued as a graduate student, and I dissected every kind of insect that came my way. My childhood training of working long hours in the grocery store prepared me for spending long periods of time in the lab.

As I dissected more and more animals, a change in my attitude toward the organisms started taking place. Now I realize that the more I knew about their internal workings, the more impressed I was by the intricacy and comprehensive nature of the functions and how closely related they were to my own.

When I was still a graduate student, I felt pressure to publish my first two papers.[1,2] I published them prematurely with faulty hypotheses and methods and erroneous conclusions. I needed more time. In my later experiments, I allowed myself the time to research the new questions brought up by results of each experiment; I tried to explore where the questions led me.

About that time, I was invited to speak at a symposium in East Germany,[3] and for the first time experienced how it felt to be in a country where personal freedom was overtly restricted. I took for granted my ready access to all kinds of literature while my hosts did not have the same privilege. Neither could they travel to the countries in the West. Upon returning to the United States, I felt fortunate and reinvigorated to finish my thesis.

What I also really enjoyed during my graduate student years was rock climbing, spelunking, and backpacking in the mountains for a month in the summers. The experience in nature inspired me to continue laboratory experiments with fresh ideas, as the physical world was luring me to seek more understanding of it. However, I stopped rock climbing and spelunking after I fell forty feet in School House Cave, West Virginia. I realized I could have been killed.

About the time of finishing our theses, a fellow student, Jim Oschman, and I decided that we would get married. We wanted to collaborate on some research as well as live together.

Immediately after my thesis defense, a member of my doctoral committee asked me where was the best place to do the work I was interested in. When I answered Cambridge, England, he urged me to go there. So with my first NIH postdoctoral fellowship, I went to work in the Zoology Department's Agricultural Research Council Unit of Insect Physiology in Cambridge. I was excited to be surrounded by so many people working on insects. Each investigator had a room, or shared a room, with lab benches and a sink, no telephone. How stimulating to spend tea time discussing some aspect of insect physiology. I reached a new understanding of how to do research and felt much more knowledgeable about insect physiology. Jim and I enjoyed being in the Zoology Department so much that between 1969 and 1976, we spent a few months there every year.

The work of Arthur Ramsay, a professor in the Zoology Department in Cambridge, was so inspiring to me that I modeled my work after his. When I got to know him we shared many humorous and enjoyable conversations and acknowledged the superior intelligence of the insects. Professor Ramsay was a pioneer and inventor who developed instruments for analyzing tiny fluid samples. He would make whatever he needed, from a flame photometer to analyze sodium and potassium in microsamples of fluid to a bell system to alert him when he needed to weigh a fecal pellet freshly expelled by a mealworm. He told me he had repeated all the experiments I had reported in a paper[4] and came to the same conclusions as I had, that salts had to be mobilized from the blood and stored in the tissues when cockroaches become dehydrated. The salts from the blood were not lost when the blood volume decreased, because when the insects drank water their blood volume quickly increased while maintaining a constant salt concentration. He was very encouraging and I felt a lot of support from him in the experiments I was doing. A few years later, some of us edited a book in his honor for his retirement from the Zoology Department in Cambridge.[5]

After Cambridge, I continued my postdoc in Cleveland, in the Biology Department at Western Reserve University. Bodil Schmidt-Nielsen, whose lab I worked in, was interested in my project, which consisted in trying to understand how water is transported across the cockroach rectum, from the rectal luminal contents, across the cells and back into the blood. Bodil had equipment for microanalysis. After much practice, I found I could handle the tiny nanoliter (10^{-9} liter) size samples as well as analyze them. For two years, I collected various samples from the spaces between the cells and surface of the cockroach rectum before I was able to make every aspect of the experiment work synchronously. The results of this experiment showed for the first time that cells are able to concentrate salt in the spaces between them, and thus create a gradient to transport water from the rectal luminal contents back into the blood.[6] This was a very exciting time.

Also during this time, Jim Oschman and I studied the structure of the cockroach rectum with regard to the way it transports fluids, and suggested that it recycles ions and water. When I drew the diagrams for that paper, stipling to the music of Vivaldi and Bach, I wondered how many people would actually see the diagrams and notice. that I had signed my Chinese name. We were pleased when, during the early 1980s, the Institute of Scientific Information (ISI) wrote to us and asked for the story behind that paper, because it had been one of the most cited papers for that year.[7] I am glad so many people found that paper interesting.

The Biology Department at Western Reserve had many postdocs and graduate students and attracted many visitors. Since there was at least one interesting lecture each day, we continually had to choose whether to stay in the lab and work or go to a lecture. Yet we did not want to miss either. We also had early morning and evening discussions about fluid and ion transport and about insect physiology and development. A few people were very competitive and were seriously after the Nobel prize, but I did not think of myself as working in competition with anyone else and I was not after any prizes. Discovering how something worked or fitted together was exciting enough. During the two years in

Cleveland, we often had international dinner parties, since there were people from many contries and it was fun to interact socially as well as scientifically.

After Cleveland, Bodil encouraged us to spend a year in Denmark, her country of origin. So, during 1970 we stayed in a guest room at the August Krogh Institute in Copenhagen and I worked in the Laboratory of Medical Physiology, headed by Christian Crone. We had a very enriching year, enjoying the science, the country, and the people. There we also got to know an American scientist, Albert Cass, since he lived in the other guest room and worked in the biophysics lab. Albert had come from New York to Copenhagen and, in addition to being an expert in membrane biophysics, he was a music scholar, pianist, fan of George Balanchine's ballets, and a jogger, but he was also obviously sad. His wife had left him suddenly five years before, and he had not talked with her since then so could not understand why she had left him.

During that year Bodil organized a symposium on fluid transport for the annual meeting of the Federation of American Societies of Experimental Biology, where I presented my findings on the cockroach rectum.[8] I spoke before an audience of a couple of thousand, which was both scary and exciting.

Jim was working in the lab of the Danish physiologist Hans Ussing. We did an experiment to see if the junctions between the cells of the frog skin would allow salts to pass from one side of the cell layer to the other. We put a barium solution on one side and a sulfate solution of the other. If the cell junctions were leaky, we would find a precipitate of barium sulfate in between the cells. When we looked at the photographs, we found tiny precipitates just inside the cell membrane, although we were looking for precipitates outside the cell. When I first saw the electron micrographs, I thought that those were calcium precipitates on the inside of the cell membrane. Neurophysiologists and cell biologists have suggested that cellular calcium in small quantities is responsible for the control of numerous functions at the membrane. Up until that time no one had shown evidence with electron microscopy that calcium was indeed located there. It took us two years before we had done enough work to substantiate that what we saw was indeed calcium[9] and two more years after that to publish the results of the microchemical analysis with the microprobe analyzer.[10]

From Copenhagen, Jim and I went to Northwestern University in Evanston, Illinois, where we taught physiology to premed majors and worked on our research projects. We also trained graduate students. I spent mornings writing papers and afternoons and evenings working in the lab. During the first year and a half, that was my work schedule seven days a week. I tested solutions of various elements similar to calcium to see if any of them could substitute for calcium. We found that a number of them could displace calcium, which may be a possible explanation for the toxic effects of certain metals, such as mercury. We were looking for precipitates along the inner surface of the cell membrane.

Only a small percentage of the results actually got published because it took so much time to get a paper written up.

Albert Cass suggested that I apply to share a lab with him and another physiologist, Alan Steinbach, at the Marine Biological Lab (MBL) at Woods Hole, Massachusetts, for the summer of 1972. I did, and that was how I began to come to Woods Hole. From then on, we spent the winter quarters in Cambridge and the summers in Woods Hole.

By the mid 1970s the technology had developed so that we could analyze thin frozen sections of intestines with the scanning electron microscope microprobe in Cambridge. We did these experiments using the rectum of the blowfly, and had to be able to dissect and manipulate the frozen tissue slices in a chamber kept cold with liquid nitrogen. I had to wear gloves to keep from freezing my hands, since I was cutting sections for so many hours. Then we would spend the next two days in the dark analyzing the frozen slices in order to get some data we could use.[11]

From the time I was a postdoc and throughout the subsequent years of research in the lab, I focused on smaller and smaller units. I was studying the relationship of structure and function of small parts of organisms.[6–12] However, I was aware that in dissecting I also destroyed the organisms' ability to continue functioning; but I accepted that as a necessary part of my work.

I enjoyed being in the lab. I liked the cockroaches. I imagined myself as a cockroach so I could understand more fully how it conserved water, and I even imagined myself inside the cockroach so I could get a picture of how the organs contributed to conserving water. As I observed the cockroach's ability to recycle water and ions, I also became more conscious of conservation and recycling in my own life. The cockroach was now teaching me.

After Jim and I had found calcium bound to the inner side of the intestinal membrane, we decided to look for calcium deposits on squid giant axons when we were in Woods Hole for the summer of 1973. Many neurophysiologists at the MBL were interested in whether calcium could be localized in the squid giant axon. When I started dissecting squid, I had to catch them and they very obviously tried to avoid being caught. The experience connected me with the pain I had felt dissecting cockroaches but had blocked out in order to be able to keep working with them. In a way this was a replay of my habitual response since childhood to emotional and physical pain in my life.

The sudden death of Albert Cass, in January 1974, was an enormous shock. He committed suicide just before he was to give his first lecture in his college course. I could not understand. I have contemplated the totality of Albert's loss of function preceding his death. I realized that it included his inability to understand why his wife suddenly left him nine years earlier. The resulting loss of self-confidence and mental anguish led to his drinking heavily to numb himself, while he pretended he was fine. Although Albert had a brilliant mind and thorough understanding of physics, that did not help him to overcome his loss and sadness. Through asking myself what would have helped him, I realized I did not know, and I wanted to find out.

I also realized it had not been a sudden decision, that he had openly discussed suicide. I never imagined he was thinking of himself when he asked what would be the fastest, most painless way to end one's life. Now, whenever I hear any talk of suicide, I listen carefully and make sure the person knows they need help, and gets help.

After spending three summers in Woods Hole, I decided that I wanted to move there because that was where I felt at home. I did not feel at home in a large city. When I told Jim I was staying in Woods Hole after the summer of 1975, he decided he would also. So we moved to Woods Hole that summer and transferred our research to the MBL.

While we had been teaching at Northwestern, we met a couple of students who told us about Rolfing,[13] a type of body work which loosens muscles and connective tissue. Jim started commuting to Boulder, Colorado, for Rolfing sessions, since he was experiencing neck and back discomfort. I said that the only way I was going to receive Rolfing work was for a Rolfer to move to Woods Hole. And then, in the summer of 1977, Jason Mixter, a Rolfer, moved to Woods Hole, and I decided to be Rolfed.

Rolfing is a process developed by Ida P. Rolf, Ph.D., born in 1896, a biochemist at Rockefeller Institute. Determined to help her son, who had physical problems, she investigated yoga, homeopathy, osteopathy, whatever was available, and synthesized them into her work. The Rolfer uses the hands and elbows to apply pressure on the body to free the structures that restrict movement. In a series of ten sessions, all the muscles in the body and the connective tissues that surround the muscles are brought closer to optimal function, from those that lie close to the surface to those that lie deeper in our bodies. As a result, our muscles and tendons are allowed to move with ease, tensions melt away, and memories are unlocked from previously frozen configurations.

Being Rolfed removed the shield of tight muscles in my chest, so that I could breathe more deeply and easier. I became aware of how I moved, sat, stood. My thinking became clearer. I could literally see more because my field of vision expanded. I also started having thoughts which hinted that I did not want to spend the rest of my life in the lab looking down a microscope.

Although life was interesting in even its smallest bits, I now realized, through being Rolfed and learning about this process, that we have the capacity to interact with, and improve, function rather than marvel at, and dissect, the structure. After I was Rolfed, I realized that the Rolfer had interacted with me in a way that improved my total functioning. My way of interacting with the organisms I worked with in the lab put a stop to their functions. This was a new frontier to me: interacting with the organism to improve its functioning.

In Woods Hole I also read many books in my friend Barbara Burwell's library, and through my discussions with her I deepened my understanding of history and of different philosophies and cultures. As a result, in 1979 I decided to go to India to study with a master. I stayed at a Tibetan Buddhist monastery, studied the Tibetan language, and listened to talks given by Kalu Rinpoche. While I was there I started writing whatever came to mind, and one of the things that I wrote was to become a Rolfer. So I decided to become a Rolfer. I also decided to get a divorce from Jim.

Often while I was in graduate school, I wondered how I would be feeling about my life by the end of the century. I did not consciously plan or expect my work and my life to go through as many changes as they have. But there was more to learn. Only this time there was no road map because the territory was in the process of being created. I went to massage school in Boulder, where my subjects were no longer insects and squid, but human beings. I reconnected with anatomy, but this time a living and moving anatomy. I volunteered for hospice, giving patients massages which comforted them during their final weeks before death, a special time of transition, of decreasing bodily function, and often of increasing wisdom.[14] I studied Rolfing and learned how to reactivate atrophied or defective functions without a lab full of equipment, but simply with my hands. I began to use my hands in a completely different way, as the instrument. I study Jin Shin Jyutsu, an ancient healing art. I went to China and studied Tuina, also known as Chinese Therapeutic Massage. While I was in China, I realized that the Chinese had been experimenting with functional interactions for thousands of years.

My studies to understand human behavior have their basis in the relationship of structure and function, from the subcellular level to the whole organism. When I am working with my hands to release tension from the body, I feel structural changes beneath my fingertips. As pressure is applied, the muscles let go and start feeling softer. Physical change is accompanied by a chemical change. Each trauma, whether emotional or physical, that the tissue has stored in its structure has the potential of being released. As the tension in the tissue is released, the structure is freer to move. The breath can reach every single cell of the body when the whole person is freed of physical tension, and the person can experience health and a sense of connectedness.

Treating the Organisms We Work with as Our Relatives

If we were to treat the organisms we work with as our relatives, we would start focusing on understanding our own behavior not as separate from other organisms but in relationship with them. We would observe the interconnectedness of all instead of each being isolated.

We would talk with the organisms as our friends. Every animal and plant, no matter how small, would have our respect as members of our family of organisms. Each of them could show us how they would like our cooperation in participating in their lives, and we would listen to them. In communicating with our animal and plant friends, not only would the concepts and practices of biology change; we would change.

Our laboratories could be places of cooperation between everyone, of sharing our knowledge and understanding, and not of competition. With respect for each other, we could all feel joined in our research in learning about the functioning of the universe of which we are a part. We would each have a commitment to ourselves to be here in such a way that all of the organisms benefit from our presence.

If dissection of animals and plants is necessary, we can ask that it be done with respect and appreciation for the lives that we are taking. We want to have respect for the total organisms we are using in our experiments. We have much in common with them. If we can think of all life as special, then we can proceed with the experiments we do in a caring, aware way. We would consider the birds as our winged friends, the cockroaches and worms as our crawling friends, the squid, the bacteria, as our cousins. For they are literally kin to us. They share many of the same physiological processes that we have.

Writing this chapter has enabled me to think through my relationship to experimental science and to cockroaches to the point that I have decided to go back into the laboratory on a part-time basis to do more experiments. I noticed, when I recently looked through the dissecting microscope, that my field of vision included more and I could see more than I saw fifteen years ago.

A Robin's Choice

The following experience with a robin happened in 1987, eight years after I had stopped working with animals in the lab.

In the seventh year of the garden my husband Rolland and I had planted in the foothills of Colorado, a robin befriended us. I think the robin had observed us for years, nesting close enough to hear us daily. One day two robins flew over my head a couple of times, almost touching me. I spoke to them, saying, "Hello, oh you want to get to know me." I was excited. A few minutes later a robin came back to where I was weeding in the garden and landed on the garden bed next to me, just about a foot away. It watched me as I was weeding, and within a couple of minutes I had dug up some earthworms. I placed an earthworm in my palm and said to the robin, "I bet you know what this is." The robin immediately jumped onto my hand and plucked the earthworm into its beak and swallowed the worm. After this, the robin continued to sit on my hand for a few minutes until I told it I wanted to continue weeding but would like for it to stay. So it sat on my shoulder and watched me weed the garden. I spent the rest of the day weeding with the robin on my shoulder. This was the beginning of my friendship with the robin.

The robin was curious about everything we did. That first night, as I was walking from our cabin to the shower house, the robin was walking along with me. I told the robin I was about to take a bath and that it could not come in with me. Well, after I finished my bath and got back to the porch, up comes the robin, all wet and hops up to the porch to start shaking and drying itself. I knew then that this robin understood what I said and wanted me to know that. I was delighted.

During that first day with the robin I realized the abundance of joy I felt, and it flashed through my mind that I would have the opposite of joy—sorrow, just as quickly.

I was willing to be with that robin as much as possible. In the mornings, I exercised and meditated in the garden. The robin would fly to the closest tree and sit on a branch within a foot of me. After I finished my meditation it would come sit on my hand, shoulder, or hat, and would accompany me around the place.

When I was weeding, it would play in the dirt digging with me, and I would feed it a few worms. When it was full, the robin would not eat any more worms.

The robin was curious about what we ate and wanted to taste our sandwiches, so I gave it a sample. It wanted to see what we did inside the cabin, and looked in every drawer, every pan, even when we were cooking. It watched me prepare food. Once when I was rolling out pie crust, with the robin watching me from outside the window directly above me, I decided to catch a couple of flies buzzing at the window for the robin. As soon as I walked out and asked the robin if it would like the flies, it flew over and ate them.

The robin gave Rolland technical assistance with a clock he was repairing. When Rolland asked, "What's wrong with the clock?" the robin jumped from Rolland's shoulder onto the bench and with its beak pointed to a loose bushing, which turned out to be the only thing wrong with the clock. Can you imagine the fun of sharing our life with a curious robin? The robin would land on our neighbors and friends who came by to visit, if they wanted. It was friendly to everyone and would sit on people's hands, shoulders, or hats, and eat earthworms from their hands.

When the robin had been with us for several weeks, Rolland packed to go to a watch show in Los Angeles. Of course, the robin inspected his bag, turned around, and left a robin dropping. At night the robin often perched under the porch roof next to the window.

The night after Rolland left for the show, I had a most awful dream. My friend Barbie was standing on the beach in front of her house telling me, "You're not coming to Woods Hole next week." In the background was the Tallulah family store in chaos. The dream woke me up and I was so disturbed by it I could not go back to sleep. I had a trip to Woods Hole planned for the following week and had just returned from a visit to my hometown, Tallulah, a few days before. At that time my older brother Skippy was running the store. When I left, I said goodbye, then I hugged him a second time to say goodbye again.

That whole day after the dream I was cleaning up the place as if I were going away on a trip. Even though I noticed my behavior, I did not think much about it. I was glad to get all the papers taken care of. Just as I was finishing the last letter, the phone rang. It was my brother Simeon telling me that Skippy had just died suddenly from a heart attack. I felt as if someone had hit me in the solar plexus.

The next morning the robin was right there waiting for me. I explained to it what had happened while I was crying and packing, and the robin listened and was watching me. I said goodbye because I knew the robin would not be there when I returned, especially since I did not know how long I would be gone. I remembered, as I was packing, the flash of sorrow I had experienced that first day of being with the robin.

After I returned to Colorado, the robin was gone. It had been a special gift. Being friends with the robin helped to sustain me through a very difficult period. I was able to put into practice forgiveness of myself and my dead brother's wife, replacing judgment, and healing my relationship with her. There was a big misunderstanding after my brother died. My mother still owned the store, but my sister-in-law thought that my brother had bought the store from my parents. My brother had talked about buying the store but had never paid them any money for it. My father's estate had never been settled after he died two years earlier. So both my father's and brother's estates had to be settled at the same time. We divided the store between my mother and my sister-in-law, but my sister-in-law was dissatisfied with that and had hard feelings toward the family. Repeatedly, I was able to visualize both myself and my sister-in-law behaving in a friendly way toward each other.[15,16] Eventually the visualization became reality, and today, a number of years later, I see much benefit from healing the relationship with my sister-in-law. Losing her as a friend would have been a real loss.

Our experience with the robin points out a very unusual instance of other parts of the biological world expressing a desire to investigate the human world. It undoubtedly exists in many instances, but we miss the opportunity when we are unwilling to become involved at this level. We must learn to open ourselves with greater sensitivity and greater participation and give up our sense of isolation as a species apart from the rest of the natural world. This change of attitude is likely to lead us to types of changes that we cannot even imagine now in the study of biological systems.

Notes

1. B. J. Wall and C. L. Ralph, "Responses of Specific Neurosecretory Cells of the Cockroach, *Blaberus giganteus*, to Dehydration," *Biological Bulletin*, 122 (1962), pp. 431–438.

2. B. J. Wall and C. L. Ralph, "Evidence for Hormonal Regulation of Malpighian Tubule Excretion in the Insect, *Periplaneta americana* L," *General and Comparative Endocrinology*, 4 (1964), pp. 452–456.

3. B. J. Wall, "Regulation of Water Metabolism by the Malpighian Tubules and Rectum in the Cockroach, *Periplaneta americana* L.," *Zoologisches Jahrbuch der Physiologie*, 71 (1965), pp. 702–709.

4. B. J. Wall, "Effects of Dehydration and Rehydration on *Periplaneta americana*," *Journal of Insect Physiology*, 16 (1970), pp. 1027–1042.

5. B. L. Gupta et al., *Transport of Ions and Water in Animals* (London: Academic Press, 1977).

6. B. J. Wall, J. L. Oschman, and B. Schmidt-Nielsen, "Fluid Transport: Concentration of the Intercellular Compartment," *Science*, 167 (1970), pp. 1497–1498.

7. J. L. Oschman and B. J. Wall, "The Structure of the Rectal Pads of *Periplaneta americana* L. with Regard to Fluid Transport," *Journal of Morphology*, 127 (1969), pp. 475–510.

8. B. J. Wall, "Local Osmotic Gradients in the Rectal Pads of an Insect," *Federation Proceedings*, 30 (1971), pp. 42–48.

9. J. L. Oschman and B. J. Wall, "Calcium Binding to Intestinal Membranes," *Journal of Cell Biology*, 55 (1972), pp. 58–73.

10. J. L. Oschman et al., "Association of Calcium with Membranes of Squid Giant Axon," *Journal of Cell Biology*, 61 (1974), pp. 156–165.

11. B. L. Gupta et al., "Direct Microprobe Evidence of Local Concentration Gradients and Recycling of Electrolytes during Fluid Absorption in Rectal Papillae of *Calliphora*," *Journal of Experimental Biology*, 88 (1980), pp. 21–47.

12. B. J. Wall and J. L. Oschman, "Structure and Function of Rectal Pads in *Blattella* and *Blaberus* with Respect to the Mechanism of Water Uptake," *Journal of Morphology*, 140 (1973), pp. 105–118. See also nn. 6–11.

13. Ida P. Rolf, *Rolfing: Re-establishing the Natural Alignment and Structural Integration of the Human Body for Vitality and Well Being* (Rochester, Vt.: Healing Arts Press, 1989).

14. E. Kübler-Ross, *To Live until We Say Goodbye* (Englewood Cliffs, N.J.: Prentice-Hall, 1978).

15. Stephen Levine, *Healing into Life and Death* (New York: Doubleday, 1987).

16. Gerald G. Jampolsky and Diane V. Cirincione, *Love Is the Answer: Creating Positive Relationships* (New York: Bantam, 1990).

PART III

Theorizing and Practice of Biology

Few practicing biologists can afford to think of our experimental subjects the way Betty Wall thought of her friend the robin, Lynda Birke thinks of her horses and dogs, or Ruth Hubbard thought of her iguana. Of course, the three of us were (or are still) living with these animals and not experimenting on them. But, being biologists, we cannot help thinking analytically about the way even our animal friends behave and live. Thus all three of us argued in the last section that our experiences of animal friendships are part of our scientific lives.

In addition, biologists who are feminists almost automatically incorporate into our work a constant effort to be aware of our own position relative to the human and nonhuman organisms with whom we live and work. At our best, we also try to imagine what it must be like to live the life of the organisms around whom we construct our theories and experiments.

Unfortunately, this is not common practice. Too many biologists try as best they can to eliminate, or at least forget, signs of character and individuality in the organisms they "use" and to think of them as reagents. At the same time, increasing numbers of biologists, anthropologists, and popularizers of science are making a genuine effort to understand how nonhuman animals experience the world. The books *Animal Minds* by Donald R. Griffin and *The Hidden Life of Dogs* by Elizabeth Thomas are examples. And biologists are going even wider afield. So, for instance, a conference was held at the National Academy of Sciences in Washington, D.C., at which biologists explored the kinds of information insects communicate by means of their chemical senses—tactile, taste, smell. According to the *New York Times* (March 29, 1994, pp. C1 and C10), researchers not only described the astounding range of abilities of these animals to secrete and receive signals, produced by means of subtly different chemicals, and the ways these capabilities affect their relationships with members of their own and other animal species. They also explored the relationships these capacities enable insects to establish with the plants that form part of their social world. Thomas Eisner, one of the conference organizers, is reported as saying that the colloquium's aim was to encourage researchers to "empathize with their subjects" by trying to imagine what it must be like to experience the world through our chemical senses, rather than through our eyes and ears.

However, when we try to get into the skins (or exoskeletons) of the animals with whom we experiment, we must not overestimate our capacities for empathy. Recognizing the inevitable differences in power between scientists and their experimental organisms, the contributors to this collection reject the dichotomies that set humans apart from the rest of nature and insist that biologists be aware of their responsibility to avoid inflicting suffering. In this vein, the contributors to this section—all of whom are research biologists—offer their thoughts about how research and theorizing in their special areas of interest would need to change if it were done, so to speak, from inside and with constant empathy and respect for our fellow organisms. However, like the editors, all of them accept that it may sometimes be necessary to use animals in experiments.

Anne Fausto-Sterling addresses the theme of respect for animals by establishing distinctions between meaningful and needless or trivial experiments that, at best, answer questions not worth asking. And in biology, pointless questions waste not only time and money; they inflict needless pain and waste lives. While granting the necessity for some animal experimentation, she takes as her example a set of experiments on sex differences in the structure of the hypothalamus, which have a shallow theoretical basis and cannot give decisive answers. Fausto-Sterling then attempts the difficult task of outlining criteria for deciding about what research is worth doing and how to do it while minimizing pain and injury to nonhuman animals. At one point, indeed, she raises the important question (mentioned also by Hilary Rose) of what would happen if we insisted on formulating our questions and designing our experiments to be uninvasive.

In that case, they could be done with people, and specifically with those people who have the power to refuse consent. But since only few, relatively privileged, people have that power, it is fairly revolutionary to suggest that experimentation be limited to them, rather than to the usual cast of human guinea pigs: prisoners, workers, poor women, and their children. However, such a radical change could have the further radical consequence that we would experiment with animals only when we wanted to know something about *them* and would not use them as cheap and convenient surrogates for humans. Fausto-Sterling does not go that far, but underlying her discussion is an insistence on reverence for life—all life—that would make taking any life a matter worthy of careful consideration.

Marianne van den Wijngaard discusses similar research that has been singularly blinkered. By focusing on sexuality and sex differences, she taps into strongly held cultural prejudices that have cast a pall over this area of animal experimentation. Again and again, scientists have turned to animals in entirely artificial settings. Yet they have interpreted the results of their narrowly constructed experiments as though these offer us insight into animal (and, by extrapolation, human) sexuality untainted by culture.

Her account would often be funny if it were not so enraging that animals are being submitted to violations for no other purpose than to reinforce cultural stereotypes of what it means to be feminine or masculine, gay or straight. Forget the fact that the animals usually are highly inbred rats, reared in deprived environments, so that their behavior does not even resemble the behavior of ordinary rats in the field. Surely, research into sex differences and sexuality is among the saddest examples of the misuse of animals by biologists just because it is so suffused with ideological commitments.

Rearing animals in laboratories and zoos limits their intelligence as well as their behavior. Lesley J. Rogers emphasizes faulty inferences about animal intelligence and thought that are often drawn from experiments done with animals who have been reared under such restricted conditions, or who have been surgically manipulated. The all too facile extrapolation of these observations to free-living animals or to humans is questionable, if not meaningless. Using a variety of examples of more carefully thought-out experiments, she shows that a good deal can be gained from investigations that do not inflict damage.

A wide range of observations challenges the frequent refusal by biologists to accept the notion that animals are conscious and self-aware and behave with forethought and intentionality, and that (to use Rogers's words) they experience "feelings of pleasure, joy, sadness, loss." By disregarding obvious manifestations of animal awareness and emotionality, biologists thus often succeed in reinforcing the unbridgeable wall Western culture and thought have erected between humans and other organisms.

Turning to evolutionary theory, another area in which there is no lack of ideological commitments, Judith C. Masters emphasizes the extent to which thinking has been structured around the notion that organisms are constantly trying to gain control—over each other and their environment. All too often evolution is interpreted as the result of a never-ending competition among organisms intent on freeing themselves from environmental constraints. This picture portrays the environment as static and dead, and ignores the fact that "the environment" is itself composed of interacting organisms who change all the time. Masters argues that attention to the co-adaptation between organisms and their surroundings would yield a more realistic account of the relationships among different organisms who constitute each others' environments.

Contradictions also arise in the conventional concept of speciation. It is hard to see how the accumulation of variations (mutations) within a species, followed by competition for static resources in preformed "niches," can give rise to the substantial differences that constitute a new species. Thus the relationship that connects mutation, via competition, to adaptation and evolutionary change, though widely accepted, is obscure.

Masters suggests that by recognizing that we are part of nature and not the apotheosis of its evolution, we would be able to develop more consistent explanations of natural processes and become more respectful toward them. This might lead us to formulate a scientific story of human origins that does not ride on the competition and violence displayed in the Darwinian and neo-Darwinian war of all against all.

7 | "Nature Is the Human Heart Made Tangible"

Anne Fausto-Sterling

> *Beauty is no threat to the wary*
> *who treat the mountain in its way,*
> *the copperhead in its way,*
> *and the deer in its way,*
> *Knowing that nature is the human heart*
> *made tangible.*

—Marilou Awiakta[1]

ONCE UPON A time, long ago, so the Cherokee Sages tell us, human hunters began to kill too many animals. If the behavior continued, life itself could disappear. So the animals met in council and the deer's chief Awi Usdi (also known as Little Deer) offered a solution:

> "We cannot stop the humans from hunting animals. . . . However, if they do not respect us and hunt us only when there is real need, they may kill us all. . . . [Therefore] whenever they wish to kill a deer, they must prepare in a ceremonial way. They must ask me for permission to kill one of us. Then, after they kill a deer, they must show respect to its spirit and ask for pardon. If the hunters do not do this, then I shall track them down. With my magic I will make their limbs crippled . . . "[2]

In thinking about the parable of Little Deer we ponder the question of human–animal interactions. While the story comes from another age and a particular culture, like any good tale, it translates through the distance of time and experience. For the Cherokee, Awi Usdi embodies what their poet Awiakta calls "the sacred law of taking and giving back with respect, the Sacred Circle of Life."[3] Awiakta's efforts to translate the wisdom of Little Deer into the atomic era have tempted me to envision one of Little Deer's woodland pathways as a road which leads toward an ethical framework for the use of animals in the biological researches of the future. I invite the reader to explore that way with me. But let the traveler beware. I'm exploring. I don't pretend, at this moment, to have definitive answers, but if we all shoulder some of the responsibility perhaps we can find the next clearing where in the leaf-filtered light we can take our bearings before setting off again in search of Awi Usdi's trail.

The question, as it comes to me, is what do we want to know and how badly do we want to know it? Or rather, just how far will we go to find out? I am a biologist who experiments on living organisms. Even before these days of high tension over the use of animals in experimentation, I hedged my bets by working with invertebrates.[4] There are no federal guidelines covering animals without fur or backbones, and not many people empathize with fruit flies or flatworms (*Planaria*). Although when friends see them up close, under a dissecting microscope, they come face to face with those exquisitely detailed antennae (fruit flies) and those totally silly-looking crossed eyes which make *Planaria* look perpetually surprised and quizzical, and a little something stirs. ("Are you *sure* they don't feel anything?" one of my companions asked. Of course I'm *not* sure, so I do a verbal shuffle.) The truth is that I can't dissociate my scientist self from the person who lives in contented commensalism with three furry cats: I draw my personal line at experimenting on mammals. On the rare occasions when I dissect vertebrates—say a frog[5]—I admire its beauty and apologize to its spirit. Yet I am willing to do it if I think I can convince my students that amphibians and their ilk are worthy of respect, that their habitats deserve preservation. (I myself was rendered speechless the first time I opened the abdominal cavity of a living frog and saw its innards laid out so neatly, pulsing vividly; the entire arrangement glowed with an awesome beauty.)

I am indeed on a slippery slope. I provide no logical justification for my own position. Rather I think it is a question of how far down the hill my psyche allows dissociation between my love of life (why else would I be a biologist?) and my willingness to take it in order to find out something new about how the world works. Nor do I presume to judge those colleagues whose ability to dissociate differs from mine. I do not feel that my co-worker who analyzes the functioning of the cat retina in the lab—while curling up with his pet feline at home—is morally less couth than I. Indeed, this piece is not about drawing moral lines in the sand; I try instead to suggest a framework for discussing a question of concern both to many working scientists and to many lay people.

Yet some animal research offends me; it is not so much that appealing animals end up dead (although we need to deal in a more honest and spiritual way with that fact). No. It's that the work wastes animal lives in an attempt to obtain information which I judge to be useless at best, socially harmful at worst— either for the furtherance of the species under study or for improving human welfare. In imagining the future of biological work I propose a mixed ethic of respect, utility, and contextual knowledge. I envision this tripartite ethic as being like a three-legged stool: it can balance properly only when each leg has been properly shaped. But we cannot cut one to the proper length without measuring each of the others. This, then, must be my method—to focus on one leg at a time, while never losing sight of the other two.

Gaining Knowledge

I've begun this balancing process by analyzing a single research article by Slimp et al.[6] It is a short report whose authors include scientists with long and well-known careers. Published in 1978, in scientific time it is also old. I have two reasons for choosing this particular piece. First, although the article is somewhat dated and today's greater restrictions on primate research make it harder to gain permission to do such work,[7] the publication continues to be cited as scientific evidence in contemporary and highly politicized debates about human sexuality.[8] Some may feel that I'm beating a dead horse by critiquing a fifteen-year-old paper, but others still use the results in current scientific arguments. Furthermore, although little work of the sort represented by Slimp et al. is done today on primates, similar studies in rodents continue.[9] Thus a careful examination of this paper can highlight issues relevant to building an ethic for animal-based biological research. Second, we can use the piece by Slimp et al. to discuss the question of contextual knowledge. What do we need to know—in this case about human sexuality—and why do we need to know it? Having looked at these questions, we can reflect on the particular mix of respect for the experimental animals, utility, and contextual knowledge represented by the piece. The reflection itself is, of course, also a matter of debate. Readers can simply accept my thoughts as their own, or they can use them to stimulate new and different responses.

I learned in the article's introduction that the rhesus was an ideal animal in which to identify the parts of the brain associated with male sexual behaviors. First, rhesus behavior had been well studied in the lab and field, and second, one could analyze the distinct components of male mating behavior: manual contact, mounting and thrusting, erection, intromission, and ejaculation. (Notice that the purpose here is to isolate the animals from their social context in hopes of establishing consistent behavior amenable to numerical analysis.) And happily, the rhesus also masturbated, both in the laboratory and in more natural settings.

What next? For their study the authors chose eleven males aged eight to twelve years. They settled on these particular animals because they could ejaculate at least 40 percent of the time (based on five or more tests) when paired with receptive females. (Female rhesus are receptive only around the time of ovulation or when scientists use hormone injections to mimic the ovulatory period. The question of female sexuality was not at issue in this study; rather, the females were experimental aids in the study of male sexual behavior. One of the unarticulated assumptions was that male sexual behavior could be rendered understandable by providing a neutral, invariant female presence. I know of no studies of this sort that examine male and female sexuality simultaneously, considering them as mutually interactive.) Slimp and his colleagues damaged small but very particularly located[10] areas of the brains of six of these males, while the other five served as undamaged controls.

Two features of studies on brain and sex in monkeys to which I will return in due time already become evident: (1) presumably because of the expense and difficulty of obtaining and rearing primates, experimenters usually use very small experimental groups; (2) all we know about the animals' past is that they were born in the wild. We have no idea how long they have been in captivity; but we do know that in the laboratory, males and females are housed separately (in the wild they live in mixed sex and age groups). Thus Slimp et al. studied the neural basis of social behavior in animals housed in highly unusual social settings.

Slimp et al. anesthetized the chosen males and, using precise instrumentation, destroyed tiny portions of the hypothalamus by passing an electrical current through a fine needle inserted into the test area. They chose this particular area of the brain because similar experiments in other mammals had implicated it in male sexual function. For sixteen months following the operation the authors examined the monkeys' sex lives; then they killed them so that they could examine microscopic slices of their brains to figure out exactly which cells they had destroyed during the original operation. As an aside, I note that, using a formula typical of such research reports, the authors never say in so many words that they have killed their experimental subjects. Instead they write that they anesthetized them and perfused them with 10 percent formalin. The cognoscenti understand that formalin is a fixative, used to curdle and coddle and preserve biological tissue so that it may be stored and studied in perpetuity. Dissembling about the fate of experimental animals has been standard practice since the 1920s, when Walter B. Cannon and F. Peyton Rous, well-known biologists and powerful editors of key scientific journals, set up editorial guidelines specifically aimed at diffusing antivivisectionist attacks on the use of animals in research.[11]

Slimp and colleagues assessed the effects of their operation on heterosexuality by watching the behavior of male-female pairs. Since the male monkeys never had to choose between a male and a female mating partner, these studies did not assess sexual preference (as sometimes suggested when the article was cited in later articles). Experimenters placed receptive females in plexiglass cages measuring five feet long, four feet wide, and two and a half feet high. The bottom was made of wire mesh. Human observers sat about thirteen feet away from the cages. Males entered such cages and had a maximum of half an hour to mate. Sometimes the authors put a male and a female on either side of a divided cage. The male could get into the female's quarters by pressing a lever, a trick learned by receiving food reinforcements. During the months following the brain operation, experimenters tested males from thirty-eight to forty-eight times. They made certain that the females would always be willing to mate by using ones in estrus (ovulating females) and ones whose ovaries they had previously removed. The latter they could bring into behavioral estrus by injecting a combination of hormones which mimicked natural heat.

Finally, the authors wished to find out about the solitary sex lives of the brain-damaged males. Well, it wasn't totally solitary. Although each animal had his own cage, all the cages were housed together in single-male-only quarters. Both pre- and post-operatively, technicians combed the drop pans beneath the wire cages looking for clumps of semen. Occasionally, they also set up video cameras and taped the monkeys, recording bouts (single events) and sessions ("collections of bouts not separated by more than 20 minutes"[12] of masturbation).

Thus we have elaborate procedures carried out for more than a year and the sacrifice of six rhesus monkeys to the cause of increased knowledge. What were the authors able to learn? First, although one of the males recovered completely, in all the others, destroying the small area of the hypothalamus under study greatly reduced the average frequency of male-female contact, mounting, intromission, and ejaculation (compared with controls, which had had sham operations with no brain damage). Ejaculation and intromission were more severely affected than initial contact and mounting. Second, the operation did not change the frequency of masturbation. Third, the males also had the chance to press a lever to gain access to estrous and nonestrous females. Before the operation the control group pressed the levers more often than the experimental group. But after the operation there was no difference, suggesting that members of the brain-damaged group actually *increased* their willingness to seek female contact. They did not, however, care whether the females were in estrus or not. Apparently they sought social contact without clear sexual intent. Finally, the authors noted that a few other behaviors changed following the operation. Compared with controls, for example, the operated males received grooming four times more frequently than before the operation. And, most interesting for interpreting the effects of the operation on monkey sexuality, females mounted operated males quite frequently. Before the operation, none of the males offered themselves for mounting ("presented") but afterwards on many of their meetings when females contacted them, they responded by presenting, and then the females mounted.

The authors concluded that damaging a particular part of the hypothalamus changed five-sixths of the males' ability to copulate but didn't alter their masturbation frequencies. In addition, male-female social interactions changed in a way that confused the authors. Here's how they discussed it:

> Presenting . . . is normally displayed to other males [but] rarely to females. The subjects of this study were not paired with other males . . . [but they] showed an increase in the number of contacts by the female and the number of presentations when contacted. The presenting by lesioned males and subsequent mounting by female partners could have been due to an increased tendency of the males to assume a feminine role. However, it is possible that these female partners were using contact and mounting behavior as a type of sexual solicitation for these relatively inactive males. (P. 119)

We learn that males normally present to other males but we don't know whether this behavior has changed, because the authors never studied male-male pairs. Furthermore, it seems that this normal male-male interaction isn't considered evidence of homosexuality. But, we are told, the willingness to present to females might indicate that the induced brain lesions increased the male's willingness to undertake a feminine role—presumably the willingness to present and be mounted. But does this make the mounting females more masculine? Nobody has tinkered with their brains, so how could that be? The authors' alternate suggestion avoids the issue by hypothesizing that female mounting might really be a female request to be mounted.

Let's look at the balance sheet. Six male rhesus monkeys which had lived in captivity for many years received brain operations followed by extensive testing before finally being killed and having their brains closely examined. In exchange for their lives, we have learned that destroying a particular portion of the hypothalamus greatly diminishes male rhesus monkeys' abilities to copulate but does not alter the frequency of masturbation. Furthermore, the operation changed male-female social interactions in ways that are uninterpretable without additional studies.

What Have We Learned?

When thinking about human sexuality we often distinguish between the physiology of the sexual response and its social meaning and context. Masters and Johnson,[13] studying humans in heterosexual, homosexual, and masturbatory activities in the laboratory, have given us the most complete information about the physiology of the human sexual response. Their work tells us little about the meanings and motivations of human sexual activity, but their conclusions about physiology are clear. To paraphrase Gertrude Stein—an orgasm is an orgasm is an orgasm.[14] There is, of course, much we don't know about the physiology of human sexuality. Exactly which brain cells become involved? Are they always the same ones? Which hormones facilitate the response and how—at the molecular level—do they work?

For the moment let's set aside the issue of whether these are useful questions to ask and put out the following query: does the work of Slimp et al. help us answer them? I suggest not. The two most obvious problems are sample size and social settings. Slimp and colleagues' work generalizes from a sample of six (of which only five showed the relevant effects). Presumably they used such a small sample size in the first place to preserve animal lives (and keep experimental costs down). Nevertheless, the use of small sample sizes to study complex behaviors presents certain problems. One of the biggest is the extent of individual variability. Like humans, other primates have individual life histories and experiences which affect their neural pathways—both physically and functionally. In Slimp and colleagues' work one of the operated males recovered complete sexual function. Why? There are two possible (and not mutually exclusive) explanations. Postmortem brain analyses revealed that the recovered monkey's brain sustained the smallest damage. Perhaps the experimenters failed to destroy the "key" sexual control tissue. On the other hand, it is also possible that the recovered monkey had had different sexual experiences before it became a lab animal. These might, in turn, have led to the development of multiple or different neural pathways enabling sexual function. The individual in question could have compensated for medial preoptic hypothalamic damage by detouring necessary brain activities along alternate routes. Finally, in the obvious problem category: the experimenters studied the sexual behavior of highly social, group-living animals in one-on-one meetings under highly restrained circumstances. Will effects on behavior observed under such circumstances tell us anything about the physiology of behavior in more social settings? The less obvious difficulties are, in my opinion, also the most damning: the experimenters do not clearly distinguish between physiological and social components of sexuality, they don't specify a theoretical framework for analyzing hetero, homo, and auto stimulation, and their comparison of male and female involves unarticulated assumptions about sexual difference.

The experiments of Slimp et al. fall into a tried and true category of neurophysiological interventions called ablation experiments. The principle is simple. Destroy something—preferably a very small and carefully localized something—and find out what's changed. The approach has its uses, but it is also littered with interpretive booby traps. A simple example will suffice. In 1968 Lawrence and Kuypers, well-known neuroanatomists, published results of studies on the functional organization of the motor system (how the brain and nerve tracts control basic movements of the arms and legs) in the rhesus monkey.[15] In their work they attempted to separate the contributions to the control of limb movement of several major nerve bundles (tracts) running from the brain to different regions of the spinal cord. To do this they systematically cut tracts—one group at a time. They then tested the effects of damage on arm and hand movements. Their major finding was that immediately after the operation to destroy the nerve bundles (technically, the pyramidal tracts), the monkeys' arms hung limp and useless at their sides. With time, however, they regained a great deal of limb use, although never the full strength or finger dexterity exhibited before the damage. The recovery suggested that when the pyramidal tracts are damaged, other pathways can take up the slack. The long-term effects of the damage implied that the pyramidal pathways may provide more uniquely for speed, agility of movement, and fine finger coordination than for use of the limb. Lawrence and Kuypers reached these conclusions after cutting both the left and right pyramidal tracts. In other monkeys, however, they cut the tracts only on one side. In this case, only one limb hung limp and useless after the operation. But when they restrained the still working arm, the monkey suddenly started to use its supposedly paralyzed limb. Something more complex was happening here. Although Lawrence and Kuypers record the result, they never discuss it.

In fact, most neurobiologists fully understand the positives and negatives of ablation experiments. While ablation experiments can be used to gain important information about how nerve cells function within specific pathways, the best experimental setups for obtaining such knowledge are those in which the microanatomy is well known and in which it is possible to remove only a small number of cells of known location.[16] Such experiments can give interpretable (and fascinating) results, but they usually concern more local neurophysiological phenomena rather than large questions concerning complex and highly variable behaviors. In addition to the hypothalamus, many other areas of the brain are needed for the expression of sexual behavior. We understand little about the integration, during the expression of sexual behaviors, of activities in these different brain regions. The argument for examining the actions of subdivisions of one brain region (while ignoring the global picture) is simple. One has to start somewhere; only after we've identified all the shapes of the puzzle pieces can we assemble them and see what the picture looks like.

Paradoxically, it may turn out that when the whole picture is included, the shapes of the individual cutouts will differ. The behaviors observed after brain damage depend on the actions of remaining brain tissue, which may either react badly to the damage or compensate for the loss in some way, thus either augmenting or decreasing the effects of ablation. A lesion can affect function in some connected area of the brain, thus leading to an overestimate of the function controlled by the damaged area. The damaged area of the brain may also normally inhibit or stimulate function in some other brain region. The effect on behavior after damage may result from a relay effect of some sort. It would then be incorrect to surmise that the damaged area "controlled" the observed behavior. To illustrate, consider the phenomenon of agraphia—the loss of ability to write after certain kinds of brain damage. Neuropsychologists Kenneth Heilman and Edward Valenstein state that "when a person is writing . . . the language and motor areas [of the brain] must be coordinated. Lesions that disconnect these areas produce agraphia even though there may be no other language or motor defect."[17]

Other work clearly demonstrates that social setting directly affects the results of both ablation and electrical stimulation experiments. In a different set of studies, experimenters destroyed groups of cells in another part of brain called the amygdala. Resulting behavioral changes included increased and "inappropriate" sexual behavior. Animals evidenced comportment changes, especially when examined in pairs. However, when the animals were housed with a larger number of unoperated companions, the behavior decreased, and it never shows up in natural settings.[18] Social setting also affects the results of experiments which focus on specific brain stimulation rather than specific neuronal destruction. Perachio notes "that sexual behaviour produced by hypothalamic stimulation is modified by . . . the social structure of the group in which the animal is stimulated, [and] the social relationship between pairs of females tested with the stimulated male. . . . "[19]

Introductory science classes usually start off by singing the praises of the scientific method. The scientist offers up a hypothesis, we are told. The hypothesis follows logically from some larger story—a theory. The goal is to submit the hypothesis to experimental test. If confirmed, the theory continues to hold. If not, the theory may be wrong. If enough derived hypotheses come up with the "wrong" answers, a general theory eventually falls into disrepute. Hopefully a new theory emerges to replace the moribund one.

It's not easy to find the hypotheses and theories in the work under discussion. In the introduction to Slimp et al.'s paper we learn first that the medial preoptic-anterior hypothalamus (MP-AH) is "critical" for "the display of male sexual behavior."[20] Rodents with damaged MP-AHs show no interest in mounting and do not ejaculate, while electrical stimulation of certain parts of the hypothalamus evokes erection in physically restrained, isolated monkeys but not in freely moving ones. Focusing us further, the authors conclude that prior literature establishes the role of the MP-AH in "the display of male sexual behavior," but they suggest that the region may also be involved "in various types of sociosexual behavior other than heterosexual copulation."[21] Does damage to this brain region, for example, affect erection and ejaculation per se or only in the context of male-female mating?

Implicitly the introduction separates questions of "pure" physiology—erection and ejaculation—from the social context. The latter seems narrowly defined as the presence or absence of a single female (with all subjects in small cages); no comments appear about the multianimal social context in which mating in the wild occurs. Apparently the hypothesis is that the MP-AH may uniquely affect male sexual response in the context of male-female interactions (defined as contact, mounting, erection, intromission, and ejaculation), but not in other contexts (defined as solitary self-manipulation and ejaculation). But what is the broader theory? With these narrow behavioral definitions, intercourse and masturbation are obviously apples and oranges. While erection and ejaculation are common physiological components, the behaviors needed to carry them out differ. The conclusion seems to be that damage to the MP-AH changes a male monkey's interest in socially but not in privately induced ejaculation. But the data make such a simple interpretation difficult. Contact and mounting—the social components of male-female interactions—are less affected than the physiological components (intromission and ejaculation). Furthermore, males remain actively interested in having company—although the company need not be in a mating mood. Even more confusing, intromission and ejaculation are interrelated, since animals presumably use intromission and thrusting as a means of achieving ejaculation. Curiously, the authors never mention thrusting—which might be analogous to manual stimulation. Thus even the basic physiological and proximate social context of male sexual response remains murky. Without a theoretical framework the results tell us little.

The authors never articulate their assumptions about heterosexuality, although they presumably have some; the passage on female-male mounting suggests they have a theory about male homosexuality as well. They propose no theoretical framework for examining female sexuality, and despite their acknowledgement that the male sexual response is deeply social, they seem to think they can profitably study it without a theory of social interaction—one in which they conceptualize the female as an active decision maker rather than a passive sexual receptacle, fully controllable by adjusting her hormone levels.

Useful Knowledge

Why do we do research? A lot of my colleagues would claim that humans have an innate need to question and understand—that we search for knowledge in all forms: knowledge for knowledge's sake. We do research because humans are curious. Others would offer more practical accounts. We do research, they would say, because we want to understand ourselves better. Greater understanding might improve management of our social world, perhaps enabling us to solve painful and still intractable problems. Although our goal is to understand humans, we use animals because it is unethical to experiment on humans. (Lawrence and Kuypers's studies on how brain–spinal cord connections control motor function, undoubtedly have practical importance for humans for the treatment of many kinds of paralysis.) Still others point to the moral imperative of fixing up the mess we have made of the earth, of averting the peril of human-induced mass extinction. To do this we must learn more.

Humans may be curious, but they always structure their knowledge seeking. We create knowledge of the natural world by following our scientific muse. Hence this paper's title, "Nature Is the Human Heart Made Tangible"; but the human heart, our scientific muse, is no more "natural" than the knowledge it helps to build. In certain historical eras, humans are fascinated by some topics but not by others. Ideas, questions, and research programs always have a history. Our scientific muse lives and develops in particular historical and social contexts. There is no doubt that research programs on mammalian sexuality are motivated by a wish to understand ourselves.[22] The experiments considered in this chapter, however, don't tell us much about monkey sexuality; and given that human sexuality is even more complex, I think we've learned little of value. We have spent resources without adequate compensation. This is the utility leg of my metaphorical three-legged stool—an ethical quid pro quo. If we take, we must give back—to humans and to the maintenance of the earth itself and the life forms living on it, or to both.

If we are to use animals to further our own knowledge systems—and especially if that use involves maiming, imprisoning, or killing the animals—we must limit ourselves to things we need to know. In a way, we already do this. Need is decided by patterns of research funding from both government and industry. The "need" in industry is often to produce saleable, profitable knowledge. Such knowledge may also benefit human health or welfare, but not necessarily. In government, "need" is arrived at through complex debate and may take into account the viewpoints of scientists, industry, the military, the health establishment, and, more rarely, the average citizen. Will the knowledge we gain through experimentation which takes the lives of animals—the knowledge we give back to the world—justify whatever destruction we undertake to obtain it? We cannot decide this without ongoing national and international debate.[23] I think that such debate already occurs in limited arenas (usually among scientists themselves), but I argue here that there must be much greater input from Jane and Joe Citizen.

We have already taken steps in that direction. No matter how poorly our democracy works, at some level the individual person-in-the-street sets some of the outside limits on the use of public funds. But more must be done. Citizen participation on local committees (in research hospitals and universities) which examine and approve research projects on vertebrate subjects is one possibility. If, as some will surely retort, the knowledge gap between the scientist and the lay citizen is too great to make a joint committee of standards work, then we scientists need to take off our white lab coats and start teaching people what they need to know and start listening to their concerns about the scientist's goals and behaviors.

How Else Can We Learn Things? Knowledge in Context

I can imagine "needing" to know about how the human brain controls sexual behavior, although the research article I've discussed makes no claims about why it is important. Properly formulated, such knowledge could, for example, help in rehabilitating victims of brain damage and perhaps eventually help us understand the origins of socially unwanted violent behavior.[24] The caveat in the previous sentence is the phrase "properly formulated." The knowledge that I want must be totally contextual. The brain grows, develops nervous pathways, and reshapes nerve functions and activities always in a social context. Individual and group experience contribute to brain function. Social history contributes to brain function. Feminist primatologists have pioneered noninvasive methods of studying animal behavior; such methods involve meticulously documented studies of individual animals—from birth until death—in their social settings. These scientists interpret behavioral patterns as parts of social networks, conditioned by individual life histories, individual personality differences, and particular social interactions. I suggest that once we understand more about contexts and individual histories, we can ask better questions about the physiology of observed behaviors. Reversing the usual order from context to specific rather than starting with the atom and building the world would enable us more plausibly to limit experiments using live animals. I do not foresee eliminating animal experimentation; I do imagine a work in which we prefer to do most of our learning first from undisturbed animals acting in their chosen social worlds.

The brain is part of a network which extends outside the individual body, just as the various components of the brain (the ones in which some scientists try to "localize" behavioral controls) form part of an interconnected neural network. Some neuroscientists have begun to think theoretically about this. In writing about past attempts to localize parts of the brain which control violent behavior, Stephen L. Chorover writes:

> By insisting that violence is *essentially* a transaction and hence a problem that needs to be addressed in interpersonal or social terms, I mean to suggest that it is not merely reducible to a social attribute . . . that it is simply not meaningful—in any scientific sense—to define or deal with violence as if it exists or could exist independently of the specific material, conceptual, and social context of which it is a part.[25]

If we need to know about the brain and behavior, then, we must figure out meaningful ways of obtaining such knowledge. Heilman and Valenstein note that while "behavioral studies in animals have . . . yielded a great deal of information . . . the applicability of this information to the study of complex human behavior is not clear-cut."[26] Animal studies would need to shift to less invasive approaches, but most important, if we develop less invasive approaches we could use human research subjects in the first place. We could derive models of human behavior from the study of human subjects.

In fact, a great deal of our knowledge of the brain and human behavior comes from the study of humans. Since the nineteenth century, scientists have studied the behavior of individuals who have suffered brain damage following illness or injury. The popularity of Oliver Sacks's contemporary accounts of such work attests to the fascination such approaches hold.[27] In addition to reporting the many case studies of this sort—including ones which localize male sexual function in the hypothalamus[28]—neurosurgeons occasionally perform electrical stimulation experiments during or after brain surgery.[29] Furthermore, even though interpretation is still a problem, modern technologies, such as PET and CAT scans, open the door to a world of new—and more holistic—information.[30] Proponents of the type of research I've critiqued in this chapter would respond that the work on humans is anecdotal and imprecise. We can study strange, Oliver Sacks–type, disruptions in behavior for years but find out where the brain lesions are only after the patient dies. By then much repair may have taken place. The brain may have compensated for the damage in ways that neuroanatomists can't detect. But the knowledge obtained by Slimp et al. is hardly less anecdotal, coming, as it does, from the study of six experimental animals.

And Finally . . .

Let's return again to my metaphor of a well-balanced three-legged chair. I've covered the legs of utility and context. What about respect? Contextual research focused on individual life experience is also more respectful. It is more accepting of what animals have to tell us about their lives and thus places animals and humans on a more equal plane. Furthermore, a relationship exists between respecting nature and respecting human difference. In her poem "Trail Warning" Awiakta states the case more succinctly than I can:

Beauty is no threat to the wary
    who treat the mountain in its way,
        the copperhead in its way,
            and the deer in its way,
Knowing that nature is the human heart
made tangible.

I can't speak for Awiakta, but here's what I see in those seemingly simple lines. Scientists spend their lives trying to understand the natural world. To do so we always work out of our own emotional, cultural, and historical contexts. We create knowledge by playing out what is in our hearts. To the extent that scientific knowledge comes to be seen as nature herself, nature is, indeed, the human heart made tangible. But this poem also suggests that nature can be dangerous if she is disrespected. Indeed, if nature is but a way of touching the human heart, then in disrespecting her we disrespect ourselves; and if we cannot respect one another, we will always construct a version of nature that endangers us now and for generations to come.

The world faces many crises. The number of AIDS victims continues to expand, with no end in sight. Ought we to use animals to develop a cure? My answer is yes, but . . . understanding more about the social body, the components of disease prevention which we can obtain only when we understand more about human sexual behavior must have equal status.[31] So, too, must the social conditions which lead to the spread of disease via drug use. Once more tuberculosis looms large in our lives. Should we use animals to help find a technological fix? Yes, but . . . not without addressing modes of disease transmission: through homeless shelters and homeless AIDS victims, and the problems of drug-resistant bacilli—often created by particular kinds of antiseptic practices in modern hospitals. Such bacilli are monsters of our own creation. The balance of social to technological disease control must change. Not only will we sacrifice fewer animals if this happens; we will also run a lower risk of damaging our environment. Many problems requiring a technological fix were created by technology in the first place.

The debate is a moral one; when is it OK to take the lives of other living beings in order to better human lives? Some say there is never justification. Others draw the line at primates.[32] On one extreme, experiments which seem not to better human or animal lives or protect the ecosystem in some fashion can be done away with. The work I dissected in this chapter falls into this category. On the other hand, most people in our culture, myself included, would justify the judicious use of animals in searching for AIDS or cancer cures.

I have tried, for better or for worse, to follow a little way along Awi Usdi's path. He is an enforcer, ensuring that humans who hunt animals maintain a proper relationship with the beings they kill. Rarely, and only to one who is deeply spiritual, a master hunter, Little Deer, will appear in person, allowing himself to be taken, his horn kept as

> a charm for the chase—
> a talisman reverently borne
> for reverence alone sustains its power
> and forestalls wrath and pain[33]

"Reverence alone sustains its power." Little Deer has much to teach us. If we take, we must do so with humility and reverence. If we take, it must be because we can use the taking to give back in kind.

The vengeance of Little Deer has many guises—the DDT-resistant mosquito, the out-of-control nuclear reactor, the desertification of the Nile basin. In the biology of the future, humans—including the scientists among us—must place ourselves *in* rather than *above* the natural world. If we can learn to respect the lessons of Awi Usdi, we will have moved in the right direction, arriving at a set of principles which can guide us in the use of animals for scientific research. Myths, stories, principles, rules—they are all guides which require interpretation. The chapters in this book attempt to set and interpret rules; so do the NIH guidelines on animal care and research, and discussions with students in our classrooms. It will take a long time to build a sturdy three-legged stool. But perhaps this chapter and the varied reactions it evokes will help us get started.

Notes

A deep thanks to Peter Sterling, James McIlwain, Evan Balaban, and William Byne for critical (and extremely helpful) comments on an earlier draft of this chapter. Of course they are in no way responsible for the final outcome. Thanks also to Ruth Hubbard and Lynda Birke for their patience about deadlines and their encouragement on the first draft. And thanks especially to Marilou Awiakta for her poetry and her comments and corrections on my first draft.

1. Marilou Awiakta, "Trail Warning," *Selu: Seeking the Corn-Mother's Wisdom* (Golden, Colo.: Fulcrum Publishing, 1993), p. 39. The chapter title also is a quotation from this book.

2. Awiakta, *Selu*, p. 29.

3. Awiakta, *Selu*, p. 32.

4. Let me make clear that I find there to be no distinct or natural dividing line between vertebrates and invertebrates. Rather, this has usually been my personal line. I do not believe it is inherently better than some other dividing line.

5. I also use chick embryos in teaching laboratories—which raises a whole other set of questions: is it more moral to work on animals before their backbone, feathers, and fur have developed than after? Emotionally, I experience far fewer problems dissecting a forty-eight-hour chick embryo than I do dissecting an adult frog. The experience, however, breeds intense discomfort for some of my students.

6. Jefferson C. Slimp, Benjamin L. Hart, and Robert W. Goy, "Heterosexual, Autosexual and Social Behavior of Adult Male Rhesus Monkeys with Medial Preoptic-Anterior Hypothalamic Lesions," *Brain Research*, 142 (1978), pp. 105–122.

7. But see more recent work by Y. Oomura et al., "Central Control of Sexual Behavior," *Brain Research Bulletin*, 20 (1988), pp. 863–870. Very similar work done on rodents rather than primates appears regularly in the current scientific literature.

8. Simon LeVay, "A Difference in Hypothalamic Structure between Heterosexual and Homosexual Men," *Science* 253 (1991), 1034–1037; Anne Fausto-Sterling, *Myths of Gender: Biological Theories about Women and Men* (New York: Basic Books, 1992).

9. See, for example, F. P. M. Kruijver et al., "Lesions of the Suprachiasmatic Nucleus Do Not Disturb Sexual Orientation of the Adult Male Rat," *Brain Research*, 624 (1993), pp. 342–346.

10. The regions were well located for the standards of the time. Today much better methods exist for localizing and studying particular neural pathways with interconnections known at the cellular level. See, for example, Patricia D. Finn and Pauline Yahr, "Projection of the Sexually Dimorphic Area of the Gerbil Hypothalamus to the Retrorubral Field Is Essential for Male Sexual Behavior: Role of A8 and Other Cells," *Behavioural Neuroscience*, 108 (1994), pp. 362–378.

11. This story is wonderfully documented in Susan E. Lederer, "Political Animals: The Shaping of Biomedical Research Literature in Twentieth-Century America," *Isis*, 83 (1992), pp. 61–79.

12. Slimp et al. (1978), p. 110.

13. William H. Masters and Virginia E. Johnson, *The Human Sexual Response* (Boston: Little, Brown, 1966).

14. Carol Ezzel reports that current sex researchers feel that while Masters and Johnson's work has informed us about the physiological preliminaries leading to orgasm, we don't really understand the essence of the event; see her "Orgasm Research: Struggling toward a Climax," *Journal of NIH Research*, 6 (1994), pp. 23–24.

15. Donald G. Lawrence and Henricus G. J. M. Kuypers, "The Functional Organization of the Motor System in the Monkey, I: The Effects of Bilateral Pyramidal Lesions," *Brain*, 91 (1968), pp. 1–14.

16. Finn and Yahr, "Projection of the Sexually Dimorphic Area."

17. Kenneth M. Heilman and Edward Valenstein, *Clinical Neuropsychology* (New York: Oxford University Press, 1993), p. 10.

18. Arthur Kling and Roslyn Mass, "Alterations of Social Behavior with Neural Lesions in Nonhuman Primates," in Ralph L. Holloway (ed.), *Primate Aggression, Territoriality and Xenophobia* (New York: Academic Press, 1974), pp. 361–386.

19. A. A. Perachio, "Hypothalamic Regulation of Behavioural and Hormonal Aspects of Aggression and Sexual Performance," in D. J. Chivers and J. Herbert (eds.), *Recent Advances in Primatology*, vol 1. (New York: Academic Press, 1978), p. 555.

20. Slimp et al., p. 105.

21. Slimp et al., p. 107.

22. This claim is thoroughly documented in Donna Haraway, *Primate Visions* (New York: Routledge, 1989).

23. Vandana Shiva (see her chapter in this volume) eloquently points out the ways in which apparently "pure" research decisions taken in developed countries adversely affect the lives and environments of people in the less industrialized nations. An international component to the establishment of ethical standards for animal research is absolutely essential.

24. Even this is, of course, difficult to define. These days, most would think I meant eliminating criminal violence. But what about socially sanctioned violence such as war? Wouldn't it be great to get rid of that, too?

25. Stephen L. Chorover, "Violence: A Localizable Problem?" in Elliot S. Valenstein (ed.), *The Psychosurgery Debate: Scientific, Legal and Ethical Perspectives* (San Francisco: Freeman, 1980), pp. 334–347.

26. Heilman and Valenstein, *Clinical Neuropsychology*, p. 13.

27. Oliver Sacks, *The Man who Mistook his Wife for a Hat and other Clinical Tales*. (Harper and Row: New York, 1970).

28. Russell Meyers, "Evidence of a Locus of the Neural Mechanisms for Libido and Penile Potency in the Septo-fornico-hypothalamic Region of the Human Brain," *Transactions of the American Neurological Association*, 86 (1961), pp. 81–85.

29. See Heilman and Valenstein. There are, of course, ethical questions about such work, but it is possible in principle to obtain informed consent before doing any such experiments.

30. See, for example, T. T. Yang et al., "Noninvasive Detection of Cerebral Plasticity in Adult Human Somatosensory Cortex," *Neuroreport*, 5 (1994), pp. 701–704.

31. I have discussed restrictions on inquiry into human sexual behavior; see Anne Fausto-Sterling, "Why Do We Know So Little about Human Sex?" *Discover*, June 1992, pp. 28–30. Restrictions on inquiry into human sexual behavior continue today; see Jennifer Steinberg, "A Sex Survey by Any Other Name . . . ," *Journal of NIH Research*, 6 (1994), pp. 25–26.

32. Paola Cavalieri and Peter Singer (eds.), *The Great Ape Project* (London: Fourth Estate, 1993).

33. Awiakta, *Selu*, p. 109.

8 | The Liberation of the Female Rodent

Marianne van den Wijngaard

Dᴜʀɪɴɢ ᴛʜᴇ ᴛᴡᴇɴᴛɪᴇᴛʜ century, sex and sexuality have increasingly become objects of scientific interest. Thousands of biologists and psychologists have studied animals with the wish to unravel the biological mechanisms of human sexual behavior. They have assumed that animals' sexual behavior, unlike that of humans, is not affected by cultural and psychological factors.

Around 1910 researchers started to ascribe differences in sexual behavior to the newly developed concept of sex hormones.[1] First, scientists ascribed a temporal function during adulthood to sex hormones in blood only. In 1959 the so-called organization theory was proposed that extended the function of sex hormones from adulthood to prenatal life and from the blood to the brain. According to this theory, prenatal hormones make the brain, and thus the later behavior, masculine or feminine. In the presence of androgens, the brain and later behavior would become male; in the absence of androgens, a female brain would come into existence. Experiments with animals formed the basis for the theory; the masculinity or femininity in the brain was registered by measuring their copulation behavior.

Scientists made clear distinctions between the laboratory rodent's male or female sexual behavior, although in nature both types occur in both sexes and in combination with either another female or male. The theory conceptualized the existence of two distinct sexes in the brain, resulting in a masculine or a feminine sexual orientation. Its popularity was due to the promising prospects of revealing biological mechanisms of human sexual behavior in general and particularly homosexuality.[2] Homosexuality was considered as a "faulty" sexual orientation caused by some abnormality in prenatal hormones before birth. Millions of rodents bred for laboratory experiments have been castrated, injected with chemicals, and operated on in the search for knowledge about sex.

In this chapter I describe how the predominantly male scientists treated laboratory rodents during the 1960s and how that treatment differed for male and female animals. With the entrance of women scientists in the 1970s, the way female animals were seen changed. I will evaluate what this has meant for the fate of laboratory animals and for the way in which we can gain knowledge about sexual behavior.

Masculinity and Femininity in Theory and Experiments

In 1959 Charles Phoenix and colleagues proposed the organization theory of brain differentiation. They ascribed an important mechanism in sexual functioning to the male hypothalamus. According to their theory, a male hypothalamus and male sexual behavior develop if androgens are present before or around birth, depending on the species. In the absence of androgens, the hypothalamus remains undifferentiated and female sexual behavior develops.[3] Scientists called male sexual behavior "mounting": one animal climbs up on another's back. They called female sexual behavior "lordosis" (an arching of the back) and regarded it as a measure of receptivity.[4]

Scientists saw male behavior, unlike female behavior, as complex. In experimental situations they distinguished a variety of component parts: mounting, intromission, ejaculation, and combinations of these acts. But they "objectified" female behavior in a different way from male behavior. Lordosis was unequivocally considered as a measure of the receptivity of the female animal for the male. In experiments the intensity of receptivity, a measure for female brain development (or better, nondevelopment, because this was seen as basic), was defined as the lordosis quotient (LQ): the number of times an animal reacted with lordosis to a mount by another animal. Thus scientists perceived female sexual behavior only in response to a "male" mount. In fact they considered male and female behavior to be opposites, corresponding with the ascribed functions of hormones in the brain. Absence of androgens was said to result in the opposite of a male, differentiated brain—a female, undifferentiated brain—causing the opposite of male sexual activity: passivity or receptivity. Scientists prenatally administered androgens to female rats to find out what dosage of androgens could make the basically "female brain" masculine. Investigations into development of the brain and behavior based on the organization theory meant that, besides adult animals, now also pregnant females and pups became experimental objects.

This theory had consequences for what could and could not be investigated. Femininity in the brain and, accordingly, active female sexual behavior did not exist within the framework of the organization theory and thus was not investigated. Androgens and masculinity in the brain ruled the investigations of the origins of sexual behavior.

How did scientists manage to neglect active female contributions to sexual interactions in experiments? And how did they match "male" hormones, male brain development, and masculinity in the behavior of laboratory rodents?

During the 1960s, the leading question in most experiments was "What makes a brain male?"[5] In their experiments, scientists focused on behavior they considered to be male and developed experimental designs in which the female laboratory rodent's contributions to sexual interactions became invisible. For example, in an article published in 1971, Roger Gorski, the director of one of the most important research institutes concerned with brain differentiation at the UCLA School of Medicine, described some problems in the field.[6] He suggested that contradictory results might sometimes be explained by the testing procedure and, at other times, by scientists' different opinions about when they considered a laboratory rat to be receptive.

In the 1960s, Gorski related, it was common practice to allow male rats a period of adjustment to the testing "arena" when performing a behavioral test. Scientists found that males removed from their home cage and placed in a test situation did not respond until they had been "adapted" to the new situation for about two hours. After this adaptation period, the test female was introduced into the cage. In 1969 the results of an experiment were published in which scientists tested "the possibility that a female rat is unusually sensitive to a novel environment." In this test scientists placed a female rat who had been treated prenatally with androgens in the test arena for two hours before introducing a male. Gorski related that "under this testing condition most females exhibited a very high LQ."[7] He did not expect this result, because according to the organization theory, the androgens should have "masculinized" the brains of these female rats, resulting in low receptivity, a low LQ. In the article Gorski did not give an explanation for this unexpected result.[8] This is typical of how scientists usually have dealt with females; they have been concerned about the effects of adaptation on male sexual behavior but have not thought of the effects of adaptation on female sexual behavior.

In discussions of the different measures of sexual behavior it is also obvious how differently scientists have regarded male and female sexual behavior in experiments. Gorski stressed that it was essential for an adequate study of brain differentiation to make a detailed analysis and proper definition of male behavior. The differences in the use of what scientists have called lordosis or receptivity, however, also gave rise to difficulties. Richard Whalen and Ronald Nadler, for example, stated that "intromission and ejaculation by the males induced a forced arching of the back in the females which did not outlast the duration of the males' response. These responses might be termed 'forced lordosis responses,' and they can be easily discriminated from the typical lordosis of the receptive female."[9]

Gorski treated females in the same way, but in opposition to Whalen and Nadler he considered them to be receptive. He used the presence of spermatozoa in the vagina as a definition of receptivity, though he considered this to be a "weaker" measure than spontaneous lordosis.[10] Whalen and Nadler agreed that if receptivity is defined by spermatozoa in the vagina, indeed the female rats were receptive. To solve the problem of definition, Whalen and Nadler suggested discriminating between forced intromission with a "passive or weakly receptive female" and mounts or intromissions which are accompanied by a normal lordosis response.

The experiments concerning adaptation time for female animals and the different definitions of female sexual behavior illustrate that researchers have seen female animals as an instrument to be used to investigate male sexual behavior.

A Cry for the Liberation of the Female Rodent

In the beginning of the sixties there was criticism of the neglect of female sexual behavior. For example, Gordon Bermant, a psychologist working with laboratory rats and J. Calhoun, an ethologist who studied wild rats, remarked on the one-sided interest in male sexual behavior.[11] In the sixties these remarks were ignored by scientists who were involved in biological research into the effects of prenatal hormones on male brain development.

In the second part of the 1970s the women's movement managed to open the masculine circle to include femininity, and the position of the female rodent changed. More women entered higher education and more female students chose the field of behavioral neuroendocrinology. They found the situation in the laboratory as I have described: interest only in male sexual behavior in experiments that did not give much opportunity for females to express other kinds of sexual behavior than lordosis.

Some of the women researchers provided historical contributions to this field of research. Monica Schoelch-Krieger, a biologist working at Rutgers University, for example, spoke at a conference in 1970 about an experiment in which she tethered male rats and gave female rats the opportunity to react to these males. With a video she demonstrated a comprehensive repertory of the females' sexual behavior, of which lordosis was only one aspect.[12] To the surprise of the audience, the female ran back and forth, darted, hopped, and moved her ears, later called "earwiggling." Researchers who had seen this behavior before had never recognized it as part of a female's sexual behavior. For the audience, this video was an eye opener.

After that congress, in many laboratories female and male scientists developed new experimental designs to investigate sexual behavior of female animals. Within a few years, several concepts, in addition to "receptivity," were in use in different laboratories, such as sexual motivation[13] and appetitive or precopulatory sexual behavior.[14] Female investigators of primate behavior also explicitly argued for the necessity to distinguish between "some parameters which reflect *initiation* of sexual interaction by the female" and those "in which the female allows interaction initiated by the male."[15]

In the 1970s, remarks like that were no longer ignored. The number of publications on the subject of female sexual behavior by both female and male researchers increased considerably.

In 1974 Richard Doty, who was involved in the research of hormonal effects on animal behavior as a postdoctoral researcher in psychology at the University of California at Berkeley, published an article entitled "A Cry for the Liberation of the Female Rodent: Courtship and Copulation in Rodentia."[16] Doty argued that "behavior of the female has been ignored in most studies of rodent copulation." According to him, a number of interacting factors probably contributed to this imbalance. The first reason he gave was that investigators of behavior traditionally had studied male rodent behavior with the idea of ultimately formulating monolithic laws of behavior; these laws were thought to apply to most members of the animal kingdom, including females. Second, he found that most laboratory studies of sexual behavior looked at copulation in small and confined test situations in which the male appeared to play the more dominant and active role. Doty argued that the earlier phases of courtship, in which mate selection, pair formation, and other behaviors related to sexual behavior occur, had not been examined. According to Doty, in the laboratory studies the male's copulatory behavior is more obvious than the female's; it can be divided easily into components which are sensitive to a number of experimental manipulations. Third, Doty stated that the estrous cycle of the female alters some of the hormonal and behavioral variables scientists use and is generally considered a nuisance by many investigators. According to him, male subjects are overrepresented numerically and conclusions drawn from experiments with males are more likely to be "overgeneralized" than are conclusions from those using females.

Doty pointed to the importance of investigations of female sex initiation. He extensively quoted Calhoun's study, published twelve years earlier, on sexual behavior of wild rats, in which the complexity of sexual behavior of males as well as females was described, in addition to the nonbehavioral aspects, such as releasing odors and sounds.[17] Doty concluded that laboratory studies may oversimplify the complexity of the sexual behavior that occurs in most natural situations. Moreover, rats used for laboratory experiments are inbred in comparison to their wild counterparts, and this probably also affects their behavior. Doty ended his article with the wish "that the present article will extend the interests of many behaviorists to courtship phenomena in wild species and will stimulate more 'equal opportunity' research on female rodents."[18]

The studies on female sexual behavior which I have mentioned so far did not deal with prenatal hormones but considered the effects of hormones administered to adult females. According to the prenatal hormone theory, females had undifferentiated brains. But this would also change.

In 1975 Frank Beach, director of the department in which Doty worked and a very influential scientist, raised the question of why there had been so much interest in male copulation and so little interest in the female part of sexual behavior. According to Beach, "this is often neglected, especially by male investigators and theorists."[19] A year later, Beach published an article on female sexual behavior that became a classic in this field of research. He wrote the article to "propose certain correctives for the existing imbalance" and to attempt to "provide a simple but heuristically useful scheme with theoretical and practical implications for the study of feminine sexuality and its behavioral expression."[20]

Beach referred to the studies mentioned above. This article, in fact, functioned as a theoretical unification of the different concepts that had been developed earlier, such as consummatory or appetitive behavior and sexual motivation of rats. Beach proposed that female sexual behavior be divided into attractivity, proceptivity, and receptivity.[21] For these three concepts he provided operational definitions in terms of stimulus-response relationships, which were susceptible to quantitative measurement.

Beach defined the new concept of attractivity for the female rat as "her stimulus value to the sexual performance of a male of the same species, or, the female's capacity to elicit ejaculation." In descriptions, attractivity covers the full range of stimulation, including bringing a male to a female, assisting males to identify the female's sex and reproductive status, synchronizing and orienting the male's coital responses, and promoting the emission of sperm while the penis is in the vagina.[22]

Proceptivity, Beach suggested, is the concept to be used for investigations into the sexual initiative of female animals: "appetitive activities shown by females in response to stimuli received from males" or "proceptive responses are parameters which reflect initiation of sexual interaction by the female."[23] Considering these definitions, however, little opportunity remained for initiative in sexual behavior for females, since Beach used the term *response* in the definition of proceptivity and in the defined behavior. Proceptivity and attractivity, in fact, were oriented toward copulation and could not be investigated as such. In Beach's words, proceptivity and attractivity "intensify the male's sexual excitement, when this is necessary and also . . . facilitate, coordinate and synchronize the bodily adjustments necessary for genital union and penile insertion." Beach's words "when this is necessary" after "intensify the male's sexual excitement" clearly indicate that he did not find it easy to ascribe these active contributions to females. In Beach's new definition, the meaning of receptivity changed from passive to active, but it remained a behavioral pattern registered only in females. For this reason Beach's definitions were not really "liberating" the female rat's sex life, but at least he contributed to correcting the imbalance in knowledge about female sexual behavior.

Beach did more. He was the one who granted female animals a female brain, whereas until then it had been regarded as undifferentiated by researchers adhering to the organization theory. Until 1975 Beach had been the only scientist who did not accept the prenatal hormone paradigm. In 1971 Beach published an article in which he criticized the research based on the organization theory severely. He even tried to persuade scientists in the field to abandon the theory on differentiation of the brain in favor of a psychological research approach which he had developed decades earlier. In Beach's model, rather than mechanisms in the brain, learning from sexual experiences was the key to understanding the development of sexual behavior.[24] Although scientists in the field took Beach's criticisms seriously, they stuck to the organization theory, and Beach became more or less isolated from his colleagues by not adhering to the prenatal hormone paradigm. In his 1976 article on active female sexual behavior, Beach finally accepted the theory. He suggested that, also for proceptive and attractive behavior, brain sites were "organized" by prenatal hormones. Especially to estrogens he ascribed a role in the developing female brain. The effect of the reclamation of the unbeliever Frank Beach was that the theory was elevated above all doubts.

Beach's 1976 article had important consequences for the study of female sexual behavior and for the localization of the control mechanisms that scientists ascribed to it. In the following years, many researchers, among them many women researchers such as Martha McClintock, Donna Emery, Mary Erskine, Anne Etgen, Kathie Olsen, Jane Stewart, and Christina Williams, focused their studies on the sexual behavior of female animals.[25] Following Beach's hypothesis, some of them studied the parts of the brain regulating the expression of proceptive behavior. The most likely candidates for the investigation of neuroendocrine regulation of female sexual behavior were brain sites thought to be involved in reproductive function, that is, various parts of the hypothalamus.[26] Scientists looked for the control mechanism of female sexual behavior in this part of the brain. They made lesions or damaged different parts of the female rat's hypothalamus.[27] Thus, after transplantations of gonads and injection of hormones, brain surgery also became the female laboratory animal's fate.

Intellectually, these changes during the 1970s had major consequences. While, until 1976, most scientists saw receptivity as a product of undifferentiated brains, now they accepted Beach's suggestion "that proceptive and receptive behavior may depend upon different anatomical or neuro-chemical systems in the brain."[28] The transformation in scientific concepts of insights originating in feminism and Beach's subscription to organization theory resulted in an enlargement of the ideas of the role of the prenatal hormones in the brain: the "second sex" was incorporated into organization theory. This meant that in the second part of the 1970s, the prenatal hormone paradigm of brain differentiation was stronger than before.[29] But certainly it did not improve the life of laboratory rats.

Environmental Factors and Behavior

During the 1970s, scientists suggested that besides sexual behavior, a number of other types of behavior were affected by the prenatal hormones.[30] Yet, at the same time, the idea that environment influences the development of behavior became stronger. In fact, the feminist "nurture" position affected neuroendocrinological research. In 1977, David Quadagno, Robert Briscoe, and Jill Quadagno of the University of Kansas published an article criticizing the research on the effects of prenatal hormones on the development of behavior of animals and humans.[31] They argued that factors other than hormones explained the observed phenomena.

In the 1960s and '70s the results on which the knowledge about effects of prenatal hormones was based derived mostly from observations of animals born with anatomical abnormalities in their sexual organs, elicited by the experimentally administered hormones. In studies of animal behavior, this factor had not been examined. Yet, according to Quadagno et al., this factor had considerably more impact on the development of behavior than researchers supposed. They suggested that rat mothers treat their pups with abnormal genitals differently than they treat normal pups. They argued that it was wrong to ascribe the outcome of this rearing to prenatal hormones.

How did scientists involved in prenatal hormone effects in animals react to the serious criticism of the Quadagnos and Briscoe? In the 1980s the social influences on behavioral development of animals became an item for research. Celia Moore found a solution which combined the nonbiological learning aspects of maternal behavior and the biological effects of hormones on the development of behavior of rats. She suggested that the maternal behavior of rats is affected by the hormonal condition of the pups and both hormones and maternal care affected their later behavior.[32] From these results we may conclude that it is at least questionable whether we can study biological mechanisms involved in sexual behavior of animals isolated from the context of the experimental settings. If this is questionable for rats, should we not ask ourselves why we torture and kill them in the belief that we can thus understand more about the nature of human sexuality?

We may wonder why rodents became so popular for the study of masculinity, femininity, and sex in the brain and in behavior. The female rat normally engages in sexual interactions only during a short interval when she is fertile. In some other species, ovulation only occurs shortly after or during copulation. However, there are also species in which females engage in copulation at any stage of the cycle, such as rhesus monkeys and chimpanzees. Do we have more ethical constraints in treating them like rats in experiments? Is it because growing and housing of rodents is cheap and easy compared with monkeys?

Moreover, it is remarkable how in these studies female animals increasingly became the victims. In the 1960s, according to the theory, the female brain was considered to be undifferentiated at birth. Therefore scientists administered androgens to female pups or their mothers to find out "how the brain becomes male." After the incorporation of the "second sex" into the prenatal hormone paradigm, scientists started to damage the female hypothalamus in order to establish the precise locus of control of sexual activity. Until now, both have been without success.

In most other experiments, scientists use male rats because they don't have a cycle that interferes with the administered hormones. In the studies described above, however, female animals were the ones that were used as "test tubes" to find a hormonal model that would resemble the natural development of the male brain.

From Celia Moore's studies we have learned that in the sexual development of rats, the behavior of the rat mother plays an important role. If we wish to understand more about such development, we cannot ignore social and cultural expectations in general, and especially of parents, even in rats. Such considerations must become integrated in research designs, even when scientists wish to define biological mechanisms affecting sex, since masculinity, femininity, and sexual behavior exist only within a context. And for animals, as well as for humans, such contexts may change with time and place.

Behavior itself is generated by developmental processes for which we have accepted that biology and environment both play a role. Lynda Birke has proposed a model in which behavior can be integrated into biological and environmental interactive processes.[33] In this model, characteristics considered to be masculine or feminine are thus not the dependent variables of an interaction between social and biological factors. Behavior is itself part of interactive processes, rather than the fixed property of an individual. Animals and humans can behave in "masculine" or "feminine" ways in different social contexts, and their preference for a male or a female in sexual interactions can change over time. In this model, "masculine" or "feminine" sexual behavior is not a fixed property due to prenatal hormone effects.[34] If we wish to understand anything about the development of behavior, we have to consider previous behavior, social context, and biological variables. This is true for animals as well as humans. It will not make research easier, but it is more likely to yield insight into the developmental processes than will brain surgery in rats. If operating on a rat's brain changes her behavior, we will never know whether this is because of what has been done to her brain or because she has a headache.

Notes

1. E. Steinach, "Feminisierung von Männchen und Maskulinisierung von Weibchen," *Zentrablatt für Physiologie*, 27 (1913), p. 49. In 1913 Steinach proposed that the differences in sexual behavior between male and female rats could be accounted for satisfactorily by the gonadal hormones secreted by the testes or ovaries in adulthood. At that time scientists performed experiments implanting ovaries in castrated males or testicular tissue in ovariectomized females.

2. Marianne van den Wijngaard, "The Acceptance of Scientific Theories and Images of Masculinity and Femininity," *Journal of the History of Biology*, 24 (1991), pp. 19–49.

3. Charles H. Phoenix et al., "Organizing Action of Prenatally Administered Testosterone Propionate on the Tissues Mediating Mating Behavior in the Female Guinea Pig," *Endocrinology*, 65 (1959), pp. 369–382.

4. In natural situations with most animals, both types of behavior occur in both sexes. In experimental situations, males and females can both be tested on display of mounting as well as on lordosis. This happens during adulthood when the prenatally "organized" center in the hypothalamus is activated by sex hormones secreted by adult gonads.

5. In the 1960s an important motivation for the research was the quest for a biological explanation for male homosexuality, traditionally considered as a femalelike sexuality. Based on the organization theory, scientists hypothesized that this could be caused by a lack of androgens before birth. See van den Wijngaard, "The Acceptance of Scientific Theories."

6. Roger A. Gorski, "Gonadal Hormones and the Perinatal Development of Neuroendocrine Function," in L. Martini and W. F. Ganong (eds.), *Frontiers in Neuroendocrinology* (New York: Oxford University Press, 1971), pp. 237–290.

7. Gorski, "Gonadal Hormones," p. 250.

8. Though the result is contrary to the expectation of the author, he did not try to explain it. Explanation of this result by suggesting that the female might have felt more comfortable after the adaptation period would have made it contradict the theory.

9. Richard E. Whalen and Ronald D. Nadler, "Modification of Spontaneous and Hormone Induced Sexual Behavior by Estrogen Administered to Neonatal Female Rats," *Journal of Comparative and Physiological Psychology*, 60 (1965), p. 151.

10. Roger A. Gorski, "Modification of Ovulatory Mechanisms by Postnatal Administration of Estrogen to the Rat," *American Journal of Physiology*, 205 (1963), pp. 842–844.

11. Gordon Bermant, "Response Latencies of Female Rats During Sexual Intercourse," *Science*, 133 (1961), pp. 1771–1773, and John B. Calhoun, *The Ecology and Sociology of the Norway Rat* (Washington, D.C.: U.S. Government Printing Office, 1962). Bermant argued that "the great majority of the studies that these and other workers have carried out focuses primary importance on the male's behavior; the female has come in for scant attention."

12. In interviews conducted with Koos Slob and Pieter van der Schoot in March 1987, they independently mentioned this lecture, which clearly impressed them, since they were able to remember it seventeen years later.

13. Bengt J. Meyerson and L. H. Lindstrom, "Sexual Motivation in the Female Rat: A Methodological Study Applied to the Investigation of the Effects of Estradiol-Benzoate," *Acta Physiologica Scandinavica*, Suppl. 389 (1973), pp. 1–80.

14. Jaroslav Madlafousek and Zdenek Hlinak, "Analysis of Factors Determining the Appetitive and Aversive Phase of Sexual Behavior in the Female Rat," *Physiologia Bohemoslovenica*, 21 (1972), pp. 416–417.

15. A. F. Dixon et al., "Hormonal and Other Determinants of Sexual Attractiveness and Receptivity in Rhesus and Talapoin Monkeys," Fourteenth International Congress of Primatology, 1973, vol. 2: *Primate Reproductive Behavior* (Basel: Karger, 1974), pp. 36–63.

16. Richard L. Doty, "A Cry for the Liberation of the Female Rodent: Courtship and Copulation in Rodentia," *Psychological Bulletin*, 31 (1974), pp. 159–173.

17. Calhoun, *The Ecology and Sociology of the Norway Rat.*

18. Doty, "A Cry," p. 169. Doty's title as well as the reference to "equal opportunity" in the last sentence referred to terms much used by the women's movement at the end of the 1960s and the beginning of the 1970s. At that time feminist thinking was mainly based on Marxist and socialist theory. Feminists saw similarities between the liberation of women and the liberation of the suppressed working class. Feminism therefore was also considered as women's liberation. These feminists aimed at achieving rights and opportunities equal to men's.

19. Frank A. Beach, "Behavioral Endocrinology: An Emerging Discipline," *American Scientist*, 63 (1975), p. 179.

20. Frank A. Beach, "Sexual Attractivity, Proceptivity and Receptivity in Female Mammals," *Hormones and Behavior*, 7 (1976), pp. 105–138.

21. Beach, "Sexual Attractivity," p. 115: "the female's tendency to display appetitive responses finds little opportunity for expression in laboratory experiments which focus exclusively upon her receptive behavior, or upon the male's execution of his coital pattern. The resulting concept of essentially passive females receiving sexually aggressive males seriously misrepresents the normal mating sequence and encourages a biased concept of feminine sexuality. Failure to recognize the importance of female initiative in the mating of laboratory rodents has been noted."

22. Beach, "Sexual Attractivity," p. 107.

23. Beach, "Sexual Attractivity," pp. 114–115.

24. Frank A. Beach, "Hormonal Factors Controlling the Differentiation, Development and Display of Copulatory Behaviors in the Ramstergig and Related Species," in Ethel Tobach, Lester R. Aronson, and Evelyn Shaw (eds.), *The Biopsychology of Development* (New York: Academic Press, 1971), pp. 249–296. The last sentences of this article read: "It is the purpose of this presentation to expose the weaknesses and contradictions that tend to weaken the theory that testosterone controls the organization of mechanisms in the brain which are destined to mediate copulatory behavior in adult males and females. It is suggested that at the present state of knowledge formal and quantitative statements of such relationships can better be made in terms of intervening variables based upon directly observable S-R relationships than in terms of hypothetical constructs such as imaginary mechanisms."

25. See Mary S. Erskine, "Solicitation Behavior in the Estrous Female Rat: A Review," *Hormones and Behavior*, 23 (1989), pp. 473–502.

26. Scientists' rationale is that the hypothalamus is connected anatomically and functionally with the pituitary gland, which in turn participates in the regulation of all the other endocrine glands in the body. These endocrine glands, such as the ovaries and the testes, in turn secrete hormones that participate in their own regulation through feedback loops acting upon the hypothalamus and the pituitary, in a manner somewhat analogous to a thermostat and a furnace heating system.

27. See, for example, Donna Emery and R. L. Moss, "Lesions Confined to the Ventromedial Hypothalamus Decrease the Frequency of Coital Contacts in Female Rats," *Hormones and Behavior*, 18 (1984), pp. 313–329.

28. Beach, "Sexual Attractivity," p. 131.

29. John Money and Anke Ehrhardt concluded that prenatal hormones also affect human brain development and behavior. For an analysis of the development in prenatal hormone studies in humans, see Marianne van den Wijngaard, "Feminism and the Biological Construction of Female and Male Behaviour," *Journal of the History of Biology*, 27 (1994), pp. 61–90.

30. See n. 29.

31. David M. Quadagno, Robert Briscoe, and Jill S. Quadagno, "Effect of Perinatal Gonadal Hormones on Selected Nonsexual Behavior Patterns: A Critical Assessment of the Nonhuman and Human Literature," *Psychological Bulletin*, 84 (1977), pp. 62–80. The first two authors, who had published before in this field of research, worked in the Department of Physiology and Cell Biology; Jill Quadagno worked in the Sociology Department. Although the character of the content of the article makes this combination plausible, it is remarkable that two researchers from a biomedical science published this article together with a sociologist. Jill Quadagno was inspired by feminist ideas concerning environmental explanations for sex differences.

32. Celia Moore, "Maternal Behavior of Rats is Affected by Hormonal Condition of Pups," *Journal of Comparative and Physiological Psychology*, 96 (1982), pp. 123–129.

33. Lynda Birke, *Women, Feminism and Biology: The Feminist Challenge* (Brighton: Harvester, 1986), esp. chap. 5.

34. In this chapter we saw how feminine sexual behavior changed in the perception of scientists. This is only one thing that changed in behavioral neuroendocrinology due to the rise of feminism. One of the other changes was the proposal of a new conceptualization of masculinity and femininity: rather than being opposites, they were proposed to be independent of each other.

9 | They Are *Only* Animals

Lesley J. Rogers

In MANY CULTURES, traditional thinking considers that it is our possession of a soul which separates us from other animal species. Scientists have used more tangible criteria to set us apart, including brain size, language, tool use, culture, adoption of an upright posture and bipedal locomotion, intellect, and consciousness. Now, ironically at a time when the survival of our closest animal relatives, the apes, in their natural environment is threatened, we are beginning to see more and more similarities between ourselves and these species. In fact, examples of tool use, cultural transmission of information, and complex social communication are being discovered even in species much more distant from us. We still see animals as being more "primitive" than ourselves, but the sharp divide between humans and other animals is disappearing.

Animal consciousness is a rapidly expanding topic of research. The aim is not simply to assume that animals lack consciousness, or indeed that they may even have it, but rather to prove it scientifically. It is an area of research which, of course, primatologists have found particularly challenging, one demanding new methods of investigation and fraught by problems of definition and interpretation. There is no clear, or at least unitary, definition of consciousness. Yet examples of self-awareness and awareness of others are being found in animals. Self-awareness, for example, has been assessed by observing the way in which an animal responds to its own image in a mirror. A chimpanzee or orangutan, seeing a mark on its forehead via a mirror, tries to rub it off its own forehead and does not reach to touch it in the mirror image (Gallup, 1977, 1983). This self-directed behavior guided by the mirror is interpreted as "self-awareness." As yet, tests of self-recognition in mirrors have not been applied to many species, but apparently elephants are unable to self-recognize, responding to the image as if it were another elephant (Povinelli, 1989). Of course, this is only one way of assessing self-awareness, and it is confined to visual awareness. Other sensory modalities also play a part in the concept of self-awareness and may be even more important in some species.

Numerous examples of tool use by animals have been reported. If we use the strict definition of a tool as an object separate from both the substrate and the user's body which is used to alter another object (Beck, 1980), the use of a rock as a hammer to crack open nuts on an anvil (a tree root or another rock) by chimpanzees qualifies as tool using (Boesch and Boesch, 1982). Moreover, Sakura and Matsuzawa (1991) have shown that chimps recognize the tool function of stones because they immediately choose the hardest spot of the anvil on which to place a nut. Similarly, the use of a rock as a hammer by sea otters to crack open shellfish (Kenyon, 1969) or the use of cactus spines by some species of finches on the Galápagos islands to probe into crevices to impale or drive out insects (Millikan and Bowman, 1967) qualifies as tool using. According to a somewhat less stringent definition of a tool, the use by orangutans of leaves arranged into a vessel from which regurgitated food is eaten represents tool using; it even involves aspects of making the tool by fashioning the leaves into the appropriate shape to hold the food (Rogers and Kaplan, 1994). Similarly, the selection of appropriate pieces of grass by chimpanzees and use of them in termite fishing is tool use (Goodall, 1968). There are other examples of tool use in captive and wild chimpanzees (Nishida and Uehara, 1980) and orangutans (Galdikas, 1982; Lethmate, 1982). A recent study reports that elephants use a variety of tools, mostly for self-directed body care (Chevalier-Skolnikoff and Liska, 1993). Over twenty types of tool use were performed by captive elephants, and nine types were observed in wild elephants.

Many still adhere to the belief that language and speech represent the last true bastion of human uniqueness and that speech first appeared suddenly around 35,000 years ago, in the Upper Palaeolithic period. As the centers in the brain which control speech and process language are, in most people, lateralized to the left hemisphere, it was until very recently commonly believed that lateralization, or asymmetry, of brain structure and function was a unique trait of the human species. However, it is now known that lateral asymmetries exist for a large number of brain functions and that they occur in a wide range of animal species (Bradshaw and Rogers, 1993). Language and tool use are associated with brain asymmetries, but humans share these characteristics with other species along an evolutionary trajectory (see Bradshaw, 1991). Moreover, recent demonstrations of proficient communication by apes using sign language or symbols are beginning to challenge the absolute uniqueness of human language (Parker and Gibson, 1990). Certainly apes are capable of symbolic communication, learning sign vocabularies, inventing new signs, and using novel combinations of signs. Researchers who have taught signing to the well-publicized chimpanzees, beginning with Washoe, and to Koko the gorilla and Chantek the orangutan have never failed to be impressed by their complexity of behavior, or intelligence. Thus Miles (1990) concluded that "the orang-utan is a viable animal model for understanding language and intelligence and within the lesser-known 'red ape' lies not only a brain but also a mind." Moreover, learning to communicate by language has not been confined to the apes. Pepperberg (1990a, 1990b) has been able to train African gray parrots to communicate, using English words to a competency equivalent to that shown by the signing chimpanzees. This rather surprising result is paralleled by other recent research demonstrating other highly developed cognitive abilities in pigeons, for symbol rotation and object classification (Delius, 1985; von Fersen and Güntürkün, 1990). These findings challenge the generally accepted perception that the cognitive abilities of birds are inferior to those of mammals (Premack, 1978), and they raise substantial questions about the importance of brain size and organization for higher cognitive abilities, the avian brain being constructed so differently from that of the mammal.

Over recent years, comparative psychologists have begun to use the term *cognition* to refer to thinking in animals. As Griffin (1984) has said, this seems to be a more scientifically respectable term than *thought*, as it does not imply consciousness, at least to the same extent. Many psychologists state that cognition implies complexity of neural processes but not consciousness, and some categorically assert that animal cognition is always unconscious. But this does not mean that all scientists take such an approach. Increasing numbers are turning their interest to the study of problem solving in animals, their ability to form concepts, to generate plans and have expectations. Some neurophysiologists also have begun to think about the possible cellular correlates of consciousness (Eccles, 1989). Yet despite the new interest in studying consciousness, in almost all of these approaches there is the underlying evolutionary concept of a gradual development of consciousness through the mammals to reach, it is assumed, its pinnacle in humankind.

Brain Size and Cognition

It is an old notion that brain size reflects brain complexity, greater ability to learn and therefore greater intelligence. There is indeed an inverse correlation between brain size and the evolutionary appearance of a species on earth, but to some extent this is due to increasing body size. If one plots brain weight against body weight on a logarithmic scale, for any given group of animals a straight line relationship is seen (Jerison, 1973); for example, smaller fish have smaller brains and larger fish have larger brains. This relationship may simply reflect the fact that a larger body has larger muscles to control and a greater surface area to monitor. However, the slope of the line thus plotted is less than one, which means that the increase in brain weight does not quite keep up with the increase in body weight. In addition, if one looks at the line plots for different groups of animals, one finds that although the slope of the line is the same in each case, the position of the line varies from group to group. Birds and mammals have relatively more brain weight with respect to body size than do fish of equivalent body size, and primates have proportionately even larger brains.

This, of course, is said to reflect the superior "intelligence" of primates compared with other animals.

The weight of the human brain relative to body weight is even greater than that of the other primates. Here, seemingly, is the basis of our superior intelligence and consciousness. Yet one may query the figures of brain and, in particular, body weight used in the "standard" calculation of the brain to body weight ratio. No mention is ever made of the details of the group from which the selected human measurements arise, even though there are marked variations in body muscle to fat ratio according to diet, exercise, and a range of other factors. Biologists, even anthropologists, seem conveniently to ignore these variables. Recognizably, these factors may cause trivial differences in the ratio compared to the magnitude of the difference of *Homo sapiens* from the great apes, but nevertheless they cannot fail to alter the size of the gap which we see as separating us from the latter.

Moreover, the same lack of consideration of group differences and dietary practices on body weight must occur when body weight is estimated for fossil hominid forms so that cranial capacity, determined from endocasts, can be plotted against body weight on a log scale (Pilbeam and Gould, 1974; also see Bonner, 1980). The estimated slope of the line so obtained is 1.73 for the *Homo* species and a mere 0.34 for the apes, suggesting that the relative increase in brain size is markedly greater for the *Homo* line than for the apes. Yet how much of this apparent uniqueness, of superior brain size in the *Homo* line, is real? If the body weights of the fossil hominids have been overestimated, the slope would be less.

Brain weight or size as an indicator of cognitive capacity has a long history of application for social and political purposes within the human species. Last century, women were thought to have smaller brains because they lacked the center for intellect. Black people, too, were considered to have smaller brains and to lack intellect. Charles Darwin (1871, p. 569) stated that "at least some of the mental traits in which women excel are traits characteristic of the lower races." He also referred to "the close connection of the negro or Australian and the gorilla" (p. 201). An era of obsessive measurement of cranium size and brain weight developed (see Gould, 1981). Not only were the brains of women and blacks considered to be smaller and lighter, but they were also considered to be underdeveloped. Thus McGrigor wrote that "the type of female skull approaches in many respects that of the infant, and still more that of the lower races" (cited in Lewontin et al., 1984).

The famous neuroanatomist Broca (1824–1880) gathered much data demonstrating that the male brain has a greater average weight than the female. The sentiments of the time were clearly expressed by one of Broca's co-workers, G. LeBon, in 1879:

In the most intelligent races, as among the Parisians, there are a large number of women where brains are closer in size to those of gorillas than to the most developed male brains. This inferiority is so obvious that no one can contest it for a moment; only its degree is worth discussion. All psychologists who have studied the intelligence of women, as well as poets and novelists, recognize today that they represent the most inferior forms of human evolution and that they are closer to children and savages than to an adult, civilized man. They excel in fickleness, inconstancy, absence of thought and logic, and an incapacity to reason. Without doubt there exist some distinguished women, very superior to the average man, but they are as exceptional as the birth of any monstrosity, as, for example, of a gorilla with two heads; consequently we may neglect them entirely. (Quoted in Gould, 1981, pp. 104–105).

It was fashionable to measure the brains of eminent men after their deaths. The brain weights of several of these men (e.g., Cuvier and Turgenev) were above the European average, and this was considered as evidence for their superior intellect. The brain weights of others (e.g., Walt Whitman and Franz Josef Gall, the founder of phrenology), were found to be embarrassingly less than the European average. Undaunted by these figures, Broca continued to claim that those with lower weight had died at an older age, were smaller in build, or that their brains had been poorly preserved. From the hindsight of our present knowledge that 80 percent of the brain's volume is extracellular space and that preserving (fixing) methods are particularly variable in the amount of shrinkage which they cause, we are able to see how extremely unreliable were all these early data on brain weight.

Also, when adjusted for body weight the differences between white male brains and those of women and blacks disappeared. Nevertheless, as late as 1903 the American anatomist E. A. Spitzka published a figure showing the brain size of a "Bushwoman" as intermediate between that of a gorilla and the mathematician Gauss, and the number of convolutions of the surface of the brain of the Bushwoman as being closer to that of the gorilla (see Gould, 1981, fig. 3.3). The lower "white" classes were also slotted into having "inferior," underdeveloped brains.

Even today some isolated attempts to correlate brain size with "intelligence" are still made, despite the crude nature of the measurement of both factors. For example, a recent report (Anderson, 1993) claiming to provide evidence for a correlation between the brain size of rats and a general score of cognitive performance used animals exposed during their embryonic development to potent drugs which affected brain development, then lumped them in with control animals for the analysis. There has been also at least one recent attempt to correlate brain size and intelligence in humans (Willerman et al., 1991).

In the final analysis, brain size, either alone or relative to body size, gives no indication of the brain's capacity to function at various levels of complexity.

This depends on how the neurones are interconnected: a brain with more neurones may not necessarily function more efficiently or more "intelligently," whatever that really means, than one with fewer neurones wired up in a different way.

It was in mammals that a completely new brain structure, the neocortex, developed, and throughout the mammalian line it progressively increased in size and complexity of organization through to humans. As Bayer and Altman (1991) state, "It is widely assumed that the evolutionary growth of mental life that reaches its zenith in humans is attributable to the progressive expansion and elaboration of the neocortex." That may be true *within* the mammalian line, but birds, which do not have a neocortex, have recently been shown to perform some cognitive tasks as well as primates, even as well as apes or young children. Thus pigeons tested in operant boxes can perform tasks which require matching of symbols rotated at various angles to an impressive degree of accuracy (Delius, 1990) and to acquire object classification skills of hundreds of different stimuli (von Fersen and Güntürkün, 1990). Added to this, as already mentioned, parrots can be trained to communicate by words to a competency equivalent to that of the signing chimpanzees (Pepperberg, 1990a and 1990b). As they do all of this without a neocortex, there must be other brain regions and mechanisms capable of analyzing complex information. In fact, because of the bird brain's difference from the mammalian brain, its study challenges assumptions that have been made about the superiority of the mammalian line of evolution, and so, the neocortex.

Pigeons have an astounding ability to perform mental rotation problems of the type included in intelligence tests for humans (Delius, 1990; Hollard and Delius, 1982). The pigeons were first trained to perform a matching-to-sample task by presenting them with stimuli on three keys. The sample (an abstract shape) was presented on the central key, and the two test stimuli on the side keys. One of the test stimuli was identical to the sample and the other was its mirror image. Pecks at the matching stimulus were rewarded with food, whereas pecks at the mirror image were punished by a brief period of darkness. Several different shapes were used in the training. In training all of the stimuli were presented at the same angle of orientation. Once the pigeons had acquired this task, they were tested with the comparison shapes rotated at various angles relative to the sample. The pigeons could still perform the task just as accurately and as rapidly as before. There was no decline in their performance when they were asked to include angular rotation in their assessment. By contrast, humans tested on the same task showed a significant decline in performance. They made more errors and needed longer to react when they were tested with the rotated stimuli. Delius (1990) said that the pigeons were geniuses in comparison to the humans!

Pigeons can also deal with perceptual concepts, such as trees, leaves, persons, water, fish, even "sphericity" (Herrnstein, 1982), and they can make mental representations of learned sequences (Terrace, 1985). Their ability to conceptualize "sphericity" was determined in an operant box which allowed presentation of solid, three-dimensional objects (pebbles, bolts, pearls, buttons, etc.) on a series of metal plates attached to an automated system. The pigeon in the box was presented at any one time with three objects on keys, either two spherical objects and one nonspherical or one spherical and two nonspherical. It had to peck spherical objects and ignore nonspherical ones. The pigeons were presented with eighteen objects of each type. Within remarkably few trials the pigeons were accurately performing the task. They were then tested to see whether they had acquired the concept of "sphericity" by presenting them with 109 novel spherical and nonspherical objects. They were able to generalize to the novel objects, recognizing them according to the abstract characteristic of "sphericity," just as humans do. They could even judge sphericity in photographs of the objects.

Performance of these tasks also requires the pigeon to have an extensive memory. Consistent with this, von Fersen and Güntürkün (1990) trained pigeons to remember over 600 stimuli of different shapes and to retain that memory over days. In fact, they still retained memory of these stimuli to an 88 percent accuracy after seven months. Similarly, Vaughan and Greene (1984) showed that pigeons can remember up to 320 slides of holiday scenes (for humans) for a period of two years. Presumably they achieve this feat by coding or labeling the information, possibly in much the same way that humans do, by using descriptive words (Delius, 1990). Cerella (1986), however, claims that rather simple mechanisms may underlie the enormously complex visual classifications attained by pigeons. Much further experimentation will be necessary to resolve these issues.

There is evidence that pigeons can learn abstract rules, such as that of oddity or difference in terms of the shape of stimuli (Lombardi et al., 1984). The pigeons learn to detect the odd stimulus in a group. Learning of the abstract rule is shown by its generalization to other types of stimuli. Earlier studies had shown oddity learning in primates, dolphins, and members of the crow family (see MacKintosh, 1983), but the data for pigeons were more equivocal. The recent work by Delius's group has confirmed that pigeons are indeed capable of oddity learning. Other researchers have shown that pigeons are capable of insight learning equivalent to that of primates (Epstein et al., 1984) and of showing awareness of their own bodies and past behavior (Epstein et al., 1981).

How do they do that with their comparatively small brains? Perhaps it is the special ability of the avian brain to make new neurones in adulthood. Nottebohm and his co-workers have shown that the nuclei in the forebrain of songbirds, such as the canary, that are involved in producing song increase in size in spring when the male canary begins to court the female (Devoogd et al., 1985;

Alvarez-Buylla et al., 1990). The testes increase in size and release more testosterone, and that triggers the neurones to divide. The male begins to develop his complex repertoire of song at the same time. Thus the neurones involved in song production multiply only when they are needed. The adult canary, for example, retains its individual song repertoire from season to season, presumably stored in those neurones which persist across seasons, but adds to it at each new season using the newly developed neuronal capacity. The growth of the song nuclei depends not only on the circulating levels of testosterone but also on the bird hearing itself sing (Bottjer et al., 1986). Auditory feedback learning and hormone levels interact to stimulate the development.

Nottebohm (1989) suggested that birds may keep their brains smaller, and so lighter in weight, because they fly. That may be why they evolved a means of modulating neurone numbers in certain brain regions, developing them only when they are required. Presumably, as the size of one brain region increases another declines, and so there may be continual jostling for, or time sharing of, brain capacity by the various functions as they need to be performed. Although new neurones have been seen in adult mammalian brains, it is an extremely rare phenomenon, and no neurogenesis has been reported for the adult primate brain (Rakic, 1985). The avian brain has solved its cognitive demands quite differently than the mammalian brain, and its small size indicates nothing of its cognitive complexity.

Environmental Enrichment Affects Brain Size and Cognition

Although the formation of new neurones in the adult mammalian brain is an extremely rare event, adult neurones can change their size and alter the number of connections which they make with each other. Diamond (1988) conducted a series of important experiments which have demonstrated that housing adult rats in an enriched environment, in which they interact with other rats and a variety of toys, leads to increased thickness of the cortical region of the brain hemispheres, compared with control rats which have been housed in isolation in standard laboratory cages without playthings. The cortical neurones of rats which experienced the enriched condition have increased branching of their dendrites and have an increased number of spines, the sites at which other neurones contact the dendrites. In fact, there was a 40 percent increase in the area of the synaptic thickenings at the points of contact between neurones. The synapses were bigger and there were more of them. In other words, the enriched condition caused increased connectivity between the neurones of the cortex. These changes occurred after as little as thirty days in the enriched environment and equally in both young adults aged four months and middle-aged rats aged two years. In another experiment, old rats (over 750 days old) were exposed to an enriched environment for 138 days, and they too had increased cortical thickness (Diamond et al., 1985).

The environment had modified or "shaped" the brain. Moreover, these cellular changes were manifest in an increased ability of the rats to solve problems in mazes: their cognitive ability had been enhanced by environmental experience. Thus these experiments have shown that brain organization is not a fixed entity, even in adulthood, but remains in dynamic interaction with the environment. It is not necessary to form new neurones to increase brain power: increasing the connections between existing neurones will suffice. But perhaps the avian brain, with its special ability to make new neurones in adulthood, is even more responsive to environmental demands. Clearly, the size of various brain regions is not a result of biological predestination but rather of the interaction between nature and nurture throughout the life span.

Interaction between biological determinants and environmental factors is paramount during early brain development. In response to environmental stimulation during early brain development, radical reorganization of the brain occurs. For example, in cats and rodents which are raised without visual stimulation, the brain regions which process auditory information expand to include those areas of the brain which would normally be given over to processing other sensory information, such as for vision (Vidyasagar, 1978; Rauschecker and Harris, 1983) and the region of the brain which processes somatosensory input from the whiskers enlarges (Rauschecker et al., 1992). Such compensatory plasticity also appears to occur in blind humans and in humans who are deaf. The developing brain adapts to maximize its cognitive abilities to the incoming information which it receives. Once again, we see that brain capacity is modified by environmental stimulation. Enrichment of the environment has its greatest effects during early life, but, as already discussed, it also modifies the structure and function of the brain in adulthood.

It is worth pointing out that the increase in cortical thickness in the adult rats which had experienced the enriched environment was not simply due to overall growth of the body. In fact, the rats from the enriched condition had lower body weights than the controls which had experienced the impoverished environment (Diamond, 1988). Therefore, if the brain measurements were scaled for body weight, the effects of enrichment would be magnified rather than diminished. The rats raised in the impoverished environment also had heavier skulls and larger jaws. The internal size of the skull did not, however, differ between the two groups. That is, the skull did not expand to encase the larger cortex of the enriched rats. If nothing more, this result shows that cranial capacity is at best a crude indicator of the neural and cognitive capacity of the brain. Presumably, the increased cortex size resulting from the enrichment (up to 16 percent in some regions) had been accommodated by a decrease in the volume of the cerebrospinal fluid, which surrounds the brain between the meninges and also occurs in the ventricles inside the brain. These aspects of the brain do not, of course, leave fossil records and therefore do not feature in general schemes for the evolution of "intelligence."

Within a species, at least, brain structure and cognitive abilities may vary markedly according to environmental factors. There needs to be more recognition of this fact when species differences in cognition are considered. For example, when the problem-solving abilities of nonhuman primates are compared with those of humans, the fact that nonhuman primates which take part in these studies have been housed in impoverished laboratory or zoo environments is never taken into account. The humans to whom they are compared have never suffered such environmental deprivations. The results, therefore, cannot fail to reinforce the assumed superiority of humans. If an intellectual gap exists between us and other primates, it is surely widened by these studies.

One might argue that it should be possible to compare the cognitive abilities of different species of nonhuman primates (e.g., chimpanzees with orangutans) without encountering these problems, as all of the experimental subjects have experienced the impoverished housing conditions. That is not so, because species vary in their adaptability to captivity and to either isolation or group living. Orangutans are, for instance, more solitary in their natural environment than are chimpanzees (Mitani et al., 1991) and so adapt differently to captivity. Furthermore, age may be a factor in adaptation to the housing and the experimental conditions. Thus Diamond (1988) suggested that old rats may fare better in isolation, whereas younger ones prefer group living. Rarely, if ever, are these species and individual preferences, or their effects on environmental interaction with brain complexity, considered in cross-species comparisons of cognitive performance or sign language use. Perhaps signing by chimps falls short of language use by humans (Rumbaugh, 1980) because of the living conditions of the animals tested.

Environmental effects on the brain also need to be considered in cross-species comparisons of brain structure and even brain size. My aim is not to overthrow the evolutionary hypotheses based on cranial or brain size but rather to add a well-considered element of doubt to the too-definite statements of these hypotheses. All too often we are presented with the argument, along with accompanying diagrams (see Eccles, 1989, diagrams 37–39A), of cranial size or brain size ranging across mammalian species, or the hominids, to illustrate the increasing size of the cortex along with more convolutions on its surface. Size is the only criterion considered. Birds, with their small brains with few convolutions of the surface, therefore fall well below cats and rats. Recent knowledge of the cognitive abilities of birds throw these conceptualizations of the evolution of "intelligence" into serious doubt.

Are There Special Neurones for Consciousness?

Eccles (1989, 1992) has developed a hypothesis for the origin of consciousness based on the evolution of the mammalian neocortex. From the outset, Eccles takes what may be called a "mammalocentric" view of consciousness. He is prepared to accept that some mammals "exhibit intelligence and a learned behavior and are moved by feelings and moods, even with emotional attachment and understanding" (1992). Thus he argues that the neocortex is the structural basis of evolving consciousness, and within the neocortex specifically the apical dendrites of the pyramidal cells arranged into bundles called dendrons. There are about forty million of these dendrons in the human cortex. The dendrons, Eccles speculates, interact with the "world of the mind," and in so doing produce units of consciousness, called psychons. Along these lines of reasoning, the evolution of consciousness can be traced to 200 million years ago when the dendrons first appeared in the primitive cerebral cortices of early mammals.

While it may not be altogether clear how this postulated mind-brain interaction really occurs, it is clear that to Eccles structure dictates brain function and so forms the basis for the scientific understanding of consciousness. He makes only limited reference to studies of animal cognition and behaviors which might signal the presence of consciousness in animals. Thus it would seem to be a case of putting the cart before the horse. Nevertheless, Eccles recognizes that birds show insightful behavior, which suggests consciousness, even though they do not have a neocortex. Therefore he calls for further examination of the Wulst region of the avian forebrain to seek structural evidence for consciousness in the organization of its neurones.

Setting these criticisms of detail aside, and even if we do not accept Eccles's theory for consciousness, at least Eccles is prepared to address the issue of consciousness in animals and to see it as something that arose gradually through the course of evolution, rather than suddenly appearing with the evolution of humans. Along with Popper (1982) he believes that consciousness is a product of evolution, undergoing natural selection because it conferred advantages to those species which acquired it (Eccles, 1989).

It seems unlikely, however, that complex thought processes are the product of one particular cell type, the pyramidal cells. Given their structural complexity and interconnectivity, pyramidal cells might well be one of the material substrates for conscious thought processes, but probably not the only one. Nor would it seem necessary to tie consciousness to a unitary anatomical substrate until a more extensive behavioral study has been made of consciousness in animal species.

Moreover, although natural selection may have acted to produce animals with larger and more complex brains, as Eccles and many others contend, it is surely not the only process which has operated. It is equally probable that brains became larger, more complex, or both as animals adapted to environmental demands, or roles. As discussed, the size of the cortex of an individual rat changes according to the environmental demands of enriched or impoverished living. Diamond (1988) has also shown that environmental enrichment during pregnancy enhances cortical size over generations. Transmission of information by learning across generations, namely culture, may have similar selective effects on brain complexity, acting quite independently of genetic factors, and so natural selection. Such cultural selection for a complex brain (and mind) is likely to have been greatest in humans, but cultural transmission of behavior patterns can also occur in other animals (Bonner, 1980).

Lateralization of Brain Function

Another aspect of brain organization which was thought to elevate the human brain above that of other animals is its lateralization. In the human brain the left cerebral hemisphere processes information differently from the right, and so is involved with a different set of functions. The first asymmetries discovered in humans were for handedness and for processing language and producing speech. In most individuals these functions are performed by regions located in the left hemisphere. The Sylvian fissure, which runs between the two major speech areas (Broca's area in the frontal lobe and Wernicke's in the temporal cortex), is longer and ends at a lower point in the left hemisphere.

Given the strong association that some writers have made between language and consciousness and the fact that some researchers even believe that consciousness cannot occur without language, the possession of conscious thought becomes associated with lateralized brain function and is seen as unique to humans. The division of functions between the hemispheres is considered to double neural capacity and, in part, explain the superior intelligence of humans. This belief in the link between consciousness and brain asymmetry is still promoted in quite recent publications (e.g., Eccles, 1989). We now know, however, that lateralization of the brain occurs in a wide range of species, including nonhuman primates, other mammals, birds, even reptiles (see Bradshaw and Rogers, 1993).

The structural asymmetry of the Sylvian fissure described for the human brain also occurs in apes, and in both Old and New World monkeys there is asymmetry in the length of the Sylvian fissure (Heilbroner and Holloway, 1988). Also, in the brains of humans and other higher primates, the anterior structures on the right side project farther forward and are wider than those on the left, whereas the reverse occurs in the back part of the brain. These structural asymmetries are paralleled by the distribution of different functions by the left and right hemispheres in humans as well as in other primates.

Pioneering studies by Sperry (1974) revealed a range of lateralized functions in humans. He examined subjects who had had the corpus callosum, the large nerve tract which connects the left and right hemispheres, severed because they suffered from intractable epilepsy. This operation prevents information from being transferred from one hemisphere to the other. Thus, when the subject visually fixates a point directly in front and visual stimuli are flashed in the extreme left visual field, the information is processed by the right hemisphere, and vice versa. When such a "split-brain" subject was presented with a picture of, say, an apple in the right field, naming of the picture was possible. By contrast, when presented with the "apple" picture to the left field, naming was impossible, because no access to the language center in the left hemisphere could be made. In the latter situation, however, the subject could choose an apple from a bowl of fruit to indicate what the right hemisphere had seen. Using such procedures, Sperry was able to show that the left hemisphere is specialized for language, speech, and forms of analytical thought including mathematical calculation, and the right for emotions, music appreciation, spatial abilities, and nonverbal ideation (Sperry, 1974). The left hemisphere also controls the right hand and so it dominates for control of many motor functions, such as writing.

It should be noted that later studies have shown involvement of the right hemisphere in some aspects of language, and the left hemisphere in analysis of music by trained musicians. The asymmetry is, therefore, relative rather than absolute.

These patients with commissurotomies added fuel to the debate about consciousness in humans. In an intensive debate Popper and Eccles (1977) considered the paradox of having two "minds" in one person. If conscious thought resided in the left and right hemispheres, then the "split-brain" subjects should have two separate minds. What would that mean for the individual identity? Or, since only the left hemisphere is involved with speech, was it the location of consciousness? If so, the right hemisphere would be unconscious. Yet the right hemisphere is capable of highly complex thought and ideas even though it cannot express them verbally. The ability for language, or specifically speech production, is not likely to be necessary for consciousness. There is no need to impute a special role for language, or for any other related processes of the left hemisphere, to consciousness. Language may simply be the vehicle by which scientists have been able to assess the presence of consciousness; that is, by which humans can communicate consciousness to one another. To assess consciousness of the right hemisphere would presumably require use of techniques similar to those which might be used to assess consciousness in animals. On the other hand, if new tests without language requirements are developed specifically to assess consciousness in the right hemisphere of humans (see

Schacter, 1992), these might well be applied to assessing conciousness in animals.

Contrary to the original belief, the possession of a lateralized brain is not unique to the human species. Rhesus monkeys with commissurotomies and also sectioned optic tracts, at the chiasma, have been tested in the same way as Sperry's "split-brain" subjects. The sectioned visual pathways further restrict visual inputs from each eye to the contralateral hemisphere, removing the need to present the stimuli in the extreme lateral field of vision. These rhesus monkeys were shown to have lateralized brain function: they discriminated faces better when they were presented in the left visual field (right hemisphere) and tilted lines when they were presented in the right visual field (left hemisphere) (Hamilton and Vermiere, 1988). In humans also, the right hemisphere is specialized for discriminating faces (Burton and Levy, 1991; Sergent and Signoret, 1992), and the same is true for chimpanzees (Morris and Hopkins, 1993). Also like humans, the rhesus monkeys showed more emotional responses to stimuli presented to the left eye (right hemisphere).

Strikingly reminiscent of the left hemisphere's involvement in language and speech in humans, the left hemisphere of the Japanese macaque is specialized for processing the species-specific vocalizations (Petersen et al., 1984). Although there has been no study of whether Japanese macaque monkeys extract meaning from their vocalizations, Seyfarth and Cheyney (1992) have shown that this is the case for the East African vervet monkey. Vervet monkeys give different calls in response to seeing different predators, such as leopards, eagles, and snakes. Each call elicits the appropriate form of defense by the other monkeys: they look up when the "eagle" alarm is sounded; stand up and look into the grass for the "snake" call; and climb trees for the "leopard" call. The monkeys certainly seem to be responding as if they knew the meaning of the calls. To add further weight to this interpretation, the researchers (Cheyney and Seyfarth, 1988) tested the monkeys using a method of habituation-dishabituation which tests speech perception in human infants. They chose two contact calls, a *wrr* which is given when the monkeys spot another group and a *chutter* which is emitted in aggressive encounters between groups. First, a subject was exposed repeatedly to the *chutter* of another individual until it no longer responded to the call. Then the *wrr* call of the same individual was played. The test subject did not respond. Given that the *wrr* is acoustically very different from the *chutter*, the test subject must have been interpreting the meaning of the calls rather than "mindlessly" responding to their acoustic content. Both calls referred to the same social situation, and habituation occurred simultaneously to both. By contrast, when the experiment was repeated using two calls which refer to very different contexts (e.g., the leopard and eagle alarm calls), habituation to one of the calls did not transfer to the other. These results indicate that the monkeys have semantic, representational communication, which is a first step toward language, although human language involves much more than referential relations between words and objects or events.

In mice, too, the left hemisphere is specialized for processing the species-specific vocalizations, such as the ultrasonic calls of the pups (Ehret, 1987). The mother mouse retrieves her pups when her left ear is blocked with wax but not when her right ear is so blocked. In gerbils the male makes courtship vocalizations, and these are produced by a nucleus in the left side of the hypothalamus (Holman and Hutchison, 1993).

Taken together, all of these findings suggest that the left side of the brain is used to process and produce the vocalizations used in communication in a large number of mammalian species. The role of the left hemisphere in human language and speech therefore had a clear evolutionary trajectory. It did not occur as a single leap with the coming of humans. In fact, it is not known when language first appeared in the hominid line. Around 35,000 years ago, in the European Upper Paleolithic, humans suddenly started to produce large quantities of art in the form of bone and ivory carvings, and some believe that this flowering of culture was accompanied by the appearance of language (Noble and Davidson, 1991). Whether or not this is so, the development of art and communication may have enhanced a brain lateralization that was already present.

Other brain functions also are lateralized in nonhuman animals (Bradshaw and Rogers, 1993). It was thought that only humans had preferred handedness (righthandedness), but this preference has now been reported for a number of primate species, and in many it is present as a population bias (MacNeilage et al., 1987). Among the lower primates, lefthandedness predominates for holding and manipulating food objects, particularly in males (Milliken et al., 1991; Ward et al., 1993, Larson et al., 1989), whereas righthandedness for manipulation is more usual in apes and humans (Bard et al., 1990; Olson et al., 1990). MacNeilage et al. (1987) have hypothesized that brain lateralization evolved to accommodate postural requirements, the right hand being stronger and used to support the body and the left being used for visually guided reaching. This was the case for food reaching in lower primates and, once bipedalism evolved, the right hand became used for manipulation.

Brain lateralization, however, evolved before the primates. It has been well documented to be present in birds and rodents, and there is increasing evidence for its presence in cats, dogs, and some marsupial species (see Bradshaw and Rogers, 1993). In rats the left hemisphere is specialized for processing information sequentially, and the right for processing it in parallel (Bianki, 1988). The right hemisphere is also used for spatial analysis and controlling emotional behavior (Denenberg, 1984; Sherman et al., 1980). This left-right arrangement is similar to that known for humans.

In dogs the right hemisphere is larger and heavier than the left (Tan and Çalişkan, 1987). One study has reported that dogs preferentially use a right paw to remove sticking tape from their eyes (Tan, 1987). There have been more studies of pawedness in cats, and overall they seem to prefer to use their left paw (Cole, 1955; Forward et al., 1962; Fabre-Thorpe et al., 1993).

Some of the best examples of brain asymmetry are known for birds. Chickens using the right eye (left hemisphere) show superior ability to discriminate food from nonfood objects (Rogers, 1991), and those using the left eye (right hemisphere) perform better on spatial tasks (Rashid and Andrew, 1989). Pigeons also recall complex visual discrimination tasks better when they use the right eye (von Fersen and Güntürkün, 1990).

Preferential foot use (footedness) is also known to occur in birds. Most species of parrots and cockatoos preferentially use their left foot for holding and manipulating food objects (Harris, 1989; Rogers, 1980). Chickens typically start with their right foot when they scratch the ground in search of food (Rogers and Workman, 1993). Other species do not show a footedness bias at the population level, but individuals show a preference to use one foot more than the other. Thus great tits show individual foot preferences for fishing up pieces of food attached to a string (Vince, 1964) and pigeons for the first foot which touches the substrate when landing from flight (Davies and Green, 1991).

Songbirds of a number of species are known to control singing by nuclei located in the left hemisphere (Nottebohm, 1977). I have already discussed the increase in the size of these nuclei when testosterone levels rise during the spring. The nuclei occur on both sides of the forebrain, but only those on the left have a functional role in song production. Thus song in birds, vocalizations in rodents and primates, and language and speech in humans are all associated with the left side of the brain. Birdsong has some of the characteristics of language communication (Nottebohm et al., 1979) and even involves learning of local dialects (Nottebohm, 1970), although it is not yet known whether the listening birds attend to the meaning of the call. Primate vocalizations do appear to be interpreted for their meaning by other members of the species. There is much need for more research of the vocal communication in all of these species, but even now we may note, with a certain amazement, that all involve brain lateralization and all the left side of the brain. There was, in fact, no major step taken by brain lateralization for language and speech in hominid species. The left side–right side differences for analyzing species-specific communications were already present long before mammals, let alone humans, evolved. Similarly, one perhaps might argue against the notion that consciousness appeared suddenly in the hominid species.

Awareness of Others

An important aspect of higher cognitive thought is to be able to attribute mental states to others. Psychologists refer to this awareness of the other as having a "theory of mind." Research on chimpanzees suggests that they possess such a theory. Premack and Woodruff (1978) showed to a captive chimpanzee a series of videotaped scenes of a human actor struggling to solve a variety of problems, such as reaching for an inaccessible bunch of bananas or getting out of a locked cage. With each videotape the chimpanzee was given a series of photographs, only one of which showed a solution to the problem, such as a stick for the bananas or a key for the lock. The chimp chose the correct photograph to solve each problem, which suggests that she understood the actor's purpose and chose accordingly. Moreover, the chimp appeared to modify her choices according to the outcome she wanted for the person appearing in the videotape. When the latter was her favorite trainer, she chose the correct solution; but when it was one she did not particularly like, she chose the incorrect solution. Of course, one might argue that she attended to the content of the videotape only when the favorite trainer appeared and therefore chose the correct solution in this case, but chimps are capable of deceit, and that may well have been her intent.

Cheyney and Seyfarth (1990) have documented many examples of deceit in apes. Deception requires awareness and assessment of the mental state or intent of others in order to trick them. Increasingly, there are reports of deception in primates and other species. For example, a wild chimpanzee was observed to lead group members away from a food source by directing their attention elsewhere and then to double back to eat the food himself. Other primates have been seen to reach out the hand to another in an appeasement gesture and, when the other responds, grab his or her arm and attack.

Play behavior often involves deception, and play is characteristic of many species. Play behavior is more common in young animals, and is commonly believed to be the way in which animals learn the skills of adulthood. For example, cats are thought to learn to hunt through play behavior, and the mother has an active role, possibly as a teacher, in this learning process. The mother cat brings prey for her offspring to play with, and she brings increasingly large prey types as their skill increases, as if she is aware of their state of knowledge (Caro and Hauser, 1992).

Teaching is also considered to involve being aware of the other, the pupil. Seyfarth and Cheyney (1986, 1992) have reported that mother vervet monkeys do not correct their offspring when they make incorrect responses to various predator alarm calls. The mothers do not appear to be aware of their offsprings' ignorance. There are other examples in which the parent would seem actively to teach the offspring.

In their review of teaching in nonhuman animals, however, Caro and Hauser (1992) argue that teaching does not necessarily involve attribution of mental states or intentionality. For example, the mother cat may change her teaching behavior according to a stereotyped time course and so be unaware of her kittens' state of knowledge. Nevertheless, the same authors present data which show that many aspects of the mother's predatory behavior was significantly negatively correlated with the increasing skills of her offspring, and there was individual variation in this pattern. That suggests adaptation to the offsprings' level of skill and possibly awareness of their state or condition.

Caro and Hauser (1992) have also used the limited cultural transmission of sign language in chimpanzees as evidence against active and aware teaching in animals. Washoe had learned sign language from human teachers. When Loulis was raised with Washoe, she acquired few of the signs and only rarely was Washoe seen to mold her hands into a sign in attempts to teach her. Yet there was at least one example of Washoe doing this; it involved the FOOD sign for a candy bar. The authors might have interpreted this example more favorably: that sign language is the preferred mode of communication only in interactions with humans and when highly motivated. Outside these situations the chimps may have relied on their own species-specific modes of communication, and that is why scientists have observed no active teaching.

Wild chimpanzees have been observed to play an active role in teaching. Boesch (1991) describes a mother reorienting the position of a nut on the anvil rock so that her infant could crack it more easily by using a second rock as a hammer. In another instance the mother reoriented the hammer in her offspring's hand so that it could be used more effectively to crack nuts. Although Caro and Hauser (1992) discount these as rare observations in the long study of Boesch, their importance cannot be denied. No matter how often active teaching occurs, the fact that it does occur indicates an apparent ability to attribute a mental state to another. The seeming rareness of occurrence of these form of teaching may depend on the opportunity for observation; it may also be due to lack of awareness in the observers. The tendency is, after all, to underestimate the abilities of animals.

While we must recognize that there is too little information for us to decide whether teaching in nonhuman species involves intentionality or the attribution of mental states to others, this is clearly an important area for future research. Much attention has been given to how and what animals learn and remember, whereas there has been relatively little interest in how, what, or why they teach. Clearly, laboratory studies of teaching in nonhuman animals will require more imaginative experimental designs than are presently used. Comprehensive knowledge of each species in its natural environment should be a basis for these new experimental designs, and there needs to be more concentration on observing how animals teach in the natural environment. In saying this, I do not mean that behavior in the natural environment is in some way superior but that more progress will be made if animals are asked to learn or teach in contexts that best suit their perceptual, motor, and information-processing abilities, rather than asking them to perform in artificial contexts. An example of the latter is the expectation that chimpanzees which have learned to sign might teach signing to other chimpanzees even though they may have their own species-specific means of communication. The fact that the chimpanzees tested so far have not taught sign language to others cannot be used as evidence that they do not have a "mind state." Similarly, the fact that chimpanzees have not learned the syntactical forms of sign language characteristic of spoken English has been taken as evidence that they lack a concept of the self (Terrace and Bever, 1980; Terrace, 1985), and the fact that signing chimpanzees have rarely asked questions using the sign language has been used to rank them as inferior to humans who use spoken English to ask a series of questions even when they are juveniles (Bronowski and Bellugi, 1970; Eccles, 1989). It would be more valid to compare the signing by chimpanzees to signing by humans who have learned the same sign language as a second language. The humans should be using sign language as their second, not their first, language, if we are prepared to begin to study the apes by assuming that their own species-specific patterns of communication are their first language. And, in fact, that is not much of an assumption to make; rather it may be considered as a more enlightened starting point for these studies. Even then we must recognize that sign languages have been developed for humans and may be less well adapted to the motor or cognitive abilities of apes; we also must recognize that the vocabularies the apes are taught may be inadequate for the sort of things they wish to express. Without considering these important issues, how can researchers claim that the differences between signing in apes and language in humans are qualitative and not a matter of more or less and therefore represent "a different type of intellectual organization"? (Chomsky, 1980, p. 439).

From the Wild to the Laboratory

I have discussed how brain organization and function adapt to enriched and impoverished environments and how many learning tasks required of animals in the laboratory may have little to do with the behaviors which they might perform in their natural environment. How useful, therefore, is the information that is gained from laboratory studies or from observation of captive animals in zoos? Behavior and brain structures will change according to the context in which the animal is living, but exactly how will they change? Could an artificial environment be enriching or is it always impoverishing? Perhaps the special environment in which Washoe was raised, in a caravan with toys, sign language, and constant contact with people, was enriching, and so she reached a potential even greater than that of wild chimpanzees. I think not. The social life of chimpanzees living in the wild is rich and complex, as also is the stimulation which they receive from that environment. Obviously, life in zoo or laboratory captivity is far less stimulating than life in the wild. Thus we cannot fail to be observing the lowest common denominator of brain development and functioning in animals living in captivity. We should begin all laboratory and zoo studies by being fully aware of the fact that we are dealing with animals in impoverished environments. Why then do scientists so often feel that it is necessary to go the extra step and painstakingly argue away any hints of consciousness that might still appear despite all of this?

We need to recognize our own cultural history, which has continually constructed animals, including apes, as different from and less than us. As Haraway (1992) so clearly describes, we have continually constructed frameworks to separate us from primates. We have created the nature-culture divide and make sure that no ape crosses it. Modern science is continuing to provide the "evidence" for this. In sexual behavior, for example, we construct apes as "uncivilized" versions of ourselves (Carpenter, 1964), as nature uninfluenced by culture and indeed by consciousness.

We are not prepared to consider that feelings of pleasure, joy, sadness, loss, and so on extend beyond the human pedestal. Some of us go so far as to say that these are emotions that can only really be felt when there is language. Language is therefore the axiom by which scientists and philosophers tend to tie consciousness to the human species. Others have no difficulty in accepting that nonhuman animals have complex thought patterns which may well rely on mental representations, but then hasten to relegate these into a category inferior to those used by humans (e.g., Terrace, 1985). The supposed superiority of the human species is thus reinforced. Yet the growing debate about animal welfare and consciousness in animals has made many of us reflect on, and sometimes change, the practices we use in animal experimentation. It is a beginning. The practice of biology must, and will, change as we recognize that the conceptual divide that Western culture has erected between humans and other animals is artificial.

References

Anderson, B. Evidence from the rat for a general factor that underlies cognitive performance and that relates to brain size: intelligence? *Neuroscience Letters*, 153 (1993): 98–102.

Alvarez-Buylla, A., Kirn, J. R., and Nottebohm, F. Birth of projection neurons in adult avian brain may be related to perceptual or motor learning. *Science*, 249 (1990): 1444–1446.

Bard, K. A., Hopkins, W. D., and Fort, C. L. Lateral bias in infant chimpanzees (*Pan troglodytes*). *Journal of Comparative Psychology*, 104 (1990): 309–321.

Bayer, S. A., and Altman, J. *Neocortical Development* (New York: Raven Press, 1991).

Beck, B. B. *Animal Tool Behavior* (New York: Garland, 1980).

Bianki, V. L. *The Right and Left Hemispheres: Cerebral Lateralization of Function*, Monographs in Neuroscience, vol. 3 (New York: Gordon and Breach, 1988).

Boesch, C. Teaching among wild chimpanzees. *Animal Behaviour*, 41 (1991): 530–532.

Boesch, C., and Boesch, H. Optimization of nut-cracking with natural hammers by wild chimpanzees. *Behaviour,* 83 (1982): 265–286.

Bonner, J. L. *The Evolution of Culture in Animals* (Princeton, N.J.: Princeton University Press, 1980).

Bottjer, S. W., Schoomaker, J. N., and Arnold, A. P. Auditory and hormonal stimulation interact to produce neural growth in adult canaries. *Journal of Neurobiology,* 17 (1986): 605–612.

Bradshaw, J. L. Animal asymmetry and human heredity: Dextrality, tool use and language in evolution—10 years after Walker (1980). *British Journal of Psychology,* 82 (1991): 39–59.

Bradshaw, J. L., and Rogers, L. J. *The Evolution of Lateral Asymmetries, Language, Tool Use and Intellect* (San Diego Academic Press, 1993).

Bronowski, J., and Bellugi, U. Language, name and concept. *Science,* 168 (1970): 669–673.

Burton, L. A., and Levy, J. Effects of processing speed on cerebral asymmetry for left- and right-oriented faces. *Brain and Cognition,* 15 (1991): 95–105.

Caro, T. M., and Hauser, M. D. Is there teaching in nonhuman animals? *Quarterly Review of Biology,* 67 (1992): 151–174.

Carpenter, C. R. *Naturalistic Behavior of Nonhuman Primates* (University Park: Pennsylvania State University Press, 1964).

Cerella, J. Pigeons and perceptions. *Pattern Recognition,* 19 (1986): 431–438.

Chevalier-Skolnikoff, S., and Liska, J. Tool use by wild and captive elephants. *Animal Behaviour,* 46 (1993): 209–219.

Cheyney, D. L., and Seyfarth, R. M. Assessment of meaning and the detection of unreliable signals by vervet monkeys. *Animal Behaviour,* 36 (1988): 477–486.

Cheyney, D. L., and Seyfarth, R. M. *How Monkeys See the World: Inside the Mind of Another Species* (Chicago: University of Chicago Press, 1990).

Chomsky, N. Human language and other semiotic systems. In T. A. Sebeok and D. J. Umiker-Sebeok (eds.), *Speaking of Apes* (New York: Plenum Press, 1980), pp. 429–440.

Cole, J. Paw preference in cats related to hand preference in animals and man. *Journal of Comparative Physiology and Psychology,* 48 (1955): 137–140.

Darwin, C. *The Descent of Man.* (London: John Murray, 1871).

Davies, M. N. O., and Green, P. R. Footedness in pigeons, or simply sleight of foot? *Animal Behaviour,* 42 (1991): 311–312.

Delius, J. D. Cognitive processes in pigeons. In G. D'Ydelvalle (ed.), *Cognition, Information Processing and Motivation* (Amsterdam: Elsevier, 1985), pp. 3–18.

Delius, J. D. Sapient sauropsids and hollering hominids. In W. A. Koch (ed.), *Geneses of Language* (Brockmeyer: Bochum, 1990), pp. 1–29.

Denenberg, V. H. Behavioral asymmetry. In N. Geschwind and A. M. Galaburda (eds.), *Cerebral Dominance: The Biological Foundation* (Cambridge: Harvard University Press, 1984), pp. 114–133.

Devoogd, T. J., Nixdorf, B., and Nottebohm, F. Synaptogenesis and changes in synaptic morphology related to acquisition of new behavior. *Brain Research,* 329 (1985): 304–308.

Diamond, M. C. *Enriching Heredity: The Impact of the Environment on the Anatomy of the Brain* (New York: Free Press, 1988).

Diamond, M. C., et al. Plasticity in the 904-day-old rat cerebral cortex. *Experimental Neurology,* 87 (1985) 309–317.

Eccles, J. C. *Evolution of the Brain: Creation of the Self.* (London: Routledge, 1989).

Eccles, J. C. Evolution of consciousness. *Proceedings of the National Academy of Science,* 89 (1992): 7320–7324.

Ehret, G. Left hemisphere advantage in the mouse brain for recognising ultrasonic communication calls. *Nature,* 325 (1987): 249–251.

Epstein, R., et al. "Insight" in the pigeon: Antecedents and determinants of an intelligent performance. *Nature*, 308 (1984): 61–62.

Epstein, R., Lanza, R. P. and Skinner, B. F. "Self-awareness" in the pigeon. *Science*, 212 (1981): 695–696.

Fabre-Thorpe, M., et al. Laterality in cats: Paw preference and performance in a visuo-motor activity. *Cortex*, 29 (1993): 15–24.

Forward, E., Warren, J. M., and Hara, K. The effects of unilateral lesions in sensorimotor cortex on manipulation by cats. *Journal of Comparative and Physiological Psychology*, 55 (1962): 1130–1135.

Galdikas, B. M. F. Orang-utan tool-use at Tanjung Puting Reserve, Central Indonesia Borneo (Kalimantan Tengah). *Journal of Human Evolution*, 10 (1982): 19–33.

Gallup, G. G. Self-recognition in primates. *American Journal of Psychology*, 32 (1977): 329–338.

Gallup, G. G. Toward a comparative psychology of mind. In R. L. Mellgren (ed.), *Animal Cognition and Behaviour* (Amsterdam: North Holland, 1983), pp. 473–510.

Goodall, J. The behaviour of free-ranging chimpanzees in the Gombe Stream Reserve. *Animal Behaviour Monographs*, 1 (1968): 161–311.

Gould, S. J. *The Mismeasure of Man.* (New York: Norton, 1981).

Griffin, D. R. *Animal Thinking* (Cambridge, Mass.: Harvard University Press, 1984).

Hamilton, C. R., and Vermiere, B. A. Complementary hemispheric specialization in monkeys. *Science* 242 (1988): 1691–1694.

Haraway, D. *Primate Visions* (London: Verso, 1992).

Harris, L. J. Footedness in parrots: three centuries of research, theory, and mere surmise. *Canadian Journal of Psychology*, 43 (1989): 369–396.

Heilboner, P. L., and Holloway, R. L. Anatomical brain asymmetries in New World and Old World monkeys: stages of temporal lobe development in primate evolution. *American Journal of Physical Anthropology*, 76 (1988): 39–48.

Herrnstein, R. J. Stimuli and the texture of experience. *Neuroscience and Behavioral Reviews*, 6 (1982): 105–117.

Hollard, V. D., and Delius, J. D. Rotational invariance in visual pattern recognition by pigeons and humans. *Science*, 218 (1982): 804–806.

Holman, S. D., and Hutchison, J. B. Lateralization of a sexually dimorphic brain area associated with steroid-sensitive behavior in the male gerbil. *Behavioral Neuroscience*, 107 (1993): 186–193.

Jerison, H. J. *Evolution of Brain and Intelligence* (New York: Academic Press, 1973).

Kenyon, K. W. The sea otter in the eastern Pacific Ocean. *North American Fauna*, no. 68 (Washington, D.C.: U.S. Bureau of Sport Fisheries and Wildlife, 1969).

Larson, C. F., Dodson, D. L., and Ward, J. P. Hand preferences and whole-body turning biases of lesser Bushbabies (*Galago senegalensis*). *Brain, Behavior and Evolution*, 33 (1989): 261–267.

Lethmate, J. Tool-using skills of orang-utans. *Journal of Human Evolution*, 11 (1982): 49–64.

Lewontin, R. C., Rose, S., and Kamin, L. J. *Not in Our Genes* (New York: Pantheon, 1984).

Lombardi, C. M., Fachinelli, C. C., and Delius, J. D. Odditiy of visual patterns conceptualized by pigeons. *Animal Learning and Behavior*, 12 (1984): 2–6.

MacKintosh, N.J. *Conditioning and Associative Learning* (Oxford: Oxford University Press, 1983).

MacNeilage, P. F., Studdert-Kennedy, M. G., and Lindblom, B. Primate handedness reconsidered. *Behavioural Brain Sciences*, 10 (1987): 247–303.

Miles, H. L. W. The cognitive foundations for reference in a signing orangutan. In S. T. Parker and K. R. Gibson (eds.), *"Language" and Intelligence in Monkeys and Apes* (Cambridge: Cambridge University Press, 1990), pp. 511–539.

Millikan, G. C., and Bowman, R. I. Observations on Galápagos tool-using finches in captivity. *Living Birds*, 6 (1967): 23–41.

Milliken, G. W., et al. Analyses of feeding lateralization in the small-eared bushbaby (*Otolemur garnettii*): a comparison with the ring-tailed lemur (*Lemur catta*). *Journal of Comparative Psychology*, 105 (1991): 274–285.

Mitani, J. C., Crether, G. F., Rodman, P. S., and Prianta, D. Associations among wild orang-utans: sociality, passive aggregations or chance? *Animal Behaviour*, 42 (1991): 33–46.

Morris, R. D., and Hopkins, W. D. Perception of human chimeric faces by chimpanzees: evidence for a right hemisphere advantage. *Brain and Cognition*, 21 (1993): 111–122.

Nishida, T., and Uehara, S. Chimpanzees, tools and termites: another example from Tanzania. *Current Anthropology*, 21 (1980): 671–672.

Noble, W., and Davidson, I. The emergence of modern human behavior: language and its archeology. *Man*, 26 (1991): 223–253.

Nottebohm, F. Ontogeny of bird song. *Science*, 167 (1970): 950–956.

Nottebohm, F. Asymmetries in neural control of vocalization in the canary. In S. Harnard et al. (eds.), *Lateralization in the Nervous System* (New York: Academic Press, 1977), pp. 23–44.

Nottebohm, F. From bird song to neurogenesis. *Scientific American* (February 1989): 56–61.

Nottebohm, F., Manning, E., and Nottebohm, M. E. Reversal of hypoglossal dominance in canaries following unilateral syringeal denervation. *Journal of Comparative Physiology*, 134 (1979): 227–240.

Olson, D. A., Ellis, J. E., and Nadler, R. D. Hand preferences in captive gorillas, orang-utans and gibbons. *American Journal of Primatology*, 20 (1990): 83–94.

Parker, S. T., and Gibson, K. R. (eds.) *"Language" and Intelligence in Monkeys and Apes* (Cambridge: Cambridge University Press, 1990).

Pepperberg, I. M. Conceptual abilities of some non-primate species, with an emphasis on an African grey parrot. In S. T. Parker and K. R. Gibson (eds.), *"Language" and Intelligence in Monkeys and Apes* (Cambridge: Cambridge University Press, 1990a), pp. 469–507.

Pepperberg, I. M. Some cognitive capacities of an African grey parrot (*Psittacus erithacus*). *Advances in the Study of Behaviour*, 19 (1990b): 357–409.

Petersen, M. R., et al. Neural lateralization of vocalizations by Japanese macaques: communicative significance is more important than acoustic structure. *Behavioural Neuroscience*, 98 (1984): 779–790.

Pilbeam, D., and Gould, S. J. Size and scaling in human evolution. *Science*, 186 (1974): 892–901.

Popper, K. R. *The Open Universe: An Argument for Indeterminism* (London: Hutchinson, 1982).

Popper, K. R., and Eccles, J. C. *The Self and Its Brain.* (Berlin: Springer-Verlag, 1977).

Povinelli, D. J. Failure to find self-recognition in Asian elephants (*Elephas maximus*) in contrast to their use of mirror cues to discover hidden food. *Journal of Comparative Psychology*, 103 (1989): 122–131.

Premack, D. On the abstractness of human concepts: why it would be difficult to talk to a pigeon. In S. H. Hulse, H. Fowler, and W. K. Honig (eds.), *Cognitive Processes in Animal Behavior* (Erlbaum, 1978), pp. 421–451.

Premack, D., and Woodruff, G. Does the chimpanzee have a theory of mind? *Behavioral and Brain Sciences*, 4 (1978): 515–526.

Rakic, P. Limits of neurogenesis in primates. *Science*, 227 (1985): 1054–1056.

Rashid, N., and Andrew, R. J. Right hemisphere advantage for topographical orientation in the domestic chick. *Neuropsychologia*, 27 (1989): 937–948.

Rauschecker, J. P., and Harris, L. R. Auditory compensation of the effects of visual deprivation in the cat's superior colliculus. *Experimental Brain Research*, 50 (1983): 69–83.

Rauschecker, J. P., et al. Crossmodal changes in the somatosensory vibrissa/barrel system of visually deprived animals. *Proceedings of the National Academy Science*, 89 (1992): 5063–5067.

Rogers, L. J. Lateralisation in the avian brain. *Bird Behaviour*, 2 (1980): 1–12.

Rogers, L. J. Development of lateralization. In R. J. Andrew (ed.), *Neural and Behavioural Plasticity: The Use of the Domestic Chick as a Model* (Oxford: Oxford University Press, 1991), pp. 507–535.

Rogers, L. J., and Kaplan, G. A new form of tool use by orang-utans in Sabah, East Malaysia. *Folia Primatologia*, 278 (1994), in press.

Rogers, L. J., and Workman, L. Footedness in birds. *Animal Behaviour*, 45 (1993): 409–411.

Rumbaugh, D. M. Language behaviors of apes. In T. A. Sebeok and D. J. Umiker-Sebeok (eds.), *Speaking of Apes* (New York: Plenum Press, 1980), pp. 231–259.

Sakura, O., and Matsuzawa, T. Flexibility of wild chimpanzee nut-cracking behaviour using stone hammers and anvils: an experimental analysis. *Ethology*, 87 (1991): 237–248.

Schacter, D. L., Implicit knowledge: new perspectives on unconscious processes. *Proceedings of the National Academmcy of Sciences*, 89 (1992): 11113–11117.

Sergent, J., and Signoret, J-L. Functional and anatomical decomposition of face processing: evidence from prosopognosia and PET study of normal subjects. *Philosophical Transactions of the Royal Society London B*, 335 (1992): 55–62.

Seyfarth, R. M., and Cheyney, D. L. Vocal development in vervet monkeys. *Animal Behaviour*, 34 (1986): 1450–1468.

Seyfarth, R. M., and Cheyney, D. L. Meaning and mind in monkeys. *Scientific American*, (December 1992): 78–84.

Sherman, G. I., et al. Brain and behavioral asymmetries for spatial preference in rats. *Brain Research*, 192 (1980): 61–67.

Sperry, R. W. Lateral specialization in the surgically separated hemispheres. In F. O. Schmitt and F. G. Worden (eds.), *The Neurosciences, Third Study Program* (Cambridge, Mass: MIT Press, 1974), pp. 5–19.

Tan, U. Paw preference in dogs. *International Journal of Neuroscience*, 32 (1987): 825–829.

Tan, U., and Çalişkan, S. Allometry and asymmetry in the dog brain: the right hemisphere is heavier regardless of paw preference. *International Journal of Neuroscience*, 35 (1987): 189–194.

Terrace, H. S. Animal cognition: thinking without language. *Philosophical Transactions of the Royal Society of London B.*, 308 (1985): 113–128.

Terrace, H. S., and Bever, T. G. What might be learned from studying language in the chimpanzee? The importance of symbolizing oneself. In T. A. Sebeok and D. J. Umiker-Sebeok (eds.), *Speaking of Apes* (New York: Plenum Press, 1980), pp. 179–189.

Vaughan, W., and Greene, S. L. Pigeon visual memory capacity. *Journal of Experimental Psychology: Animal Behavior Processes*, 10 (1984): 256–271.

Vidyasagar, T. R. Possible plasticity of the rat superior colliculus. *Nature*, 275 (1978): 140–141.

Vince, M. A. Use of the feet in feeding by the Great Tit, *Parus major. Ibis*, 106 (1964): 500–529.

von Fersen, L., and Güntürkün, O. Visual memory lateralisation in pigeons. *Neuropsychologia*, 28 (1990): 1–7.

Ward, J. P., Milliken, G. W., and Stafford, D. K. Patterns of lateralized behavior in prosimians. In J. P. Ward and W. D. Hopkins (eds.), *Primate Laterality: Current Behavioral Evidence of Primate Asymmetries* (New York: Springer-Verlag, 1993), pp. 42–72.

Willerman, L., et al. In vivo brain size and intelligence. *Intelligence*, 15 (1991): 223–228.

10 | rEvolutionary Theory
Reinventing Our Origin Myths
Judith C. Masters

... we are, as evolutionary biologists, indirectly working on *nothing less than an important part of our culture's very own creation myth.* Is the combination of the pointlessness of chance with the tyranny of necessity, competitive exclusion, expedience, and obedience to external forces what we really want to think of as the sources of our origins?

—Stanley N. Salthe[1]

M**ANY SCIENTISTS STRONGLY** resent the notion that the science they practice, in terms of the problems identified and questions asked, the methodologies employed, and the interpretations advanced, is in large part informed by the broader ideologies prevalent in the societies in which their science is embedded, as well as by their own particular ideological agendas. For an illustration of this resentment, one has only to note the defensive tone of the majority of reviews by scientists of Haraway's recent sociohistorical study of twentieth-century anthropology, *Primate Visions.*[2] Cachel described the book as "a jeremiad against science" "replete with antiscience sentiments."[3] Cartmill complained that "Haraway's approach to science in general and to primatology in particular is an unfriendly one, which makes no effort to understand or sympathize with the intentions of scientists."[4] The most common response was to excommunicate Haraway from the scientific community, despite her doctoral degree in biology. Landau stated authoritatively: "What sets [Haraway] apart from the primatologists she studies is not only her intellectual commitments (and perhaps her political ones as well) but also her background (she is a historian of science) and approach."[5] Dunbar's review provides enough material for a postmodernist essay in itself. Not only did he refer to Haraway as "Ms. Haraway" throughout and impugn her scientific professionalism, he devalued the opinion of anyone who might be injudicious enough to accord the book value: "The less knowledgeable may greet it with enthusiasm; those with firsthand experience are likely to find it, at best, frustrating."[6] The message for scientists is clear: turn your critical gaze inward at your peril.

Despite resistance and even denial by scientists, science is as marked by prevailing ideologies as are other human activities. The most conspicuous ideological commitment of Western science is its portrayal as a means of controlling and dominating nature.[7] This ideology of domination demands the separation of science and the people who practice it from a conquered, exploited nature, which is of course construed as female. Nowhere is this ideology more explicit than in evolutionary theory, where our scientific convictions are enmeshed in our beliefs regarding how we came to be the way we are—i.e., our creation myths.[8]

Ideologies act like topographic landscapes to direct ideas and interpretations along channels that characterize particular research traditions. We all experience such constraints, and the effects are not necessarily negative. Collaboration and peer assessment would be impossible without the development of research traditions and their associated modes of expression, or paradigms.[9] On the other hand, when a particular approach has been unsuccessful or has reached the limits of its potential in explaining a suite of problems, extricating one's mind from the well-worn pathways in order to try a new approach can be very difficult. I see this situation prevailing in several areas of evolutionary theory and believe it has become necessary to backtrack and to dissect the ideology that has brought evolutionary theory to this impasse. In this chapter, therefore, I contrast the ideology of "being in control of and separated from nature" with one of "being part of nature" as they relate to our interpretations of evolutionary change.

Direction and progress are highly controversial ideas among evolutionary biologists (see 3B below). Nevertheless, because of their intimate association with the domination/separation ideology, they continue to reappear in evolutionary arguments. Julian Huxley, one of the major players in the consolidation of the neo-Darwinian synthesis, was strongly criticized for his forthright espousal of evolutionary progress; yet few contemporary neo-Darwinists would quibble with his description of evolutionary advance:

> The distinguishing characteristics of dominant groups all fall into one or other of two types—those making for greater control over the environment, and those making for greater independence of the environment. Thus advance in these respects may provisionally be taken as the criterion of biological progress.[10]

It is on these grounds that lungs are considered to be advances over gills and the development of a shelled egg containing its own aqueous environment within an amniotic membrane to constitute a "breakthrough into a new adaptive zone." Adaptive zones are constructs of neo-Darwinism—spaces and associated ways of life that are systematically conquered and exploited with the assistance of "key adaptations." Evolution of the amniote egg allowed vertebrates to become fully terrestrial, and—as any textbook will attest—independent of water for their reproduction. Since no land vertebrate is truly independent of water, this assessment of independence is equivocal.[11]

The tale of human evolution, in particular, is often told in these value-laden terms, echoing Adam's god-given dominion over nature. See, for example, Johanson and Shreeve's statement:

> The whole human career is a story of a species never relenting in its effort to pull itself free, by the sheer force of intellect, from the natural constraints that bind all other species to their biological fates.[12]

Griffin has argued cogently that this ideology of the cultural control of nature stems from a desire to deny and conquer what we construe as nature within ourselves: mortality, sexuality, compassion, emotions, feelings.[13] The parallels between this ideology and masculine socialization in our own society are obvious.

The extent of our culture's terror of nature is vividly captured in Eiseley's description of the psychological and emotional consequences of the general acceptance of our own evolutionary roots:

> [Man] has been convinced of his rise from a late Tertiary anthropoid stock. Through neurological and psychological research he is conscious that the human brain is an imperfect instrument built up through long geological periods. Some of its levels of operation are more primitive and archaic than others. Our heads, modern man has learned, may contain weird and irrational shadows out of the subhuman past—shadows that under stress can sometimes elongate and fall darkly across the threshold of our rational lives. . . . We have frightened ourselves with our own black nature and instead of thinking "We are men now, not beasts, and must live like men," we have eyed each other with wary suspicion and whispered in our hearts, "We will trust no one. Man is evil. Man is an animal. He has come from the dark wood and the caves."[14]

In a similar vein, Huxley posed the question of whether "some of [man's] inherited impulses and his simpler irrational satisfactions may not stand in the way of higher values and fuller enjoyment." In his view, the future evolutionary progress of humanity lay in the suppression of "the mammal . . . within us," in favor of the development of "more complex satisfactions, in the realm of morality, pure intellect, aesthetics, and creative activity."[15]

Domination/separation ideology is manifest at three levels of generality in the construction of evolutionary theory: (1) the preferred scientific methodology; (2) the processes and forces deemed to be responsible for evolutionary change; (3) the interpretations of particular patterns in extinct and extant nature. I shall investigate two examples at each of these levels.

1. The Preferred Scientific Methodology

Evolutionary biology, like many other Western sciences, has opted for a methodology which fragments the world into manageable pieces and then attempts to understand the world by analyzing the subsystems that comprise it.

The link between this methodology and the ideology I have outlined above has been eloquently described by Ho:

> To many great civilizations past and present, the unity of nature is simply a fact of immediate experience that needs no special pleading. In the West, however, much of the history of science is concerned with separating and reducing this unity into ever smaller and smaller fragments out of which nature has somehow to be glued together again. It is a history not only of fragmentation but of our own alienation from nature.[16]

This fragmentation has two main manifestations in evolutionary theory. One is the rigid separation that has been created between organisms and their environment; the second is a form of reductionism which has elevated the base sequences of the genes of living organisms to a level of significance far higher than that occupied by any other organizational level in nature.

A. The Organism-Environment Dichotomy

Because their favored mechanism for evolutionary change, natural selection, is mediated by the environment, neo-Darwinists operate on a fundamental dichotomy between organisms and the world around them (organic and inorganic).[17] This has given rise to a strangely static view of natural organization, whereby the world is seen as having been partitioned into niches which exist prior to the organisms that evolve to fill them.[18] Where no organism has evolved to follow a particular lifestyle, the niche is described as being "vacant" or "empty," and this hypothetical space is held to be a potent force for evolutionary change (see 2B below).

Lewontin has contrasted this view with a dynamic one whereby organisms construct their own niches to a large extent; they alter and are altered by their environments, biotic and abiotic, which also evolve. Lewontin's approach is the basis of ecological succession, calling into question simplistic models of the evolution of optimal "solutions" to static environmental "problems."[19] It also challenges one of the fundamental premises of evolutionary ecology, i.e., that no two species can share the same ecological niche. If niches can only be defined relative to the organisms which create them, then this premise is meaningless. Countless person hours have been expended in tracing the subtle differences in substrate use, seasonality, activity periods, etc. that "allow" taxa with similar resource requirements to coexist sympatrically, without violating this rule. These subtle discrepancies are then accorded a theoretical significance that may bear little or no relation to their meaning in the organisms' lives.

Reconstructing the organism-environment relationship as suggested by Lewontin means abandoning notions of niche priority and focusing on organismal activities and interactions for their own sakes, rather than as support for an idealized model of how the world should be ordered. This perspective of trying to understand nature from inside it, as opposed to attempting to impose order from the outside, will pave the way for an urgently needed revision of theories regarding competition and coexistence in nature.

B. Genetic Reductionism

Neo-Darwinism has always accorded the genetic level of organization far greater significance than any other, from the early days when evolution was described as a "change in gene frequencies"[20] to the situation that prevails today, when it seems to many that the only questions deemed worth asking involve the comparison of base sequences. To some biologists, this obsession with genetics has a strongly social dimension. Ho has argued that heredity has been reified by neo-Darwinism, conceived as the transmission of some material entity—the germ plasm—from parents to offspring.[21]

In a similar vein, Webster has explicated the hidden analogy between the inheritance of property in the social domain and concepts of genetic "inheritance."[22] In the social realm, inheritance functions not so much to maintain sameness as to perpetuate difference and inequality in wealth and power. The same ideology has been at the base of genetic theories, where the existence of genes devolves from the observation of difference.

Ho's alternative vision is one of heredity as a process, consisting of a multilevel system of interactions and feedbacks involving both organisms and environments.[23] It is the interdependent nature of the systems that is responsible for the reproduction of similar forms in successive generations, not a rigid set of deterministic rules imposed by a single component of the system—the DNA molecules. Thus, according to Ho, stability is based on fluidity, not on rigid, unchanging sets of instructions. This interpretation is more appropriate in a world of fluid genomes constantly under rearrangement owing to the activities of moveable elements such as transposons or to processes such as concerted evolution and molecular drive, which involve rapid changes in base sequence at disparate parts of the genome almost simultaneously.[24]

It is also more in line with the ideas put forward by the neutralists, who suggested that the majority of mutations occurring at the level of nucleotide and amino acid substitutions will have no noticeable phenotypic effect.[25] If heredity is viewed as a set of interacting and mutually responsive processes rather than as a fixed blueprint, it is easier to understand how changes in one part of the system may be buffered or accommodated elsewhere.[26] Ho's reformulation of heredity and understanding of cellular organization is explicitly informed by her rejection of the domination paradigm. In her words, "The 'control' of development is, therefore, not localized in the genes, but is diffused and distributed over the whole system."[27]

2. The Processes and Forces Deemed to Be Responsible for Evolutionary Change

A. Natural Selection

An interpretation of evolutionary advance based on dominance requires a compatible mechanism for its realization. Natural selection, with its Malthusian "struggle for existence," its competitive battles for ecological and reproductive dominance, fits the bill ideally. Darwinism in general and natural selection in particular have attracted criticism for as long as they have been in existence. This criticism has intensified over the past decade with the publication of several book-length critiques and essay collections highlighting the mistakes and omissions in the old evolutionary paradigm and heralding a range of new alternatives.[28] The question is not whether natural selection occurs; as Lewontin has pointed out, selection is inevitable among any natural entities that show variation, reproduction, and heritability.[29] The question is whether natural selection can really account for the origin of evolutionary novelties or is just a useful model for the maintenance and spread of evolutionary novelties once they have arisen.

To many critics of neo-Darwinism, our current evolutionary paradigm has chosen to place its major emphasis on a theory of mechanism that relegates the crucial problems surrounding the origin of diversity to the domain of the insoluble.[30] Briefly, mutations occur at random (with respect to desired effect), and every now and again one appears that, by sheer chance, aids in the construction of a phenotype that fares better than its competitors in the survival and reproduction stakes.[31] Through a series of such largely unpredictable occurrences, new morphological features are somehow built up. In the words of Eden, this is akin to

> the probability of typing at random a meaningful library of one thousand volumes using the following procedure: Begin with a meaningful phrase, retype it with a few mistakes, make it longer by adding letters; then examine the result to see if the new phrase is meaningful. Repeat this process until the library is complete.[32]

As an interesting corollary to this unsatisfactory situation and a good example of the pervasive nature of our need to control the unknown, Thompson, in his introduction to the Everyman edition of *On the Origin of Species*, had the following to say concerning the appeal of Darwin's proposition:

> because of the extreme simplicity of the Darwinian explanation, the reader may be completely ignorant of biological processes yet he feels he really understands and in a sense dominates the machinery by which the marvellous living forms have been produced.[33]

An evolutionary theory deriving from an ideology which eschewed control over nature would be free to investigate the dialectical relationship that exists between organisms and their environments.[34] Moreover, questions surrounding the origin of evolutionary novelties would be released from their magical black box and become the focus of evolutionary research. For example, if the nature of mutations and the forces generating them are not regarded as side issues in the study of adaptation, this opens up an important area of research into genome-environment interactions. We need to know more about the genomic and systemic effects of prolonged inbreeding and substantial climatic change that characterize many speciation models. We also need to know much more about how genetic differentiation is translated into changes of form. Some interesting proposals regarding the latter issue have been put forward by Newman, who holds that form changes are based on intrinsic physical properties of living tissues. They can therefore occur prior to genomic changes in evolutionary time and become stabilized or canalized by the appropriate mutations. The results of his experimental program in developmental biology support this model.[35] More studies of this nature would mean that investigations of the *Origin of Species* could indeed be directed to this end instead of being forays into the fine-tuning of populations to their local environments.

B. Competition

The fossil record is not simply a chronicle of the independent originations and extinctions of taxa through time. Rather, it is characterized by successions of entire floras and faunas that appear and disappear in a relatively coordinated fashion.[36] The traditional explanation for these patterns of biotic replacement involves competition on a grand scale and in two different guises. First, competition is held by many to be an important cause of extinction.[37] Gould and Calloway and Benton have queried the evidence for this.[38] The second major role envisaged for competition is to curb diversification: while one group of organisms is dominant, others are prevented from entering new "niches" in the economy of nature.[39] Eventually, either through competition or through major environmental shifts, the incumbent group becomes dislodged from its exclusive position and another group is then released to radiate into the vacated space.

Like the "explanation" discussed above in 2A this one sidesteps the most crucial aspects of evolutionary diversification.[40] How do taxa perceive "empty ecological space," and why do they radiate into it? Why do some groups radiate more extensively than others? In short, what is the real motor behind evolutionary change?

An evolutionary theory in which competition was a minor component in the totality of forces impinging on a biota would seek more direct explanations for these phenomena. Its research program would involve exploring correlations between speciation and extinction rates across taxa, as well as potential associations between these events and earth history. Indeed, macroevolutionary theorists have begun to make inroads into this previously neglected area of research, and patterns are beginning to emerge that were invisible at the level of isolated taxa or populations. This will undoubtedly be one of the growth points of evolutionary theory in the future.

3. The Interpretations of Particular Patterns in Extinct and Extant Nature

A. The Nature of Species

A model of evolutionary change based on the gradual accumulation of minute changes to the phenotype does not fit easily with observations that variation in nature tends to be clumped and discontinuous, i.e., ordered within and among species. In fact, to Darwin's contemporary critics the existence of species provided the main evidence that natural selection was *not* the primary motor behind evolutionary change.[41]

Darwin's response to this apparent conundrum was to deny the existence of species altogether. Species, in his view, were simply strongly marked varieties, having no more consistent distinguishing characteristic than that they had been so deemed in "the opinion of naturalists having sound judgement and wide experience."[42] This line is not a popular one among present-day evolutionists working with extant organisms, although it is often invoked in paleontological studies when lineages are described as diverging gradually and seamlessly over time.[43]

The approach taken by the chief architects of the neo-Darwinian synthesis was entirely different: they interpreted species as advantageous states so that natural selection could indeed be responsible for their evolution.[44] In the biological species concept (BSC), species are comprised of particularly harmonious or superior gene complexes and have evolved devices (called isolating mechanisms) to protect them against harmful gene flow from other gene pools. The earmark of a species is reproductive isolation, and the barriers to hybridization which make that isolation possible are the fundamental organizing principles in nature.

Critics of the BSC have demonstrated through the use of population genetic models that despite any advantages they may confer, barriers to interbreeding cannot arise by means of natural selection.[45] Defenders have argued that the nature of the force responsible for the evolution of isolating mechanisms is irrelevant; what counts is their presence.[46] But what this rebuttal fails to recognize is that if natural selection is *not* the force behind the origin of the essential criteria for the delimitation of species, then its role as the major force in the origin of diversity must be questioned. Neo-Darwinism therefore carries at its heart a fundamental contradiction that apparently has not disturbed many of its adherents: either natural selection is the major force behind diversity and species do not exist or species do exist and natural selection is not the major force behind the evolution of diversity.

The approaches taken by Darwin and the neo-Darwinians to unraveling the nature of species can both be understood as attempts to impose order on the world from a position outside or above it, in accordance with one's preconceptions as to how the world ought to be ordered. The neo-Darwinian idea of species as harmoniously coadapted gene pools with an integrity to be protected against pollutants from other gene pools has a long history. "Hybrids," "mongrels," and "bastard races" have been seen as a threat to the natural order for as long as organisms have been classified.[47] And this, no matter what the professed (and even heartfelt) ideology of its proponents, is the premise at base of the BSC.

An alternative approach, which was formulated to provide a workable population genetic basis for the origin and maintenance of species, was proposed by Paterson and his co-workers.[48] A species, in this view, can best be defined as "that most inclusive population of individual biparental organisms which share a common fertilization system".[49] Species have coadapted systems of mate attraction, recognition (chemical or behavioral), and fertilization which, because of their bipartite nature (having male and female components), will be subject to strong pressures of stabilizing selection. Any aberrant signal or response on behalf of either mating partner may terminate the fertilization process and hence be selected against. Even though Paterson's recognition concept (RC) still sees natural selection as providing the major force for change, the concept allows organisms—specifically sexual organisms—to define their own systems of order in nature.

The RC was devised within an essentially neo-Darwinian context. Paterson viewed the fertilization system as being closely fitted to the environment within which speciation occurred. The primary motor for change was held to be environmental perturbation. Should the environment alter so radically that the fertilization and recognition systems of a population of organisms could no longer function adequately, there would be strong directional selection to evolve new systems. However, Paterson never viewed selection as the only source of such change. A drastic shift in environmental conditions would fragment populations and reduce population sizes, according greater significance to the role of genetic drift or the random fixation of genes. Pleiotropic (indirect) effects arising from genetic changes at loci concerned not with mate recognition and fertilization but simply with survival in the new habitat could influence these systems. Finally, and most importantly, altered environmental conditions could influence ontogenetic (developmental) pathways, setting in motion the mutually responsive dialectical relationship for the coevolution of organism and environment envisaged by Lewontin.[50] Thus Paterson's species concept can be seen as a bridge between the two paradigms in evolutionary theory. While it invokes the negative pressures of stabilizing selection predicted by neo-Darwinism for the maintenance of species, it allows for processes other than selection to be the major causes behind the origin of species. And indeed this may ultimately be the solution to the neo-Darwinian "species problem."

B. *Evolutionary Progress*

As I mentioned earlier, the concept of "evolutionary advance" or "progress" is a highly controversial one, and the official neo-Darwinian view is far from clear. Darwin denied repeatedly that his theory was progressive; natural selection is a model to explain adaptation to a specific set of environmental conditions, not incremental improvement on some grand, cosmic scale. However, Løvtrup has pointed out that progress is a logical consequence of a theory in which the only forms that survive are those that outcompete "inferior" taxa.[51] In his view, the issue is one of organizational levels: intraspecific competition will lead to perfection of local adaptations, as Darwin predicted, but interspecific competition will lead to a succession of dominant taxa, each superior to the last.

Huxley, who was of a similar opinion, dedicated a chapter of his book to the definition of "evolutionary progress" and "dominant types."[52] Stebbins, who was responsible for bringing botany into the neo-Darwinian synthesis, entitled one of his books *The Basis of Progressive Evolution*.[53] Gould has argued that the idea of progress "continues to obsess evolutionists, despite the fact that there is little evidence for it."[54]

An evolutionary theory in which competition was de-emphasized, particularly at the macroevolutionary level, would contain no such vector for progress. Such instances of increase in structural, behavioral or ecological complexity as were observed, would require explanations based on the structures, behaviors, or ecological relationships themselves, rather than on some hypothetical force operating outside of them.[55] Once again, investigation would be brought down from the level of pseudo-explanation and speculation on how the universe ought to be ordered to the level of actual forces impinging on real organisms and the mechanics of structural change.

Adherents of the structuralist school of biology have made inroads in this field.[56] For example, Saunders and Ho have devised a "Principle of Minimum Increase in Complexity" which makes predictions regarding the most likely structural changes that any given form will undergo. These predictions provide null hypotheses against which phylogenies can be tested, as well as indicators of the potential mechanisms involved.[57]

The ideology of "being in control of nature" has given us a partial explanation of natural history. It has provided us with an evolutionary paradigm that accounts well for the maintenance of useful structures in organisms. For example, cave fish go blind and lose their pigmentation when no longer subjected to pressures that promote the upkeep of such physiological features as eyes and pigment cells. What this paradigm has sidestepped consistently, however, is a theory to account for the origin of evolutionary novelty.

An alternative ideology, that of "being part of nature," has opened up vast new areas of research that promise to revolutionize our understanding of natural organization. However, old habits die hard, and the resistance of many evolutionists even to the contemplation of some of the alternatives offered has been remarkable. A recent collection of essays on evolution, entitled *Evolutionary Biology at the Crossroads* and compiled to answer questions raised against the neo-Darwinian synthesis, made no mention of any of these alternatives;[58] and one may search the pages of the major evolutionary journals in vain for signs of their contribution to the field.

But the rumblings against neo-Darwinism continue. One of the most vociferous critiques has come from systematics, where proponents of "transformed" or "pattern cladistics" have become so disenchanted with the uninformative nature of neo-Darwinian "explanations" for evolutionary change that they have advocated a complete break between interpretations of pattern (systematics) and process (evolutionary theory).[59] However, one of the major themes of the new ideology is the dialectical and dynamic nature of these two approaches to studying natural organization, and I do not believe that they can be either meaningfully or usefully separated.

Finally, the new ideology grants us the opportunity to reinvent our origin myths. For too long scientists and laypeople alike have contributed to a vision of the violent birth of humanity. Inter- and intraspecific competition have been portrayed repeatedly as the forces that drove us out of the forests and onto the savannahs, where we had only our wits to defend us against the vicious predators that lurked there. It is a theme much beloved both of wildlife television shows and of undergraduate biology courses that the "survival of the fittest" is the natural mechanism whereby a population grows stronger and better adapted to its environment. Our "fitness" has enabled us to endure and prosper and to become a "dominant fauna" in our own right. Our position of evolutionary and ecological dominance has been taken as a license for us to exploit, plunder, and destroy as we please. A reckless adherence to the old ideology has brought our world to the brink of destruction. It's time to change.

Notes

I am grateful to Meg Conkey for placing at my disposal her information regarding the more than sixty reviews of Donna Haraway's book. I thank Chris Green, Stuart Newman, Dick Rayner, Kaye Reed, Stanley Salthe, Peter Saunders, and the editors for their many helpful criticisms of earlier versions of the manuscript. I conducted this research as a Research Fellow of the South African Foundation for Research Development, Pretoria, and I thank Robin Crewe and Shirley Hanrahan for making the facilities of the Zoology Department at the University of the Witwatersrand available to me.

1. Stanley N. Salthe, Commentary in *Evolutionary Biology at the Crossroads*, ed. Max K. Hecht (Flushing, N.Y.: Queens College Press, 1989), p. 175.

2. Donna Haraway, *Primate Visions: Gender, Race, and Nature in the World of Modern Science* (New York: Routledge, 1989).

3. Susan Cachel, "Partisan Primatology," *American Journal of Primatology*, 22 (1990), pp. 139–142.

4. Matt Cartmill, Review of *Primate Visions*, *International Journal of Primatology*, 12 (1991), pp. 67–75.

5. Misia Landau, Review of *Primate Visions*, *Journal of Human Evolution*, 20 (1991), pp. 433–437.

6. Robin Dunbar, "The Apes as We Want to See Them," *New York Times Book Review*, January 10, 1990, p. 30.

7. See William Leiss, *The Domination of Nature* (New York: George Braziller, 1972); Susan Griffin, *Women and Nature: The Roaring Inside Her* (New York: Harper & Row, 1978); Carolyn Merchant, *The Death of Nature: Women, Ecology and the Scientific Revolution* (New York: Harper & Row, 1980); Evelyn Fox Keller, *Reflections on Gender and Science* (New Haven, Conn.: Yale University Press, 1985); Hilary Rose, "Nothing Less than Half the Labs," in *Science and Beyond*, ed. Steven Rose and Lisa Appignanesi (Oxford: Blackwell, 1986), pp. 179–196; and Ruth Bleier, *Feminist Approaches to Science* (New York: Pergamon, 1988).

8. Salthe, Commentary. See also Stanley N. Salthe, *Development and Evolution: Complexity and Change in Biology* (Cambridge, Mass: MIT Press, 1993), and Peter T. Saunders, "Evolution Theory as a Creation Myth," *World Futures* 38 (1993), pp. 89–96.

9. See Thomas S. Kuhn, *The Structure of Scientific Revolutions*, 2nd ed. (Chicago: Chicago University Press, 1970).

10. Julian Huxley, *Evolution: The Modern Synthesis* (London: George Allen & Unwin, 1942), p. 562.

11. I am not attempting to deny that some kind of individuation and independence of entities within nature is necessary in order for change to be possible. I am simply pointing out that estimates of evolutionary advance are value laden and that the values derive essentially from Victorian and post-Victorian views as to what constitutes manhood.

12. Donald Johanson and James Shreeve, *Lucy's Child: The Discovery of a Human Ancestor* (New York: Avon, 1989), pp. 289–290.

13. Susan Griffin, *Pornography and Silence: Culture's Revenge against Nature* (London: Women's Press, 1981).

14. Loren Eiseley, *Darwin's Century* (London: Victor Gollancz, 1959), p. 345.

15. Huxley, *Evolution*, p. 575.

16. Mae-Wan Ho, "On Not Holding Nature Still: Evolution by Process, Not by Consequence," in *Evolutionary Processes and Metaphors*, ed. Mae-Wan Ho and Sidney W. Fox (Chichester: Wiley, 1988), pp. 117–118.

17. See Richard C. Lewontin, "Gene, Organism and Environment," in *Evolution from Molecules to Men*, ed. D. S. Bendall (Cambridge: Cambridge University Press, 1983), pp. 273–285, and "The Organism as the Subject and Object of Evolution," *Scientia*, 118 (1983), pp. 65–82.

18. Stanley Salthe has pointed out that I have used several niche definitions interchangeably here. I agree, but since all hold the niche to be prior to the organism, this does not cloud the argument.

19. Dynamic environment models, like Van Valen's Red Queen hypothesis (see Leigh N. Van Valen, "A New Evolutionary Law," *Evolutionary Theory*, 1 [1973], pp. 1–30), and selection models, like frequency dependent selection, are attempts to combat this

static view from within the neo-Darwinian paradigm. Their continued emphasis of the organism/environment dichotomy, however, has prevented them from introducing any radically new ways of thinking into the theory structure.

20. See, for example, G. Ledyard Stebbins, *Processes of Organic Evolution* (Englewood Cliffs, N.J.: Prentice-Hall, 1966).

21. Mae-Wan Ho, "Heredity as Process: Toward a Radical Reformulation of Heredity," *Rivista di Biologia*, 79 (1986), pp. 407–447.

22. Gerry Webster, "Structuralism and Darwinism: Concepts for the Study of Form," in *Dynamic Structures in Biology*, ed. Brian Goodwin, Atuhiro Sibatani, and Gerry Webster (Edinburgh: Edinburgh University Press, 1989), pp. 1–15. See also Salthe, *Development and Evolution*, figure 6.1.

23. Ho, "Heredity as Process" and "On Not Holding Nature Still."

24. See Gabriel A. Dover, "Molecular Drive in Multigene Families: How Biological Novelties Arise, Spread and Are Assimilated," *Trends in Genetics*, 2 (1986), pp. 159–164.

25. Motoo Kimura, *The Neutral Theory of Molecular Evolution* (Cambridge: Cambridge University Press, 1983).

26. Many neo-Darwinists have accounted for the observations of the neutralists by assuming that DNA comes in two kinds: gene-coding DNA, which is subject to selection, and "junk" DNA, where most of the mutations accumulate. This assumption has set up much acrimonious debate within the neo-Darwinist school as to whether or not selection would allow for the continued existence of a substantial component of useless DNA in the genome (e.g., see commentaries by Walter J. Bock and Lee Ehrman in *Evolutionary Biology at the Crossroads*, ed. Max K. Hecht [Flushing, N.Y.: Queens College Press, 1989]). A systems approach to genome function renders this argument superfluous. I thank Peter Saunders for this insight.

27. Ho, "Heredity as Process," p. 426.

28. See Mae-Wan Ho and Peter T. Saunders, *Beyond Neo-Darwinism: An Introduction to the New Evolutionary Paradigm* (London: Academic Press, 1984); Jeffrey W. Pollard, *Evolutionary Theory: Paths into the Future* (London: Wiley, 1984); David J. Depew and Bruce H. Weber, *Evolution at the Crossroads: The New Biology and the New Philosophy of Science* (Cambridge, Mass: MIT Press, 1985); Niles Eldredge, *Unfinished Synthesis: Biological Hierarchies and Modern Evolutionary Thought* (Oxford: Oxford University Press, 1985); Søren Løvtrup, *Darwinism: The Refutation of a Myth* (London: Croom Helm, 1987); Daniel R. Brooks and Edward O. Wiley, *Evolution as Entropy: Toward a Unified Theory of Biology*, 2nd ed. (Chicago: University of Chicago Press, 1988); and Salthe, *Development and Evolution*.

29. Richard C. Lewontin, "The Units of Selection," *Annual Review of Ecology and Systematics*, 1 (1970), pp. 1–18.

30. See Mae-Wan Ho and Peter T. Saunders, "Adaptation and Natural Selection: Mechanism and Teleology," in *Towards a Liberatory Biology*, ed. Steven Rose (London: Allison & Busby, 1982), pp. 85–102.

31. Peter Saunders has pointed out that "while neo-Darwinists claim that they assume that variations (not just mutations) are random in the sense of not being connected with need, they actually assume they are 'really' random. They have to, because otherwise the nature of the variations would have an effect on evolution and natural selection would no longer be the sole creative force" (personal communication).

32. Murray Eden, "Inadequacies of Neo-Darwinian Evolution as a Scientific Theory," in *Mathematical Challenges to the Neo-Darwinian Interpretation of Evolution*, ed. P. S. Moorhead and M. M. Kaplan (New York: Alan R. Liss, 1967), pp. 109–111.

33. W. R. Thompson, Introduction to *On the Origin of Species* by Charles Darwin (London: Everyman, 1956).

34. See Richard Levins and Richard Lewontin, *The Dialectical Biologist* (Cambridge, Mass: Harvard University Press, 1985).

35. See Stuart A. Newman, "Generic Physical Mechanisms of Morphogenesis and Pattern Formation as Determinants in the Evolution of Multicellular Organization," *Journal of Biosciences (Bangalore)*, 17 (1992), pp. 193–215.

36. See Anthony Hallam, "What Can the Fossil Record Tell Us about Macroevolution?" in *Evolutionary Biology at the Crossroads*, ed. Max K. Hecht (Flushing, N.Y.: Queens College Press, 1989), pp. 59–73.

37. See Karl J. Niklas, Bruce H. Tiffney, and Andrew H. Knoll, "Patterns in Vascular Land Plant Diversification," *Nature*, 303 (1983), pp. 614–16; Løvtrup, *Darwinism*; Geerat J. Vermeij, *Evolution and Escalation: An Ecological History of Life* (Princeton, N.J.: Princeton University Press, 1987); and David M. Raup, *Extinction: Bad Genes or Bad Luck?* (New York: Norton, 1991).

38. See Stephen J. Gould and C. Bradford Calloway, "Clams and Brachiopods—Ships That Pass in the Night," *Paleobiology*, 6 (1980), pp. 383–396, and Michael J. Benton, "Progress and Competition in Macroevolution," *Biological Reviews*, 62 (1987), pp. 305–338.

39. See Brooks and Wiley, *Evolution as Entropy*.

40. See Judith C. Masters and Richard J. Rayner, "Competition and Macroevolution: The Ghost of Competition Yet to Come?" *Biological Journal of the Linnean Society*, 49 (1993), pp. 87–98.

41. See Georges J. Romanes, "Physiological Selection: An Additional Suggestion on the Origin of Species," *Zoological Journal of the Linnean Society*, 19 (1886), pp. 337–411, and John T. Gulick, "Divergent Evolution through Cumulative Segregation," *Zoological Journal of the Linnean Society*, 20 (1887), pp. 189–274.

42. Charles Darwin, *On the Origin of Species*, 6th ed. (London: Wiley, 1872), p. 56.

43. See Ernst Mayr, *Animal Species and Evolution* (Cambridge, Mass: Harvard University Press, 1963), and Bock, Commentary in *Evolutionary Biology at the Crossroads*.

44. See Theodosius Dobzhansky, *The Biology of Ultimate Concern* (New York: New American Library, 1967), and Ernst Mayr, *Populations, Species and Evolution* (Cambridge, Mass: Harvard University Press, 1970).

45. See Hugh E. H. Paterson, "More Evidence against Speciation by Reinforcement," *South African Journal of Science*, 74 (1978), pp. 369–71; David M. Lambert, Marc R. Centner, and Hugh E. H. Paterson, "Simulation of the Conditions Necessary for the Evolution of Species by Reinforcement," *South African Journal of Science*, 80 (1984), pp, 308–311; Hamish G. Spencer, Brian H. McArdle, and David M. Lambert, "A Theoretical Investigation of Speciation by Reinforcement," *American Naturalist*, 128 (1986), pp. 241–62; and Hamish G. Spencer, David M. Lambert, and Brian H. McArdle, "Reinforcement, Species, and Speciation: A Reply to Butlin," *American Naturalist*, 130 (1987), pp. 958–962.

46. See Jerry A. Coyne, H. Allen Orr, and Douglas J. Futuyma, "Do We Need a New Species Concept?" *Systematic Zoology*, 37 (1988), pp. 190–200; C. Ray Chandler and Mark H. Gromko, "On the Relationship Between Species Concepts and Speciation Processes," *Systematic Zoology*, 38 (1989), pp. 116–25; and Jerry A. Coyne, "Recognizing Species," *Nature*, 364 (1993), p. 298.

47. See Hugh E. H. Paterson, "Darwin and the origin of species," *South African Journal of Science*, 78 (1982), pp. 272–275, and Judith Masters, David Lambert, and Hugh Paterson, "Scientific Prejudice, Reproductive Isolation and Apartheid," *Perspectives in Biology and Medicine*, 28 (1984), pp. 107–116. I should point out here that while I believe that our current concept of reproductive isolation has direct links to Victorian xenophobia, this does not mean that I see no danger in transferring genes across species lines, as is done regularly today in biotechnology laboratories (see Stuart Newman's contribution to this volume).

48. Hugh E. H. Paterson, "The Recognition Concept of species," in *Species and Speciation*, ed. Elisabeth S. Vrba (Pretoria: Transvaal Museum, 1985), pp. 21–29, and *Evolution and the Recognition Concept of Species* (Baltimore: Johns Hopkins University Press, 1993).

49. Paterson, "The Recognition Concept of species," p. 25.

50. Lewontin, "Gene, Organism and Environment," and "The Organism as Subject and Object."

51. Løvtrup, *Darwinism*.

52. Huxley, *Evolution*. As noted, Huxley was strongly criticized by his peers for his views on evolutionary progress even though his views embodied the ideals of the society within which Darwinism was developed. Stanley Salthe has suggested that "one might suppose it was the blatant exposure of these views like this, revealing them in a mythological framework, that bugged neo-Darwinian scholars intent upon 'scientific objectivity'" (personal communication). I agree.

53. G. Ledyard Stebbins, *The Basis of Progressive Evolution* (Chapel Hill: University of North Carolina Press, 1969).

54. Stephen J. Gould, "The Paradox of the First Tier: An Agenda for Paleontology," *Paleobiology*, 11 (1985), pp. 2–12.

55. See Salthe, *Development and Evolution*, for thermodynamic and information theoretic illustrations of this approach.

56. See Ho and Saunders, *Beyond Neo-Darwinism*; Mae-Wan Ho and Sidney W. Fox, *Evolutionary Processes and Metaphors* (Chichester: Wiley, 1988); and Brian Goodwin, Atuhiro Sibatani, and Gerry Webster, *Dynamic Structures in Biology* (Edinburgh: Edinburgh University Press, 1989).

57. Peter T. Saunders and Mae-Wan Ho, "The complexity of organisms," in *Evolutionary Theory: Paths into the Future*, ed. Jeffrey W. Pollard (London: John Wiley and Sons, 1984), pp. 121–39.

58. Max K. Hecht, *Evolutionary Biology at the Crossroads: A Symposium at Queens College* (Flushing, N.Y.: Queens College Press, 1989).

59. See, for example, Colin Patterson, "The Impact of Evolutionary Theories on Systematics," in *Prospects in Systematics*, ed. David L. Hawksworth (Oxford: Clarendon Press, 1988), pp. 59–91.

PART IV

Border Crossings:
Human/Animal, Live/Inanimate

THOUGH THE THREE chapters in this final part are very different in concept and content, they all are concerned with the landmarks that cultures and scientists within them use to erect and police boundaries—between human beings and other organisms, between different species, and between organisms and non-living matter. They relate to the basic program around which this book is organized in that all three explore rules used to establish or question an organism's or a species' identity and thereby its right to exist.

Stuart A. Newman reflects on metaphors and explanations that have dominated Western thinking about the separation and merging of species and of individuals within them. Tracing the theological and philosophical antecedents of scientific thinking, he highlights two divergent strands. One is epitomized by the Darwinian and present-day assumption that the differences we observe among organisms are consequences of accidents ("mutations"), unguided by the intrinsic properties of the material substances of which different organisms are composed. The other assumes that various intrinsic properties of living matter ("flesh" is the word Newman uses) in fact constrain structural possibilities and thus guide differences among bodily forms and functions that can arise and become established over time.

The former, currently dominant, view lends itself to the triumphalism of limitless "genetic engineering," constrained only by limitations imposed by human ingenuity, need, and a desire for knowledge or profits. The latter, which has its own long-standing tradition (of which Newman is an exponent), implies that there are dangers in believing that biological matter can be mixed and matched at will. Such dangers arise not only from the practical excesses to which such exploitation readily leads but also from neglect or misunderstanding of the potential for misguided, indeed dangerous, transformations.

Newman thus advocates ecological caution, based on the need to recognize and understand the character of fleshly limitations and constraints that should not be transgressed. This approach offers an interesting counterpoint to Vandana Shiva's eloquent plea for ecological justice, voiced in part I, and to her political rebellion against the profit-driven Western project of limitless subjugation and domination of "nature."

Arnold Arluke and Boria Sax describe a different form of subjugation and domination as they explore the ways in which Nazi ideologues of the 1930s and '40s drew their boundaries between "wild" and "tamed" nature. Humans and animals could fall into either category, depending on their position in the Nazi scale of being. Thus the ideal German—the Nordic "blond beast"—ranked high, along with wild animals and beasts of prey. By contrast, Jews, stereotyped as overly domesticated and sophisticated, were classed as inferior, tamed "counterfeits," along with pigs, cattle, and dogs.

Contradictions abound in the way boundaries and hierarchies were constructed. They perhaps climax in the fact that Hitler and his closest associates gladly murdered millions of human beings while condemning the hunting of hares and deer and adhering to a strictly vegetarian diet.

This chapter raises more troubling questions than any other. To come face to face with Nazi theorizing about humans and other animals is like looking into a distorting mirror that reflects gibbering caricatures of the ways we ourselves construct boundaries.

Quite different rules and contradictions are explored by Emily Martin, who looks at current attempts by anthropologists and sociologists of science to explore human/other boundaries, where the "others" can be nonhuman animals or, indeed, inanimate devices, such as machines or computers of different levels of sophistication.

She shows that whereas the human and nonhuman actors are sometimes described as seemingly equal partners in the boundary crossings, the terms of the transactions are set in such a way that the nonhuman "agents"—be they living or inanimate—get incorporated into the world view of their human "partners" as though by their own choice.

This apparent equalization of activities and intentions obscures the hierarchies of wealth and power between people of different nationalities, classes, genders, and races. In the end, because the opportunities to set the terms of such boundary crossings exist only for the investigator-entrepreneurs who oversee them, the seemingly critical descriptions of how science operates in fact reinforce traditional, politically constructed differences. In Martin's account, boundary crossings can never be unproblematic because they can be initiated only by people with the power to decide where and how the boundaries are erected, breached, or leveled.

All three of these very different contributions are about power—the power to construct and enforce boundaries, to have them accepted as "natural," and to give them practical and political meaning.

11 | Carnal Boundaries
The Commingling of Flesh in Theory and Practice
Stuart A. Newman

H UMAN FLESH IS on the cultural menu. Cannibalism as the last resort of ordinary people under duress has received sympathetic treatment in recent films that recount events that took place two decades[1] and a century and a half[2] ago. Cannibalism as the characterological organizing principle of a fictional mad genius was the hook of the Academy Award–winning best picture of 1991.[3] Persistent rumors about the ceremonial consumption of the livers and hearts of "class enemies," initially by zealots and then by ordinary villagers, in the Guangxi Autonomous Region of China during the Cultural Revolution of the late 1960s were kept under wraps for more than twenty years. The dissident journalist Liu Binyan, questioned about these stories in 1984, said that he had avoided writing about them "because the subject was so nasty." But the American public is now presumably ready: two nonfiction books on the Guangxi cannibalism incidents by the recently expatriated novelist Zheng Yi, although not yet available in English, have been widely discussed in the U.S. press by, among others, Liu Binyan himself.[4] And during the same period the emergence and trial of an individual in Milwaukee who had been eating or freezing portions of his murder victims' bodies played to a fascinated public and defined new outer limits of social pathology.[5]

How can we account for the recent interest in violation of one of the most fundamental of social taboos?[6] I suggest that cannibalism is a potent symbol of the erasure of traditionally conceived boundaries between different kinds of flesh. Such crossing of biological borders—previously the stuff of art and mythology—is, in the late twentieth century, increasingly a technological reality. This is evident in the recent profusion of research, medical, and commercial ventures involving transplantation of human embryonic tissues, human gene modification, and the production of "transgenic" animals.[7] The clash of this "carnal pragmatism" with traditional ideas of the mainstream culture inevitably fascinates and unsettles, providing a social context for entertainment technologies such as "morphing" (the computer-simulated transformation of one body into another seen in Arnold Schwarzenegger movies and Michael Jackson videos), the obsession in certain sectors of society with body building and cosmetic surgery, as well as the morbid interest in cannibalism.

The commingling of flesh is a central principle of animal existence, and its representation in the minds of humans in the form of metaphor and myth is a fountainhead of drives and taboos. Indeed, various senses of "commingling"— the mixing of the substance of one kind of individual with another, as in procreation, the transformation of the substance of one individual into that of another, as in pregnancy, and the incorporation of the substance of an individual into that of another, as in consumption, have been conflated in the conceptual frameworks of different cultures.[8] Functionally, animal procreation is only possible through the merging of cells produced by what are arguably the most biologically *distinct* members (i.e., the two sexes) of the most biologically *uniform* of populations (i.e., a species). Human sexual relations do not necessarily correspond to biologically defined roles, but all societies have sexual prohibitions (e.g., incest taboos) formulated in relation to the "facts of life" as they are understood. And the fact that women's bodies can grow babies makes the maintenance of carnal boundaries a qualitatively different issue for women than it is for men.

In Freud's view, the organization of the human psyche itself is initiated by the infant's recognition that its body and that of the mother are *not* confluent.[9] Freud's notion, as he acknowledged, is only a modern form of Plato's famous speculation in the *Symposium* that human love is based on a longing for a primeval state in which the lovers' bodies were a single entity. The special terror of cancer among all the ailments to which the human body is subject is related to its inversion of the ecstatic mixing of flesh. In this disease the body's own flesh produces an alien tissue that malignantly invades and may eventually suffocate its host.

Consumption of meat is the most common mode by which one individual's flesh is commingled with another's, and few if any cultures are without some restrictions in this area. Human flesh is, of course, subject to the most severe taboos, and when it has been eaten, the purpose almost invariably has been ceremonial rather than nutritive. Moreover, most if not all human groups have religious or aesthetic prohibitions against eating the flesh of certain kinds of animals, or concerning when or in what form animal flesh may be consumed. Maintenance of carnal boundaries thus appears to be a constant of human social organization and mental life, a point made repeatedly, on the basis of wide-ranging ethnographic evidence, by the anthropologist Claude Lévi-Strauss.[10]

Apart from procreation and diet, the mixing of the flesh of different individuals or species has traditionally been a strictly theoretical, rather than practical, possibility. And on a speculative plane the connotations of chimeras, organisms made up of mixtures of naturally occurring types, have not been always negative. The gods of ancient Egypt and India and the ancestral figures of Native American and Aboriginal Australian legends, for example, frequently partake of a combination of human and animal characteristics. Representations of human-animal hybrids in Mediterranean and Northern European pagan cultures could similarly convey positive values. However, with the rising domination of Judeo-Christian ideas in Europe, the specter of human-animal, or even purely human, chimeras became increasingly disquieting, as can be seen in medieval monster allegories such as *Beowulf* (ca. 1000), seventeenth-century were-wolf hysteria,[11] and *Frankenstein* (1818).

The mixing of flesh and production of chimeras is no longer purely theoretical. Human organ transplantation began in the 1950s, first with kidneys, followed by livers (1963), hearts (1967), and lungs (1981).[12] By the late 1980s, transplantations of fetal tissue into adult brains and pancreases were being performed, with the goal of alleviating Parkinsonism and diabetes. More recently, attempts have even been made to cross species lines, with transplantations of baboon hearts and livers to human recipients.

In recent animal experiments, mixture of the flesh of different individuals and biological types has progressed even further, yielding chimeric "geeps" (animals formed by the jumbling together of cells from goat and sheep embryos)[13] and such curiosities as mice with four biological parents.[14] Most significant from the point of view of human biology is the new capacity to produce transgenic cells and animals, which originate in one individual or species but contain genes derived from another. If such alterations are made in "somatic" or body cells which are returned to the body of the cell donor, the individual thus reconstituted may thereby produce substances that he or she was previously incapable of making. This procedure has already been used by medical scientists as an experimental palliative for certain life-threatening human diseases. If the alterations are made in "germ line" or reproductive cells, then individual animals or humans will develop which are genetically chimeric in every cell of their bodies and can pass this condition to their progeny—and hence into their species' gene pool.

Beginning with organ transplantation forty years ago and continuing into the contemporary "brave new world" of gene manipulation, critics have raised ethical questions about the commercialization of body parts, the patenting of transplantable cells, and other means by which human tissues have been desacralized and introduced into the material culture. Simultaneously, the rise of militant ecological preservation and animal rights movements has challenged traditional attitudes concerning species integrity and value and the cultural implications of producing transgenic animals.

In what follows I will examine the historic bases of attitudes concerning the crossing of carnal boundaries in the Western societies in which the contemporary scientific culture originated and has achieved its most dominant form. I intend to demonstrate that the prevailing Judeo-Christian conceptual framework affected the development of scientific ideas concerning matter and flesh to such an extent that modern biology retains a profound but covert affinity with religious ideology. Furthermore, since the elements carried over from the earlier belief systems are generally unacknowledged by the scientific mainstream, so are those which have been discarded. I will attempt to show that powerful concepts that are useful in understanding the generation and maintenance of biological form and the nature of species differences have been suppressed as a quasi-mystical notion of genetic determinism (related, as I will show, to older religious ideas) became dominant.

Finally, the traditional religious and philosophical concepts of nature considered here carried specific ethical and moral implications which, to varying extents, continue to inform value systems of contemporary technological cultures. I will therefore explore how certain Western views of the relation between flesh and matter have promoted and other such views have provided a basis for resistance to increasing pressures to commodify tissues, trivialize species identity, and generally bring all flesh into the realm of commerce and manufacturing.

Kashruth and Eucharist

The ancient Hebrews, whose laws are one of the foundations of European-American moral codes, committed themselves to a set of precepts about the preparation and consumption of meat that embodied strict notions of boundaries between different forms of life. Indeed, Hebrew myth stated that the diet of humans before the Fall was vegetarian, as decreed by God in one of the earliest passages of the Bible (Gen. 1:29). In the interpretation of the anthropologist Jean Soler,[15] the fundamental difference between human and God is thus expressed by the difference in their foods. God's food is animal sacrifices, which serve as his "nourishment" according to the Bible, and that of humans is the edible plants. Only after the Flood, which humans brought upon themselves by their violence, was it permitted for them to eat meat: "Every moving thing that lives shall be food for you; as I gave you the green plants, I give you everything" (Gen. 9:3). Yet even then a distinction continued to be made between God's portion and that of humans, with the added injunction: "Only you shall not eat flesh with its life, that is, its blood" (Gen. 9:4). According to Soler,

> Blood becomes the signifier of the vital principle, so that it becomes possible to maintain the distance between man and God by expressing it in a different way with respect to food. Instead of the initial opposition between the eating of meat and the eating of plants, a distinction is henceforth made between flesh and blood. Once the blood (which is God's) is set apart, meat becomes desacralized—and permissible.[16]

This new dietary regime signifies a covenant between God and Noah's descendants, e.g., all human beings, but one that acquiesces to human corruption. The Lord says, after the Flood: "I will never again curse the ground because of man, for the imagination of man's heart is evil from his youth" (Gen. 8:21). Only when Moses appears is a third dietary regime instituted, and the purpose of this one is to make a distinction between the Hebrews and other peoples: "I am the Lord your God, who have separated you from the peoples. You shall therefore make a distinction between the clean beasts and the unclean; and between the unclean bird and the clean; you shall not make yourselves abominable by beast or by bird or by anything with which the ground teems, which I have set apart for you to hold unclean" (Lev. 20:20–25).

Close examination of the dietary regime prescribed by Leviticus and Deuteronomy shows it to embody a theory of boundaries between living things that goes beyond issues of mere food. These laws have been considered in detail by the anthropologist Mary Douglas in *Purity and Danger* (1966)[17] and by Jean Soler in his 1973 article "The Semiotics of Food in the Bible." The following discussion is based in large part on their analyses.

It is clear that the concept of holiness from which the dietary laws flow is tied as well to nondietary prescriptions. As Douglas notes, "Hybrids and other confusions are abominated."[18] Leviticus states that "you shall not lie with any beast and defile yourself with it, neither shall any woman give herself to a beast to lie with it: it is perversion" (Lev. 18:23) and that "you shall not let your cattle breed with a different kind; you shall not sow your field with two kinds of seed; nor shall there come upon you a garment of cloth made of two kinds of stuff" (Lev. 19:19). Deuteronomy 22:11 contains its own version of the last proscription: "You shall not wear a mingled stuff, wool and linen together."

Both Douglas and Soler note that the conceptual framework of the Hebrews is continually referred back to the conditions that prevailed at the creation. The dietary laws themselves contain as a necessary (but not sufficient) condition that no flesh of carnivorous animals be eaten. And indeed, such animals were not included in the plan of the creation (Gen. 1:29–30). This is why hoofed animals, which have no means of seizing a prey, and cud-chewing animals are the only class of mammals from which the "clean" varieties are selected and birds of prey, such as the eagle, are specifically listed as "unclean." In Soler's interpretation, "Carnivorous animals are unclean. If man were to eat them he would be doubly unclean."[19] Along with this back reference to the earlier vegetarian regime of Paradise, the postdiluvian blood taboo is carried over into the new regime: "You may slaughter and eat flesh within any of your towns . . . as of the gazelle and as of the hart. Only you shall not eat the blood; you shall pour it out upon the earth like water" (Deut. 12:15–16). This is the condition for expiation of the "blood guilt" (Lev. 17:4) that attaches to anyone who kills a living being, the major prohibition of the Bible.

A further connection of the dietary regime to the plan of Creation is the stricture that animals to be eaten "shall conform fully to their class."[20] This echoes the prohibition against sacrificing to the Lord any animal with a "blemish" or "defect" (Lev. 22:21, Deut. 17:1). As Soler notes, "A fundamental trait of the Hebrews' mental structures is uncovered here. There are societies in which impaired creatures are considered divine."[21]

The relationship of cleanliness to the boundaries established at the creation is enforced by imposing a set of "ecological" precepts. In Genesis a division is made into three elements: the waters, the earth, and the firmament. Living creatures are brought forth in a specific relation to each of these elements: "Let the waters bring forth swarms of living creatures, and let birds fly above the earth across the firmament of the heavens" (Gen. 1:20); "Let the earth bring forth living creatures according to their kinds, cattle and creeping things and beasts of the earth according to their kinds" (Gen. 1:24).

The classification into creatures of the earth, water, and air is reiterated in the chapters of Leviticus and Deuteronomy that establish the dietary laws. "Any class of creatures which is not equipped for the right kind of locomotion in its element is contrary to holiness," Douglas writes.[22] Thus edible creatures from the water must have fins; those that do not move about (mollusks) and those that have legs and can walk (arthropods) are unclean. The qualifying trait for a bird is that it "flies in the air" (Deut. 4:17). Thus the ostrich is specifically prohibited, as are birds which spend most of their time in the water, such as the swan, the pelican, the heron, and all stilted birds. Moreover, "Every swarming thing that swarms upon the earth is an abomination; it shall not be eaten" (Lev. 11:41). The Hebrew word *shérec*, translated as "swarming" or "teeming,"[23] denotes appropriate locomotion for creatures of the water but not for those of the earth. So most insects are prohibited, but not all. Leviticus (11:21) permits as food insects such as grasshoppers and locusts, which have "legs above their feet, with which to leap on the earth."

Douglas and Soler consider the dietary laws of the Hebrews and the conceptual framework from which they derive as one example of the kinds of social-mental structures by which, Douglas writes, cultures "create unity in experience."[24] Although the specific items designated as significant and their relationships will vary among different cultures, Soler adds, "man knows that the food he ingests in order to live will become assimilated into his being, will become himself. There must be therefore, a relationship between the idea he has formed of specific items of food and the image he has of himself and his place in the universe."[25]

What goes for food must certainly also go for other commingling of tissues where this becomes feasible. Most relevant with regard to the structure of the Jewish dietary laws and the world view from which they flow is their pervasive influence (mainly through the authority of the Hebrew scriptures, Christianity's Old Testament) on the intellectual and moral development of European-American societies. For not only is a theory of biological boundaries and ecological order built into this perspective; these concepts are also closely tied to attitudes toward corruption and killing and the guilt that attaches to them and its expiation. These remain embedded (though perhaps to a diminishing extent) in contemporary culture, irrespective of how any of its biological notions may have been challenged by scientific activities.

Of course the view of nature embodied in the Hebrew scriptures is not the only ancient influence on contemporary notions of carnal boundaries. Christianity, from its inception, broke with a number of tenets of the Hebraic world view in a very decisive fashion. Soler, from his analysis of the Jewish dietary laws and the underlying conceptual structures, concludes that it is understandable that the Hebrews did not accept the divine nature of Jesus: "A God-man, or a God become man, was bound to offend their logic more than anything else."[26] Christ is the ultimate hybrid.

Christianity, in its mission to convert the Gentiles, drew a new line of demarcation. The new covenant sought to place all peoples and God to one side and the rest of Creation to the other side while simultaneously erasing the structures that separated the Hebrews from the other peoples. The New Testament is explicit about this. Mark (7:15) quotes Jesus as saying that "nothing that goes into a man from outside can defile him; no, it is the things that come out of him that defile a man" and goes on to comment: "Thus he declared all foods clean" (7:20). In Acts, Peter has a vision of a great sheet being lowered to the ground containing "creatures of every kind, whatever walks or crawls or flies." A voice says to him: "Rise, Peter, kill and eat." When Peter refuses, saying, "No Lord, no: I have never eaten anything profane or unclean," the voice replies, "It is not for you to call profane what God counts clean" (Acts 10:11-16). Later Paul instructs the Corinthians: "You may eat anything sold in the meat market without raising questions of conscience; for the earth is the Lord's and everything in it" (1 Corin. 10:25-26). And while the Hebrew scriptures assign guilt to the shedding of nonhuman as well as human blood and contain several passages condemning cruelty to animals, the New Testament is completely lacking in any such injunctions.

The Greco-Roman world in which Christianity first established itself was one in which human-animal hybrids were a commonplace of legends and gods were routinely assumed to take human form. This culture, unlike that of the Hebrews, was receptive to the concept of species blends and chimeras, especially a God-man. The unification of communicants with God in the Eucharist, an act of ritualistic consumption of the blood and body of Christ, represents the most extreme break possible with a Jewish moral order based on radical separation of the human from the divine, ritual extraction of blood from meat, and sanctification of living boundaries. But the Christian notion that humans, of all

beings living on earth, have an immortal soul and are therefore the only creatures qualified for fusion with God initiated an alternative moral order which,
while opening the way for glorification of human action (particularly in the
service of religion), also fostered an alienation from, and reductive manipulation of, nature.

The hegemony of Christian thought during most of European history ensured the persistent influence of the Old Testament doctrine of genuine boundaries between life forms. However, the New Testament emphasis on the demarcation of humans from the rest of creation undercut the regulative role of this
doctrine as a moral foundation and guide to right behavior. When an attack on
the principle of species integrity was eventually mounted in the late nineteenth
century by Darwinism, it beset an ancient framework of attitudes about the living world which, under the Christian regime, had already become rickety, its
main pillar being a mystical notion of human uniqueness. And as this article of
faith increasingly came into question in the present century, those elements of
the traditional moral order grounded in respect for the inviolability of distinctions among living things began to dissolve.

Flesh and Matter in Pre-Enlightenment Europe

Partly because of its deliberate desanctification of biological boundaries,
early Christianity had a tumultuous encounter with the question of the relation
of living flesh to nonliving matter and in the process incorporated and rejected
various views that had been inherited from non-Christian sources. The influential pagan idea that the earth itself is a benign female organism continually fell
afoul of the patriarchal ideology of the Roman Catholic Church, which encouraged the persecution of animistic witches. It would nonetheless be invoked pe-
riodically by utopian and communitarian protestors against the alienation of
land and resources from serfs and peasants. Polemicists such as Agrippa (1486–
1535) and poets such as Spenser (1552–1599) and Milton (1608–1674) sought support in this idea for their resistance to the despoliation of nature caused by mining, which they likened to rape.[27]

In general, the Church fathers promulgated the Aristotelian notion that during the conception and development of all living beings the matter is provided
by the female but remains inert without the animating principle supplied by
the male. Manichaeanism, which flourished widely in Europe in the fourth century, took this notion further, asserting that spirit (identified with Good and the
male principle) and matter (identified with Evil and the female principle) were
substances that originally existed in radical opposition to one another. The mixture of the two substances in the contemporary world was held to be the basis
of profound corruption and degradation, and the Manicheans promoted an as-
cetic ideal that included abstinence from sex, procreation, meat eating, and cul-

tivation and harvesting. Although Manichaeanism was branded a heresy by the Church, its major precepts reemerged in the form of later Christian sects such as the Cathari of the twelfth and thirteenth centuries (also celibate and vegetarian), who went so far as to reject the doctrine of the Incarnation, i.e., that God could become flesh.

While treading the line between the organicist and dualist heresies to either side of it, the mainstream Church and its congregants struggled incessantly with the ambiguities of flesh and matter. In a 1989 essay,[28] Caroline Walker Bynum describes the vision of Christ received by Colette of Corbie, a fifteenth-century Franciscan reformer, as she prayed to the Virgin Mary: Christ appeared to Colette as a dish completely filled with "carved-up flesh like that of a child," while the voice of God warned her that it was human sin that minced his son into such tiny pieces. Bynum also writes about the cult of the eucharistic host of the late Middle Ages, involving miracles in which the bread of the eucharist turned into bloody flesh on the paten or in the recipient's mouth and related visions, such as that of the Viennese Beguine Agnes Blannbekin (died in 1315), who reported receiving Christ's foreskin in her mouth and finding it to taste as sweet as honey.[29] Around the same time theologians were debating whether, at the Second Coming, God would have to reassemble the same bits of matter that had before been animated by a particular soul.[30]

In these examples the transformation of matter into flesh and flesh into matter is always imbued with mystery and the supernatural. There is no implication, as there would be for later biological science, that the material commonality of all flesh represents the medium by which different kinds of beings could turn into one another in the normal course of things. Indeed, even the Devil could not change one kind of body into another. This was put quite explicitly in the influential late ninth-century Christian text known as the *Canon Episcopi*:

> Whoever therefore believes that anything can be made, or that any creature can be changed to better or worse or be transformed into another species or likeness, except by God himself who made everything and through whom all things were made, is beyond doubt an infidel.[31]

This doctrinal position, solidly grounded as it was in the Old Testament view that each living thing was created "according to its own kind," often ran up against the widespread medieval belief in werewolves, humans who involuntarily assumed the forms of animals, and in the presumed ability of witches to so transform themselves for specific evil purposes. Saint Augustine proffered a mystical explanation for these alleged phenomena, stating that such transformations affect neither the body nor the soul but a third part of the human called the *phantasticum*, a ghostlike double that is given a deceptive visible appearance by a demon. Saint Thomas Aquinas also invoked the Devil, who was said to be able to mold air around a body to make it appear transformed without being physically altered.[32]

Jean Bodin (1530–1596), a French lawyer and political philosopher, was an interesting transitional figure in these matters. Usually regarded as a proto-Enlightenment thinker for his advocacy of constitutional monarchy and religious toleration, he was also a misogynist who promoted the persecution of witches. His controversial views on werewolves anticipated, in certain ways, later "scientific" thinking on organismal boundaries. Bodin rejected the doctrine of the *Canon Episcopi*, holding that the commonality of humans and animals on the material plane made it reasonable that Satan could actually transform the body of one species into that of another. He used the analogies of men's ability to transform iron into steel and create hybrid plants to demonstrate that qualitative changes could be brought about in both nonliving and living materials by clever artifice. But since the real essence of the human being, according to Bodin, was not the physical form but rather the rational faculty, transformation into a werewolf (which was typically reported to occur without impairing reason) would leave the "true" human form unchanged.[33]

The line of thought initiated by Bodin was carried to its most extreme form in the following century by the French philosopher and mathematician René Descartes (1596–1650). Descartes was a pious Christian who combined his religious beliefs with the distinctively modern mode of philosophical and scientific thought which he is credited with founding. According to Descartes the animal body was essentially a machine, functioning according to mechanical laws like a clockwork or a highly perfected automaton. In the case of humans this machine was inhabited by an immortal soul, which, apart from commanding the voluntary motions, had no connection with the body's operations. Referring to animals, which he considered to lack a soul and therefore consciousness, Descartes stated: "I assume their body to be but a statue, an earthen machine formed intentionally by God to be as much as possible like us . . . "[34]

Descartes's views were clearly influenced by the great progress being made in the physical and engineering sciences by contemporaries such as Galileo and Kepler. In turn, they brought the study of the animal and the human bodies under a common scientific regime. Although the separate creation of the different species had yet to be questioned, the biological boundaries between them were being blurred. Indeed, as noted by the philosopher Hans Jonas, the concept of body as machine raised the question of why different species were created by God in the first place, "especially since mere complexity of arrangement does not create new quality and thus add something to the unrelieved sameness of the simple substratum that might enrich the spectrum of being."[35] In any case, the dissection of animals in order to advance knowledge of human anatomy gained impetus from the Cartesian theory. Moreover, the Christian theological division of the human and the divine to one side and the rest of creation to the other, which was integral to Descartes's model, actually relieved his adherents of qualms that might otherwise have stood in the way of their scientific activities. An account of experimenters working at the Jansenist seminary of Port-Royal in the late seventeenth century evokes the extremes of this mind-set:

> They administered beatings to dogs with perfect indifference, and made fun of those who pitied the creatures as if they felt pain. They said the animals were clocks; that the cries they emitted when struck were only the noise of a little spring that had been touched, but that the whole body was without feeling. They nailed poor animals up on boards by their four paws to vivisect them and see the circulation of the blood which was a great subject of conversation.[36]

The theoretical possibilities unleashed by the notion that living substance was just a variety of matter can be seen in the utopian writings of Descartes's contemporary, Francis Bacon (1561–1626). Bacon was the first great ideologue of modern science, touting its force in overcoming ancient superstition and its capacity to materially benefit society. Near the end of his life he began to set down his vision of human activity organized according to his progressive view of science in the uncompleted fragment known as *New Atlantis*.[37] In this work Bacon describes a scientific community organized very much along the lines of a modern research institute, a social entity that would in fact not be realized for another three centuries. In this institute, called Salomon's House, research is conducted on the physical and chemical sciences and mathematics as well as into the development of technologies, such as desalinization and the harnessing of wind power. But it is the vision of biological research presented by Bacon that most concerns us here.

Allowing himself free rein to speculate on the capacity of living materials to yield to scientific manipulation in a fashion well beyond anything found in the writings of the more mechanically minded Descartes, Bacon envisions botanical gardens in which the resident scientists practice

> all conclusions of grafting and inoculating, as well of wild-trees and fruit trees, which produceth many effects. And we make (by art) in the same orchards and gardens, trees and flowers to come earlier or later than their seasons; and to come up and bear more speedily than by their natural course they do. We make them also by art greater much than their nature; and their fruit greater and sweeter and of differing taste, smell, color, and figure, from their nature. And many of them we so order, as they become of medical use. We have also means to make divers plants rise by mixtures of earths without seeds; and likewise to make diverse new plants, differing from the vulgar; and to make one tree or plant turn into another.[38]

Salomon's House also has its own *Jurassic Park*-like facilities, but their uses go even further than those envisioned in the late twentieth-century novel, encompassing many of the touted prospects of modern transgenic biotechnology and its proposed application to human biology:

> We have also parks and inclosures of all sorts of beasts and birds, which we use not only for view or rareness but likewise for dissection and trials, that thereby we may take light what may be wrought upon the body of man. . . . By art likewise, we make them greater or taller than their kind is; and contrariwise dwarf them, and stay their growth: we make them more fruitful and bearing than their kind is; and contrariwise barren and not generative. Also we make them differ in color, shape, activity, many ways. . . . We find means to make commixtures and copulations of different kinds; which have produced many new kinds, and them not barren, as the general opinion is.[39]

The optimism of Bacon's utopian vision is exhibited in an overconfidence that exceeds that of most practioners of modern biotechnology, though perhaps not by much: "Neither do we do this by chance, but we know beforehand of what matter and commixture what kind of those creatures will arise."[40]

Carolyn Merchant, in *The Death of Nature*,[41] considers these passages in the *New Atlantis* to constitute an explicit rejection of respect for the natural world and the "beauty of existing organisms." And in a discussion of this work, Leonard Isaacs, who is more credulous than Merchant of Bacon's frequent avowals of religious and ethical sentiments, is nonetheless also troubled about the attitude toward nature implied by the goal of (in Bacon's words) the "effecting of all things possible." This program, according to Isaacs, constitutes "one of the most corrosive conceptions ever developed; and it has been eating away at the bedrock of religion for at least 3 centuries."[42] As the social ideology of modern science emerged, it was thus yoked to a technological imperative for which biological boundaries, whatever their significance in previous systems of thought, were just obstacles to be overcome.

Biological Types and the Chain of Being

Irrespective of the theories of Descartes and the speculations of Bacon, by the seventeenth century, flesh was still too solid to melt. Before the entry of evolutionary ideas into the mainstream of European thought, notions of the natural relationships between the various types of organisms were dominated by the concept of the *Scala Naturae*, or Great Chain of Being.[43] Taken over largely from the Greeks, this idea held that all natural entities, ranging from the inanimate through the animate, were unique and separate, occupying singular positions in sequences of forms of ascending complexity or perfection.

Perhaps the most widely diffused statement of this doctrine was that of the poet Alexander Pope, in his *Essay on Man* (1733):

Vast chain of being! which from God began,
Natures aethereal, human, angel, man,
Beast, bird, fish, insect, what no eye can see,
No glass can reach; from Infinite to thee,
From thee to nothing.—On superior pow'rs
Were we to press, inferior might on ours;
Or in the full creation leave a void,
Where, one step broken, the great scale's destroy'd;
From Nature's chain whatever link you strike,
Tenth, or ten thousandth, breaks the chain alike.

In its Platonic version, exemplified in Pope's verse, the discrete essences of the successive members of the chain were emphasized. The Aristotelian version, in contrast, stressed the principle of continuity and shading off of the properties of one class into those of the next. Scholars from medieval times on debated whether the perfection of the universe was manifested in the multiplicity and variety of things as they are normally encountered in the real world or whether qualitative gaps between types would represent imperfections and therefore must be a function of incomplete knowledge. However, in no case was the notion of the Chain of Being taken to imply that the created essences were actually transformable into one another.

In a world in which the boundaries between living things were supposed to be static, the transgression of these boundaries implied by the sudden appearance of biological novelties could be a source of public fascination. And given the right socioeconomic setting, such fascination could be genuinely disruptive. This was the case with the speculative episode known as "tulipomania," which brought Holland to the brink of bankruptcy in the early seventeenth century.[44] Tulips had been introduced into Western Europe from Turkey in the 1550s and became widely cultivated throughout the continent. Rare tulips, which exhibited variegated patterns unlike their parental strains, were particularly valued. At the peak of the speculative fever in the 1630s, individual bulbs sold for as much as several thousand florins. (For comparison, a thousand pounds of cheese sold for about 120 florins during the same period.) The sudden change in flower pattern seen in these tulips was known as "breaking." This phenomenon, which we now know to be caused by sporadic viral infection of the plants, was unpredictable and uncontrollable with seventeenth-century technology, making the often beautiful results a matter of mystery and luck. Hybridization, the technique responsible for most of the unusual varieties of tulips available today, requires raising the plants from seeds, which takes five to seven years for tulips. Such systematic manipulations could not have been undertaken without an understanding of the sexual nature of plants, which only became available with the publication of *De Sexu Plantarum Epistola* by the German botanist Rudolf Jakob Camerarius in 1694.

Although scientific plant breeding began in earnest in the eighteenth century, affording a more tractable experimental medium than animal breeding for testing the reality of the biological boundaries, the results of such studies had anything but a uniform effect on thinking about biological novelty and integrity. This can be seen in conflicting tendencies in the evolution of thought of two major figures of eighteenth-century biology, Carl Linnaeus (1707–1778) and George-Louis de Buffon (1707–1788).

Linnaeus, renowned for his system of classification of living organisms into branching family trees still in use today, was insistent in his early writings on the constancy and sharp delimitation of species from one another. In his *Fundamenta botanica* (1736) he stated: "Species are as numerous as there were created different forms in the beginning."[45] However, doubts about this proposition began to emerge several years later when he observed an abrupt morphological transformation in the plant *Linaria*:

> Nothing can be more wonderful than what has happened to our plant. The deformed offspring of a plant that used to produce flowers of an irregular form have now reverted to a regular form. This is not merely a variation with regard to the maternal genus, but an aberration in terms of the whole class; it provides an example unequaled in the whole of botany, which may now no longer be thought of in terms of the differences between flowers. What has happened is indeed no less wonderful than had a cow given birth to a calf with the head of a wolf.[46]

Since this *Linaria* was fertile and bred true, Linnaeus's notions of species integrity were shaken. He crossed out the words *"Natura non facit saltus"* (Nature does not make leaps) from his own copy of his *Philosophia botanica* (1751).[47] Further studies with interspecific hybrids led Linnaeus to the ultimately erroneous but intellectually courageous speculation that crossbreeding within a genus was a means for producing new species. In a 1764 letter he wrote: "We may assume that God made one thing before making two, two things before making four ... first a single species from a genus, and then mixed the different genera so that a new species would form." The statement *"nullae species novae"* (no new species) was removed from the last edition of his major work, the *Systema Natura* (1766).[48]

By contrast, Buffon, in his early writings on the species question, dismissed the idea of sharp boundaries between types of organisms and rejected the Linnaean taxonomic scheme:

> Nature progresses by unknown gradations and consequently does not submit to our absolute divisions when passing by imperceptible nuances from one species to another and often from one genus to another. Inevitably there are a

great number of doubtful species and intermediate specimens which one does not know where to place.[49]

To this way of thinking only individuals have a true existence, and a species is just a convention of human thought, a position explicitly stated by Buffon in the first volume of his major work *Histoire naturelle* (1749). But on the basis of evidence that he obtained himself and from correspondents during the following decade that whereas some distinct varieties of plants or animals could form fertile hybrids, others, like the donkey and the horse, produced sterile offspring, Buffon revised his views. The hybridization results appeared to provide proof for the objective reality of species as "the sole essences of Nature." A species, according to the thirteenth volume of *Histoire naturelle* (1765), was "a whole independent of number, independent of time; a whole always living, always the same; a whole which was counted as one among the works of the creation, and therefore constitutes a single unit of the creation."[50]

Just as different machines, e.g., clocks and water pumps, are built from similar parts and materials according to different plans, so could the different types of organisms have a common material basis. His eventual conclusion that species identities were discrete thus did not require Buffon, who was originally a physicist, to relinquish the eighteenth-century mechanistic world view in which he had been steeped. In reconciling his materialism with the idea of the separateness of species, Buffon proposed the radical concept of the "interior mold," a kind of three-dimensional template that determined the organizational properties of an organism's matter.

By this concept Buffon purported to solve one of the longest-standing controversies among natural philosophers: preformation versus epigenesis. The preformationists were advocates of the ancient idea that organisms successfully reproduced their kind by virtue of the presence of a miniature individual of the same type in either the sperm or the egg which enlarges but does not change its form during development. This doctrine had been criticized for numerous logical inconsistencies. One of the most compelling points was Buffon's own argument that since each "homunculus" would have to have a proportionately smaller one nested in its own germ cells for the production of the subsequent generation, the miniature being of the sixth generation would be smaller than the smallest possible atom.[51] The epigeneticists, led by the embryologist Caspar Friedrich Wolff (1733–1794), thought of organization, in contrast, as reemerging anew during embryonic development. One analogy that they used was the curdling of milk during the formation of cheese. Unlike the preformationists, however, Wolff and his followers had no notion of how such a process could be reproducible from generation to generation. In Buffon's view, this was the function and purpose of the interior mold.

This concept, of course, had to extend beyond the analogy of the sculptor's mold into which wax or plaster is poured, since this kind of template can only reproducibly render surface characteristics. Buffon recognized that embryonic development cannot be achieved by mere addition of molecules to surfaces. He therefore hypothesized the interior mold as "an intussusception that penetrates the mass,"[52] a hidden structure that organizes matter during embryonic development so as to produce a child in the image of its parents, and to provide "a general prototype in each species upon which all individuals are moulded."[53] Not being a philosophical idealist, he readily acknowledged that this prototype could be "altered or improved, depending on the circumstances, in the process of realization."[54]

Buffon's notion of the interior mold was not well received during his lifetime, partly because few scientific thinkers of the eighteenth century were willing to consider any view of matter other than the "corpuscular" theory then in fashion. This view attributed the qualities of materials much more to the properties of their irreducible atoms (which were completely hypothetical at that time) than to any interrelationships among these basic units. Buffon's great insight that complex molecular organization could be transmitted from one parcel of living matter to other parcels derived from it continued to be derided long after his death, most recently by the evolutionary biologist Ernst Mayr, who, as an unwavering proponent of "particulate" inheritance, considered the interior mold to be a Platonic idea.[55] The molecular biologist François Jacob, who was more appreciative of Buffon's notion, nonetheless considered its validity as a biological principle to be limited to the one-dimensional "mold" or template represented by the DNA molecule.[56] However, recent biological research has demonstrated that the egg and all the tissues subsequently derived from it contain "cytoskeletons," "nuclear scaffolds," and "extracellular matrices," all consisting of highly articulated three-dimensional networks of molecular fibers that are partitioned between daughter cells at each cell division, along with and in addition to the DNA. These findings underline the prescience of Buffon's concept of the interior mold.

Like staunch eighteenth-century citizens, Buffon couched his materialistic concepts of biological organization and Linnaeus his notion that new species could continually arise over time in terms of special creation. The implications of their insights pointed in a different direction, however. The embryologist Wolff, from his studies of birth defects in animals, was, like Linnaeus, impressed with the abruptness with which new forms could appear and with their subsequent stability.[57] In unpublished manuscripts he concluded that not every biological structure or species was a primordial product of nature which had received its existence directly from the hand of God. Like Buffon, he believed that the constancy of species and genera was derived from the specificity of a structured substance that reproduced itself, but he also entertained the possibility that external factors could modify this substance and cause hereditary changes.[58]

Thus, while the discreteness of biological boundaries had yet to be called into question, the possibility of their being breached in the normal course of events was raised by the new observations and the materialistic explanations offered for them. Among the boundaries increasingly characterized as permeable during the eighteenth century was the moral distinction between humans and nonhuman animals, the absoluteness of which had been a fundamental tenet of Christian dogma and Cartesian dualism. Alexander Pope, in addition to being a popularizer of the Chain of Being, was also a strong advocate of the humane treatment of animals and one of the earliest public opponents of scientific vivisection. Accordingly, he included mental faculties among the ascending chain of graded qualities in his *Essay on Man* and entertained the idea that animals, like humans, had immortal souls.[59] It is also significant, in light of the present discussion, that when Pope sought religious justification for the humane treatment of animals he invoked the authority of the Old Testament.[60]

During the next century the spiritual argument for the unity of creation would be replaced in the writings of philosophers and natural scientists by concepts such as "laws of form," "functional adaptation," and "community of descent." Debates about these ideas ultimately resolved into the Darwinian doctrine that biological boundaries exist not as a matter of principle but as a matter of contingency or historical accident. Since, as we shall see, the purported scientific foundation for this doctrine is debatable, it is of interest to examine the varying degrees to which this outcome was driven by evidence on one hand and by ideology on the other.

Nineteenth-Century Theories of Biological Transformation

Flesh is matter, but the Newtonian concept of matter that continued to prevail throughout most of the nineteenth century could not account for the distinctive properties of flesh. In the classical picture, matter is inert. Although its motion is governed by mathematically precise laws, the outcome of this motion is entirely dependent on the initial preparation of the system—the arbitrarily given position and velocity of each particle. In order for the matter in a multicomponent system to become organized in a complex fashion, it would have to be "set up" in an appropriate way. That is why Descartes, Newton, and the other founders of the mechanistic world view could simultaneously be physical determinists and religious believers: God was in the initial conditions.

The German philosopher Immanuel Kant (1724–1804) was a critic of the notion that the existence of God could be derived from the design of the natural world. But he was equally dismissive of the hope that the principles upon which organisms were constructed could be derived from causal analysis based on physical science. In this regard he was affirming the independence of the laws of motion from the initial conditions, noted above. Kant argued that characterizing the functional relationships among parts of a complex structure such as a clock, a painting, or an organism was not the same as understanding the principle or purpose of its organization. While the purpose of a human-made artifact derives from the concept that led to its production, the "concept" behind a living thing cannot be discerned by scientific experimentation. Moreover, not only do organisms, like machines, exhibit a high degree of functional integration among their parts, but they are also self-generating and self-renewing; as Kant stated it, "every part is reciprocally purpose and means."[61] The principles behind the arrangement of matter in a living organism are even more opaque to causal analysis than those of machines.

Although science and religion were progressively diverging in the domains they sought to explain during the nineteenth century, they remained joined to one another by virtue of the recognition, epitomized in Kant's analysis, that organizational principle could not be derived from mechanism. And while progress in chemistry had undermined earlier beliefs that living matter consisted of elements other than those found in nonliving matter, for the generation of biologists that followed and were influenced by Kant (referred to as the "teleomechanist" school by the historian Timothy Lenoir),[62] the chemical mixtures that occurred in living tissues were organized in ways that were irreducibly different from anything in the inorganic realm.

Some of these scientists went beyond Kant's skepticism about arriving at conclusions about principles of organization from causal analysis and actually postulated the existence of vital principles or forces in living tissues. For example, the Swedish chemist Jöns J. Berzelius (1779–1848), writing about the "catalytic force" exhibited by certain biological molecules, indicated that this was not to be understood as "a capacity independent of the electrochemical relationships of matter" but rather as "a special sort of expression of those relationships . . . that remains hidden from us."[63] But if living materials are organized according to principles beyond our abilities to discern scientifically, then the different species or life forms also may be organized according to distinct suprascientific principles. Thus the materialist paradigm, now well established in physiological circles, could be maintained side by side and quite consistently with a firm belief in special creation.

The range of differing opinion on the principles upon which organisms were constructed and the stage at which the hand of God exerted its direct influence on organismal form can be seen in the debates between the "structural" or "transcendental" morphologists, who dominated on the Continent, and the "natural theologians," who held sway in Britain. Georges Cuvier (1769–1832), the French founder of paleontology and comparative anatomy, was a Christian believer who brought a holistic materialism to bear on the question of the separateness of creatures. The notion of a structural basis for biological uniqueness, which Buffon had proposed to underlie heredity and embryonic development, was generalized to the organismal level by Cuvier in his theory of the correlation of parts. He held that all the functions of an organism are interrelated by a "necessity equal to that of metaphysical or mathematical laws." He stated, moreover, that "if one of these functions were modified in a manner incomparable with the modification of the others, the creature could no longer continue to exist."[64]

Cuvier provided a set of examples that interestingly echoes the biological classification scheme underlying the dietary prohibitions of Leviticus:

> An animal that digests only flesh must be able to see its prey, follow it and tear it apart. Consequently, it must have a piercing eye, a keen sense of smell, a swift gait, agility and strength of leg and jaw. For this reason, cutting teeth for tearing through flesh are never found in the same species with a foot encased in horn that can only support the weight of the animal and cannot be used for grasping.[65]

His conclusion from such arguments was that abrupt discontinuities in the organization of the animal kingdom and gaps in the fossil record represented the absence of transitional forms that are, in fact, biologically impossible.

Cuvier was a strong opponent of the notion that the different types of organisms were derived from common ancestors by a process of "organic change," a theory proposed by his countryman Jean-Baptiste Lamarck (1744–1829) in his *Philosophie zoologique* (1809).[66] Indeed, Cuvier's ridicule of Lamarck's ideas and his caricature of the role of volition in the selection by organisms of their environments (the famous "giraffe" example) served to deny Lamarck his rightful place in the history of biology as the first systematic proponent of evolution.[67] Etienne Geoffroy Saint-Hilaire (1772–1844), Cuvier's colleague and fellow structural morphologist, believed that a single set of geometrical-topological "laws of form" were responsible for generating all animal types.[68] That led him to be more sympathetic to evolutionary ideas than was Cuvier. For while Cuvier was content to consider that the main branches of the animal kingdom and the various species within them had been separately created to occupy distinct functional niches, Geoffroy's emphasis on the common principles that underlay the generation of all animal types suggested that structure determined rather than reflected function (e.g., birds fly because they have wings, not the other way around). Moreover, it raised the question of why the laws of form had led to a variety of outcomes rather than a single type. In his later writings he proposed the idea that the environment, acting during embryogenesis, could modify "organized bodies," leading to new biological types.[69] However, Geoffroy was a religious conservative; while his laws of form might generate the occasional new type of organism as conditions changed, he believed that all living things conformed to a Unity of Plan that was unchanged since the time of creation.

Natural theology represented an alternative framework to transcendental morphology in accounting for the design of the living world. Like that doctrine, it was a blend of science and belief, albeit one that violated the Kantian precept of the impossibility of deriving knowledge of a creator from the nature of the "created." The commissioning by the Earl of Bridgewater, by a legacy upon his death in 1829, of a series of works "on the power, wisdom, and goodness of God, as manifested in the Creation" is representative of the flavor of the movement. Where natural theology most differed from transcendental morphology was in its emphasis on "adaptations," the suitability of biological structures for the functions they performed. Although Cuvier also stressed the interrelatedness of structure and function, his principle of the correlation of parts placed constraints on the possible activities that could be served by anatomical variation. Certainly the primacy given by Geoffroy and his successors to material properties of living systems and the concomitant rule-generated structures in determining biological function was bound to conflict with the natural theologians' view that the hand of God was manifested in even the smallest detail of each being's construction.

The nature of the debate between the structural morphologists and the natural theologians can be seen in the neurologist Sir Charles Bell's *Bridgewater Treatise* of 1833. At issue was the proposal by the structural morphologists that the *incus* (anvil) bone, one of the chain of three bones constituting the sound transmission system of the mammalian middle ear, was absent in birds because it had been "transformed" into the quadrate bone of the upper jaw joint. Bell's alternative account was that the bird has an articular structure in its upper jaw because it requires the extra mobility in order to catch insects. He comments:

> It is above all, surprising with what perverse ingenuity men seek to obscure the conception of a Divine Author, an intelligent, designing, and benevolent Being—rather clinging to the greatest absurdities, or imposing the cold and inanimate influence of the mere "elements," in a manner to extinguish all feelings of dependence in our minds, and all emotions of gratitude.[70]

Out of this tradition of natural theology came the British naturalist Charles Darwin (1809–1882) and his theory of evolution by natural selection. The central doctrine of this theory—that given the small morphological, physiological, or behavioral variations encountered in any natural population of a single kind of organism, the competition of marginally different individuals for limited resources has been sufficient to generate the entire array of biologically distinct types seen on the face of planet—is too familiar to require detailed discussion here. Because the mechanism that Darwin proposed for achieving organismal form and the interrelationships among parts in a living system is, on the face of it, so "unguided," the congruence between the structure of his theory and that of natural theology, while occasionally noted, is not considered by contemporary Darwinians to detract from the theory's scientific standing. Thus Ernst Mayr makes what he considers to be the "rather paradoxical claim" that much of the intellectual structure of the *Origin of Species* can be accounted for by the fact that the leading paleontologists and biologists of Darwin's day were natural theologians whose descriptions were filled with "what we would now call adaptations." He goes on to state that "when 'the hand of the creator' was replaced in the explanatory scheme by 'natural selection,' it permitted incorporating most of the natural theology literature on living organisms almost unchanged into evolutionary biology."[71]

While modern Darwinians thus concede the formal similarities between the "perfectionism" of natural selection and that of natural theology, the proposal that Darwin's mechanism of organismal change also shares some of the teleological assumptions of theistic metaphysics is much more controversial. Nevertheless, the historian of science John Cornell, studying Darwin's notebooks from the period in which he was formulating the theory of natural selection, concludes that

> the power he attributed to his new mechanism depended specifically on the assumption of a divine Being, intelligent like man but superior and utterly lawful. This assumption underlay both Darwin's idea of natural species' "perfect adaptation" and his stunning analogy of selective breeding to describe nature in terms of this mechanism.[72]

And the historian Robert J. Richards has inferred from Darwin's embryological writings that he believed that evolution was a progressive force,[73] leading to a situation in which (according to the last edition of the *Origin of Species*), "The inhabitants of each successive period in the world's history [are] higher in the scale of nature" than their predecesors.[74]

It should not be surprising that Darwin incorporated the theistic ideas of his cultural milieu into his biological theory. Given Kant's Newtonian judgment that knowledge of mechanism carried no implication concerning knowledge of organizational principle or "purpose," there were only two possible pathways to a more naturalistic understanding of the living world. One route was the incorporation of a dynamic conception of matter into biology, a conception that differed from the static view of matter of classical mechanics. If matter itself has "self-organizing" properties that can lead to the formation of structures relatively independently of the initial conditions of its preparation, then the goal-directedness of organismal physiology, development, even evolution could potentially find interpretations in the physical properties of biological materials.

We know that distinct kinds of nonliving matter assume preferred forms and patterns: liquids flow, taut strings vibrate as a whole and in discrete segments along their lengths, and soapy solutions form bubbles and foams. If, as I have suggested elsewhere,[75] there were analogous tendencies for primitive living tissues to assume preferred forms and patterns—hollow, multilayered, segmented, or jointed structures, for example—then the appearance of a particular set of body plans and organ forms over life's history would represent the inevitable emergence of stereotypical morphologies. A subset of these forms might meet with differential success under different circumstances—natural selection would still be possible—but the array of possible "types" would be intrinsic to fleshly matter and limited rather than open-ended.

Glimmerings of a dynamic view of matter were emerging in the early nineteenth century with the new science of thermodynamics, and our own century has seen sustained exploration of the forms and patterns generated by the dynamic behavior of fluids and of "excitable media." Geoffroy Saint-Hilaire's organismal "laws of form" were a prescient application of this view of matter to the living world.

The other pathway out of Kant's impasse, the one taken by Darwin and his followers, was to retain the classical view of matter and with it the formal structure of natural theology. Newtonian matter can be molded into any form or pattern, subject only to the constraints of the initial conditions, which are entirely arbitrary. Correspondingly, Darwinism makes no a priori statements about why organisms have the appearance and characteristics that they do. As the philosopher Thomas Nagel has noted, the theory of natural selection "explains the selection among those organic possibilities that have been generated, but it does not explain the possibilities themselves."[76]

Some of Darwin's most enthusiastic adherents, such as the Christian Darwinists Asa Gray (1810–1888) and George Frederick Wright (1838–1921), took comfort in the fact that Darwin's theory made no attempt to explain the origin of the variations preserved by natural selection, since, as Wright stated, it "left God's hands as free as could be desired for contrivances of whatever sort he pleased."[77] While the theistic metaphysical context from which Darwin's theory emerged was rapidly disavowed by most of his immediate successors, who emphasized the role of impersonal forces in the generation of biological types, it is difficult to avoid the conclusion that the Darwinian "mechanical" materialism that emerged as the scientific mainstream was a doctrine less threatening to the dominant social classes that espoused it than the alternative of a Geoffroyan "dynamical" materialism would have been. If fleshly matter was inexhaustibly malleable, there would be few limitations on the forms it could assume in the course of evolution. Whether through the workings of the hand of God or, as later became fashionable to believe, through "chance" coupled to "survival of the fittest," those who came out on top could well imagine themselves the products of a perfecting process. In contrast, if we all, humans and other species, were, in all our complexities, truly products of common natural forces, the biological world would have to be seen in more pluralistic terms and the human species would no longer be, in the words of Julian Huxley, one of the leading Darwinians of this century, "the highest form of life produced by the evolutionary process on this planet."[78]

In any case, the secular Darwinists, in order to render natural selection acceptably materialistic, needed to address Kant's precept that the organizing principle of a complex whole could not be derived from analysis of the functional interactions among its parts. But for the reasons already discussed, the solution to this problem could not appeal to the self-organizing properties of biological matter without undermining the Darwinian view of the external determination of the direction of evolutionary change. Specifically, the Darwinians needed an independent guiding force, analogous to the God of the natural theologians, to give form and reproducibility to the inert materials of their biological world. This force was soon provided by the theory of the isolation and continuity of the germ plasm proposed by August Weismann (1834–1914) and its eventual recasting into the scientifically questionable modern idea of the "genetic program."[79]

The Apotheosis of the Gene

According to standard accounts, Darwin's theory of evolution by natural selection was so intellectually compelling that it required only a plausible theory of inheritance to gain general acceptance among all but the most obstinate of the scientifically disposed. The "rediscovery" of Mendel's laws by several independent investigators around 1900 supposedly satisfied this requirement. There are certain historical distortions in this scenario; for example, Mendel's findings were never really "lost" to the scientific community,[80] and the nature of the mechanisms of heredity was never central to the most cogent scientific criticisms of Darwinism before or after 1900.[81] But more important, examination of the logic of both Darwin's and Mendel's contributions in light of present knowledge of the nature of biological stability and variation and of the mode and tempo of evolutionary change suggests that the neo-Darwinian "modern synthesis," as it became defined by the mid–twentieth century, incorporated uncritically held beliefs to an extent comparable to any of the earlier theories of nature discussed in this chapter.

In keeping with the requirement of his theory that biological organizing principles must reside in a medium independent of fleshly matter itself, Darwin put forward "the hypothesis of pangenesis" a decade after he first published the *Origin of Species*. This theory held that each cell in the body produced, and was represented by, invisible particles called "gemmules" which circulated freely throughout the system and accumulated in the reproductive organs. The mixing of the gemmules of two individuals would result in offspring that were in part similar to the parents and in part novel. Darwin, like most nineteenth-century biologists, also believed in the inheritance of acquired characteristics. He therefore proposed that the tissues of the body, upon being affected by "changed conditions" such as use and disuse, will "consequently throw off modified gemmules, which are transmitted with their newly acquired peculiarities to their offspring."[82]

In its proposal that vanishingly small, structurally unspecified particles could embody and regenerate specific living qualities, Darwin's theory verged on philosophical idealism. The idea that the traits of an organism could be "represented" in distinct and independent particles, which upon mixing would produce new versions of those same traits, is notably deficient in explanatory power. This is particularly evident when compared with the tentative attempts by Buffon and the members of the "teleomechanist" and "structural morphology" schools to conceptualize how the properties of "organized bodies" could be transmitted from generation to generation.

The fact that Darwin's theory of evolution could incorporate a virtually mystical notion of the transmission of biological qualities suggests that it lacked the specificity that might be expected from a purported explanation of organismic form and function. The substitution of Mendelism for pangenesis did little to correct this problem. Indeed, the version of Mendel's concept of inheritance that was incorporated into the Darwinian paradigm was much closer conceptually to pangenesis than to what Mendel in fact deduced from his studies of the transmission of traits in peas and hawkweeds.

Gregor Mendel (born 1822) entered the Augustinian monastery in Brno, Moravia, in 1843, and died there as abbot in 1884. As a Catholic monk he was steeped in Church doctrine and Aristotelian philosophy. Much of Mendel's success in classifying the qualities of plants and conceptualizing the regularities of their combination and transmission is attributable to his Scholastic intellectual background, according to the geneticist H. Kalmus.[83] Despite the efforts of the eminent population geneticist R. A. Fisher[84] and some later Darwinians to portray Mendel as a convert to evolutionism, Mendel's writings contain no evidence of affinity with this concept. Moreover, L. A. Callender has persuasively argued that Mendel was, rather, an adherent of the Linnaean version of the doctrine of special creation, discussed above, which left open the possibility that new forms could arise through hybridization.[85] Mendel's work with peas addressed the inheritance of alternative versions of the same characteristic within a given species (i.e., flower color, seed shape, or texture), while his studies of hawkweeds suggested that hybrids formed between preexisting species could exhibit some long-term stability. None of his results implied that major transformations between biological forms (that is, species and, ultimately, more divergent groups such as classes and phyla) could result from successive alterations in heritable determinants or "elements" (what we now call "genes"). But this, of course, would be a necessary condition if Mendel's factors were to provide the herditary basis for Darwin's doctrine that new biological types originated by sequences of gradual modifications.

The botanist Hugo de Vries (1848–1935) interpreted his own and Mendel's findings within a conceptual framework similar to that of Darwin's pangenesis, proposing that differences among individuals could be dissected into "unit characters," each with its own hereditary basis. For example, de Vries stated in his 1889 book, *Intracellular Pangenesis*: "If one considers the species characters in the light of the doctrine of descent, it then quickly appears that they are composed of separate more or less independent factors."[86] He saw no reason to change his opinion in 1900 after he became aware of Mendel's work.[87] But Mendel himself came to no such universal conclusions, even referring to the only law that he ever enunciated, that concerning the nature of the progeny of hybrids with two alternative characters, as the "Law Valid for *Pisum* (peas)." There is no implication that all traits, let alone all species, conform to this law. Moreover, in considering the class of features that were inherited in this fashion, Mendel wrote that "the distinguishing traits of two plants can, after all, be caused only by differences in the composition and grouping of the elements existing in dynamical interaction in their primordial cells."[88] Mendel's notion of how these "elements" affected the production of traits was therefore, from this limited evidence, a developmental one. That is, those factors of heredity which can exist in alternative states influenced the outcome of a generative process in a complex system rather than "representing" distinct traits, as they did for Darwin and for de Vries.

Eventually some biologists rejected the naive notion, embodied in Darwin's and de Vries's pangenesis theories, that each heritable characteristic is carried by an independent factor. For example, the Danish botanist Wilhelm Ludwig Johannsen (1857–1927) wrote in 1909:

> By no means have we the right to define the gene as a morphological structure in the sense of Darwin's gemmules or biophores or determinants or other speculative morphological concepts of that kind. Nor have we any right to conceive that each special gene (or a special kind of genes) corresponds to a particular phenotypic unit-character or (as morphologists like to say) a "trait" of the developed organism.[89]

Thomas Hunt Morgan, an embryologist turned geneticist and a central figure in the discovery that genes are parts of chromosomes, made the same point even more forcefully several years later:

> Failure to realize the importance of these two points, namely, that a single factor may have several effects, and that a single character may depend on many factors, has led to much confusion between factors and characters, and at times

to the abuse of the term "unit-character." It cannot, therefore, be too strongly insisted upon that the real unit in heredity is the factor, while the character is the product of a number of genetic factors and of environmental conditions. . . . So much misunderstanding has arisen among geneticists themselves through the careless use of the term "unit character" that the term deserves the disrepute into which it is falling.[90]

Six decades of subsequent research have led to the generally accepted view that in functional terms, a gene constitutes nothing more than the cell's replicable record of the primary sequence of an RNA molecule, or indirectly, a protein. With the recognition that the molecules specified by genes can influence one another's synthesis and physiological activity, the modern view is seen to be in full accord with Mendel's notion of a "dynamical interaction" of elements causing the production of distinguishing traits.

The subtleties involved in relating genes to traits can be seen in the example of sickle cell disease, which is the classic case of an association of a gene mutation with impairment of health in humans. Persons who have this condition produce only mutated versions of a hemoglobin protein ("hemoglobin-S") in their red blood cells. The severity of the condition depends on the proportion of cells which are "sickled" and because of their shape may clog small blood vessels; severity therefore varies from asymptomatic in some individuals to life threatening in others. This variability exists because the degree to which the blood cells are made abnormal by the presence of hemoglobin-S is controlled by physiological factors quite independent of the hemoglobin gene; these factors differ from individual to individual and even vary in a given individual under different conditions.[91] Despite the ubiquity of such complexities, in the course of the development of the modern gene concept the warnings of Johannsen and Morgan have been often ignored.

The growing influence of Darwin's theory of natural selection required a formal disconnection between an organism's "plan" and its fleshly matter. Otherwise evolution of form (i.e., alterations of the plan) would be driven as much by the intrinsic propensities of such matter to generate certain structures as by transformations resulting from adaptation to changed conditions. To draw another example from sickle cell disease, one may look at a red blood cell as a partially deformable object which can take on a small number of different shapes depending on variations in its constituent molecules and in its local environment. Both genetic and environmental changes can induce a transformation from the standard biconcave disk shape of red blood cells to a sickle shape. Moreover, a given genetic change (i.e., mutation of hemoglobin to hemoglobin-S) can have the effect that a greater proportion of cells are sickled under standard conditions. But no plausible sequence of genetic changes can turn a red blood cell into a five-pointed star. Analogous constraints on the ability to generate form must have pertained to the multicellular parcels of matter that provided the raw material for the elaboration of the earliest plants and animals.[92] In other words, throughout the history of life on earth the relationship between genes and traits must inevitably have been mediated by what Geoffroy referred to as "laws of form."

However, as a result of the general acceptance of the Darwinian-Weismannian version of inheritance by the mid-twentieth century, the genome—an organism's collection of genes—came to be typically characterized as embodying or determining all of the organism's traits. Although the naive picture that each trait was "represented" by a particular gene or set of genes was no longer fully subscribed to, the notion of the presumed role of genes in determining all aspects of every biological feature and their collective autonomy in executing their functions was carried over from the pangenesis of Darwin and de Vries into the modern view. Thus the physicist Erwin Schrödinger, in his influential book, *What Is Life?* (1945), summarized the emerging consensus of many of his biologist colleagues and sounded the keynote for the new field of molecular biology when he wrote: "The chromosome structures are instrumental in bringing about the development they foreshadow. They are the law-code and executive power—or to use another simile, they are the architect's plan and builder's craft in one."[93] That this passage did not simply represent the musings of a nonspecialist prior to the discovery of the chemical nature of the gene is evident from a 1976 symposium paper by Max Delbrück, a founding figure of molecular biology. Under the wry but telling title "How Aristotle Discovered DNA," Delbrück stated: "It is my contention that Aristotle's principle of the unmoved mover [i.e., God] originated in his biological studies. . . . [U]nmoved mover perfectly describes DNA. DNA acts, creates form in development, and it does not change in the process."[94]

Therefore, while molecular biologists in their day-to-day work followed a Cartesian functionalist paradigm that permitted them very effectively to determine the manner in which the various molecules and cells of organisms were organized, some theorizers convinced themselves that they had also solved the problem posed by Kant of identifying the principle that guided this organization. This guiding principle was considered to be contained in the base sequences of DNA, a substance to which were attributed the powers of "self-reproduction" and the ability to "act" and "create."

The Modern Ideology of Flesh

The invention in the mid-twentieth century of the digital computer depended on developing a conceptual framework in which data and the programs to manipulate them were entirely separable from the machine itself. For such a machine, what it was actually made of (e.g., vacuum tubes, transistors, integrated circuits) had little bearing on the tasks it performed. This provided an apt metaphor for the presumed separation of instructions from material in what was becoming the standard description of biological systems. Thus the molecular biologists Alexander Rich and S. H. Kim wrote that "it is now widely known that the instructions for the assembly and organization of a living system are embodied in the DNA molecules contained within the living cell"[95] and the physicist Freeman Dyson, discussing the origins of life, declared: "Hardware processes information; software embodies information. These two components have their exact analogues in a living cell; protein is hardware and nucleic acid is software."[96]

What organisms are actually made of is entirely irrelevant from this viewpoint. The biologist Richard Dawkins is quite explicit about this in the following remarkable passage from his book *The Blind Watchmaker*:

> [Molecules of living things] are put together in much more complicated patterns than the molecules of nonliving things, and this putting together is done following programs, sets of instructions for how to develop, which the organisms carry around inside themselves. Maybe they do vibrate and throb and pulsate with "irritability," and glow with "living" warmth, but these properties all emerge *incidentally*. What lies at the heart of every living thing is not a fire, not warm breath, not a "spark of life." It is information, words, instructions. If you want a metaphor, don't think of fire and sparks and breath. Think instead of a billion discrete digital characters carved in tablets of crystal. If you want to understand life, don't think about vibrant, throbbing gels and oozes, think about information technology.[97] (Emphasis added.)

The philosopher Mary Midgley calls this mode of thinking about life "getting away from the organic." In her book *Science as Salvation* she notes that much speculative writing by male scientists during the twentieth century contains "quasi-scientific dreams and prophesies" involving visions of escape from the body coupled with "self-indulgent, uncontrolled power-fantasies."[98] An example is this fascinating statement from an article by Freeman Dyson:

> It is impossible to set any limit to the variety of physical forms that life may assume. . . . It is conceivable that in another 10^{10} years, life could evolve away from flesh and blood and become embodied in an interstellar black cloud . . . or in a sentient computer.[99]

The history of ideas developed in the present chapter provides a context for interpreting these denials of specificity to the material substratum of life, coupled with an exaggerated notion of human agency—what Midgley calls "predictions of the indefinitely increasing future glory of the human race, and perhaps its immortality."[100] One could infer, for example, that by opting so completely for a Hellenistic–New Testament concept of nature, with its de-emphasis on the reality and moral significance of biological boundaries, over the biologically pluralistic Hebraic–Old Testament concept, the Darwinian mainstream reinforced a cultural paradigm that valued power to transform the world over a respect for variety, balance, and limits. But the previous discussion also suggests that the Darwinian model to a significant degree represents a projection of scientifically unsubstantiated beliefs onto the natural world.

An alternative to the Darwinian-Weismannian view, referred to above,[101] attributes the bodily forms assumed by complex multicellular organisms to the intrinsic properties of the semisolid materials that constituted flesh at early stages of its evolution. The array of biological forms that populate the world are thus considered to be limited and stereotypical, not an open-ended set of structures whose particular characteristics depend mainly on the vagaries of extrinsically imposed functional adaptation. This alternative view is clearly closer to the "laws of form" perspective of Cuvier and Geoffroy than it is to the "descent with modification" paradigm of Darwin. In particular, by treating biological types as intrinsic to the matter from which organisms are made, the alternative view explicitly rejects the dualism inherent in Darwinism.

In these two views of the evolutionary process, which I will refer to as the "externalist" and "internalist" models, we can distinguish two very different notions of biological boundaries. In the externalist model, the organism is continually evolving into something different from itself. According to the philosopher Hans Jonas, for Darwinism "the emergence of forms falls wholly to the random play of aberrations from pattern, which as aberrations are by themselves indifferently 'freaks,' and on which the distinction between deformity and improvement is superimposed by entirely extraneous criteria." He continues:

> [I]f the gene system is the transmitter of heredity, stability—the condition of faithful transmission—is its essential virtue. Since [a mutation] is a mishap to the steering system of a future organism, it will result in something which from the point of view of the original pattern can only be termed a deformity. However "useful" it happens to be, as a deviation from the norm it is "pathological." As similar mishaps continue to befall the same gene system in succeeding generations, an accumulation of such deformities under the premium system of selection may result in a thoroughly novel and enriched pattern: but the "enrichment" would still be an excrescence on the original simplicity, a slipping of the discipline of form multiplied over and over again under the licensing of selection; and thus the high organization of any animal or of man would appear a gigantic monstrosity into which the original amoeba has grown through a long history of disease.[102]

For the externalist, transgression of biological boundaries is thus the neverending evolutionary norm. In the internalist view, in contrast, almost all overt biological diversification occurs early on, when primitive organisms, because of the physical contribution to the determination of their forms, are to a certain extent mutually transformable. Through subsequent evolution the disparate kinds of organisms, by accumulating mechanisms which promote their capacity to develop "true to type" despite genetic mutation ("morphological stasis") and to maintain their phenotypic character in the face of changing conditions ("physiological homeostasis"), turn more and more into "themselves." According to the internalist view, then, the intensification of uniqueness, rather than the open-ended production of overt difference, may thus be the hallmark of organismal evolution once it has left its early, "physical" stage. This view implies, furthermore, that mixing and matching the biochemical capabilities of modern organisms by transgenic manipulations could be profoundly disruptive of species and individual identity and integrity in a fashion different from anything encountered during evolution.

Only time will tell whether the internalist or externalist model better accounts for the facts of organic evolution. But can adherence to one or another of these paradigms possibly make a difference to our practical interaction with the natural world? After all, even in the Darwinian picture organisms are unique over any time scale in which human action can have any consequence. Indeed, the arch-Darwinist E. O. Wilson is one of the most eloquent scientific advocates of the preservation of the diversity of life. He argues that the expected generation of truly novel forms would be much too slow to compensate for any losses by extinction or ecological disturbances resulting from destructive policies with respect to the biosphere.[103]

But when Wilson's rationale for his valuation of diversity is examined closely, it is seen to embody a view of biological boundaries as disparaging as that implicitly held by the ecological despoilers whom he deplores. True to his practice of attributing all important organismal features to Darwinian processes, Wilson postulates that a long process of random genetic change, along with competition among individuals and groups, has produced humans with a set of "impulses and biased forms of learning loosely characterized as biophilia,"[104] by which he means a propensity to protect and cherish life in the particular array of forms that have co-evolved with us. Respect for nature is therefore in the genes. The world's flora and fauna, which, with Jonas, Wilson would have to admit are just so many Darwinian "freaks" and "monstrosities," are to be valued because they are *our* freaks and monstrosities.

But like many such "reified traits"—abstractions endowed with biological concreteness—Wilson's "biophilia" founders on the shoals of arbitrariness. As Stephen Jay Gould asks in his review of Wilson's *Diversity of Life*,

> Why should these valued features be deeper, more innate, more definitive of our nature, than our rapacity? Was it any less natural to kill all the moas of New Zealand, all the mammoths of North America? Surely for each biophile in the United States there are ten who would kill a deer for sheer sport, rather than for needed food; ten who will build the suburban shopping mall for each cry of "woodman, spare that tree."[105]

The sociologist Howard Kaye, in *The Social Meaning of Modern Biology*, places Wilson firmly within the tradition of natural theology in the attempt to biologize ethics and communal purpose.[106] Along with other critics he finds deep affinities between Wilson's view of the ideal society as a "social organism" and that of nineteenth-century pre-Darwinian social evolutionists. The philosophical dualism that connects Darwinism to natural theology also permeates Wilson's sense of the significance of biological boundaries, which, in this world view, must be seen as ultimately arbitrary, or at least extrinsically imposed. On one hand Wilson rhapsodizes that "[t]he flower in the crannied wall—it *is* a miracle. . . . Every kind of organism has reached this moment in time by threading one needle after another, throwing up brilliant artifices to survive and reproduce against impossible odds."[107] But on the other hand he approvingly quotes the entomologist Thomas Eisner:

> As a consequence of recent advances in genetic engineering, [a biological species] must be viewed . . . as a depository of genes that are potentially transferable. A species is not merely a hard-bound volume of the library of nature. It is also a loose-leaf book, whose individual pages, the genes, might be available for selective transfer and modification of other species.[108]

An ultimate celebration of genetic manipulation can be found in the following passage, reminiscent of Bacon's *New Atlantis*, in Dyson's *Disturbing the Universe*:

> Imagine a solar energy system based on green technology, after we have learned to read and write the language of DNA so that we can reprogram the growth and metabolism of a tree. All that is visible above ground is a valley filled with redwood trees, as quiet and shady as the Muir Woods below Mount Tamalpais in California. These trees do not grow as fast as natural redwoods. Instead of mainly synthesizing cellulose, their cells make pure alcohol or octane or whatever other chemical we find convenient. While their sap rises through one set of vessels, the fuel they synthesize flows downwards through another set of vessels in their roots. Underground, the roots form a living network of pipelines transporting fuel down the valley. The living pipelines connect at widely separated points to a nonliving pipeline that takes the fuel out of the valley to wherever it is needed. When we have mastered the technology of reprogramming trees, we shall be able to grow such plantations wherever there is land that can support natural forests.[109]

Such schemes must be considered overweening and dangerous fantasies, not only because of what might "go wrong" but also for the continued negative cultural impact of this Procrustean view of living beings.

The dualist conception of living organisms as program plus its execution, which has prevailed with the ascendancy of the Darwinian world view, has provided a notion of biological boundaries that corresponds perfectly with the requirements of modern commercial biotechnology in its drive to generate products such as experimental mice that contract cancer at high rates,[110] tomatoes that remain ripe-looking despite being weeks off the vine,[111] pigs that have leaner meat,[112] and, ultimately, children with enhanced athletic or social skills. Indeed, the molecular biologist and editor of *Science* magazine Daniel Koshland, Jr., contemplates the possibility that prenatal gene modification of humans could be perceived to meet future "needs" to design individuals "better at computers, better as musicians, better physically."[113] With the advent of human in vitro fertilization and embryo cloning, proposed applications of transgenic technology now threaten to bring human individuals into the realm of manufactured items.[114]

According to a report called *Patenting Life* by the U.S. Office of Technology Assessment, the nature of a species is "rooted in the identity of the genetic material carried by the species," although "how a species might be defined genetically is not yet apparent". It therefore follows that since mammals may contain 50,000 to 100,000 or more genes, "[w]hatever it is in the organization and coordination of activity between these genes that is fundamental to their identity as species, it is not likely to be disrupted by the simple insertion or manipulation of the small number of genes (fewer than 20) that transgenic animal research will involve for the forseeable future."[115]

These assertions are based on the erroneous assumption that there is a straightforward relationship between genetic difference and the "distance" between organisms in a typological sense.[116] Since biological boundaries are, in this view, historically contingent products of gradually accumulated genetic change, they can be slightly breached with only slight consequences. The government report on patenting life thus provides false reassurance to Congress and the public that the freaks and monstrosities almost certain to arise from transgenic research will be no different from the Darwinian garden variety that have supposedly spurred evolution on its way.

But as we have seen, there is an alternative to the scientific view on which this analysis is based. With roots, in Western culture, in the Hebrew creation myth, and with scientific branches represented by the thought of the naturalist Buffon, the morphologists Cuvier and Geoffroy, the embryologist K. E. von Baer (1792–1876), and in our own century the morphologist D'Arcy W. Thompson, the alternative view holds that the various types of organisms that populate the biosphere are the virtually inevitable formations of living matter, much as the elements of the periodic table are inevitable formations of subatomic particles. A consequence of this view is that as the different biological types began to emerge, evolution's effect would have been to sharpen rather than blur biological boundaries. In this nondualistic view, the properties of flesh define a range of organic possibilities to which any evolved genetic "programs" must necessarily conform.

I have argued in this chapter that biological dualism has emerged in European-American culture in concert with a value system that gives automatic preference to the drive to manipulate the living world over a more ancient stance—what William Wordsworth called "natural piety." But far from representing a scientific rejection of obsolete concepts of nature, the dualistic view of living beings, I have suggested, is at odds with reality. The public, which pays for and bears the consequences of technological change, has been sold a view of organisms as entities lacking in self-definition that are entirely malleable and programmable. But as we have seen, the acceptance of this view (among both scientists and the lay public) has often had less to do with scientific evidence than resonance with mythic and religious traditions embedded in the culture. Unearthing the multifarious intellectual pathways that have led to the currently dominant view has revealed that the culture contains, as well, alternative theories and traditions that could foster receptivity to a new respect for and scientific understanding of carnal limitations and biological uniqueness.

Notes

1. *Alive*, Touchstone Pictures; Frank Marshall, director; 1992.
2. *The Donner Party*, Steeplechase Films; Ric Burns, director; 1992.
3. *The Silence of the Lambs*, Orion Pictures; Jonathan Demme, director; 1991.
4. Barbara Rudolph, "Unspeakable Crimes," *Time*, January 18, 1993, p. 35; Liu Bin-yan, "An Unnatural Disaster," *New York Review of Books*, April 8, 1993, pp. 3–6.
5. Tom Mathews, "Secrets of a Serial Killer," *Newsweek*, February 3, 1992, pp. 44–49, 50–51.
6. I am assuming here that there is a continuity between a culture's socially sanctioned preoccupations and its "peculiar forms of pathology" (Christopher Lasch's term). This notion is developed, for example, in books by Michel Foucault (e.g., *Madness and Civilization*, R. Howard, trans., New York: Pantheon, 1965), Christopher Lasch (e.g., *The Culture of Narcissism*, New York: Warner Books, 1979), and Susan Bordo (e.g., *Unbearable Weight: Feminism, Western Culture, and the Body*, Berkeley: University of California Press, 1993).
7. See Andrew Kimbrell, *The Human Body Shop*, San Francisco: Harper Collins, 1993.
8. See, for example, Claude Lévi-Strauss, *The Raw and the Cooked*, J. and D. Weightman, trans., University of Chicago Press, 1969.
9. Sigmund Freud, *Beyond the Pleasure Principle*, J. Strachey, trans., New York: Liveright, 1928.
10. Lévi-Strauss, *Raw and Cooked*.
11. Caroline Oates, "Metamorphosis and Lycanthropy in Franche-Compté, 1521–1643," in M. Feher, R. Naddaff, and N. Tazi, eds., *Fragments for a History of the Human Body*, part 1, New York: Zone, 1989; Adam Douglas, *The Beast Within: A History of the Werewolf*, London: Chapmans, 1992.
12. Reneé C. Fox and Judith P. Swazey, *Spare Parts: Organ Replacement in American Society*, New York: Oxford University Press, 1992.
13. C. B. Fehilly, S. M. Willadsen, and E. M. Tucker, "Interspecific Chimerism between Sheep and Goat," *Nature*, 307 (1984), pp. 634–636; S. Meinecke-Tillmann, "Experi-

mental Chimeras—Removal of Reproductive Barrier between Sheep and Goat," *Nature*, 307 (1984), pp. 637–638.

14. B. Mintz, "Clonal Expression in Allophenic Mice," *Symposium of the International Society of Cell Biology*, 9 (1970).

15. Jean Soler, "The Semiotics of Food in the Bible," in R. Forster and O. Ranum, eds., *Food and Drink in History*, Baltimore: John Hopkins University Press, 1979, pp. 126–138.

16. Soler, "Semiotics," p. 128.

17. Mary Douglas, *Purity and Danger*, London: Routledge and Kegan Paul, 1966.

18. Douglas, *Purity*, p. 53.

19. Soler, "Semiotics," p. 131.

20. Douglas, *Purity*, p. 55.

21. Soler, "Semiotics," pp. 132–133.

22. Douglas, *Purity*, p. 55.

23. Douglas, *Purity*, p. 56.

24. Douglas, *Purity*, p. 2.

25. Soler, "Semiotics," p. 126.

26. Soler, "Semiotics," p. 136.

27. Carolyn Merchant, *The Death of Nature: Women, Ecology and the Scientific Revolution*, San Francisco: Harper and Row, 1980.

28. Caroline Walker Bynum, "The Female Body and Religious Practice in the Later Middle Ages," in M. Feher, R. Naddaff, and N. Tazi, eds., *Fragments for a History of the Human Body*, part 1, New York: Zone, 1989, pp. 161–219.

29. Bynum, "Female," p. 164.

30. Bynum, "Female," p. 192.

31. Quoted in Oates, "Metamorphosis," p. 319.

32. Oates, "Metamorphosis," pp. 317–318.

33. Oates, "Metamorphosis," pp. 319–320.

34. René Descartes, *Treatise of Man*, 1662.

35. Hans Jonas, *The Phenomenon of Life*, New York: Harper and Row, 1966, p. 58.

36. Quoted in Peter Singer, *Animal Liberation*, New York: New York Review of Books, 1975, p. 220.

37. Francis Bacon, *The Advancement of Learning* and *New Atlantis*, Oxford: Oxford University Press, 1979.

38. Bacon, *New Atlantis*, p. 241.

39. Bacon, *New Atlantis*, p. 241.

40. Bacon, *New Atlantis*, p. 241.

41. Merchant, *Death*, p. 183.

42. Leonard N. Isaacs, "The Effecting of All Things Possible: Molecular Biology and Bacon's Vision," *Perspectives in Biology and Medicine*, 30 (1987), pp. 402–432.

43. Arthur O. Lovejoy, *The Great Chain of Being*, Cambridge, Mass.: Harvard University Press, 1936.

44. Terry Berger, " 'Tulipomania' Was no Dutch Treat to Gambling Burghers," *Smithsonian*, April 1977, pp. 70–77.

45. Quoted in Hans Stubbe, *History of Genetics*, T. R. W. Waters, trans., Cambridge, Mass.: MIT Press, 1972, p. 97.

46. Quoted in Stubbe, *History*, p. 97.

47. N. Hofsten, "Linnaeus's Conception of Nature," *Kungliga Vetenskaps-Societeten Arsbok* (1957), pp. 65–105; cited in Ernst Mayr, *The Growth of Biological Thought*, Cambridge, Mass.: Harvard University Press, 1982, p. 259.

48. Mayr, *Growth*, p. 259.

49. Mayr, *Growth*, p. 260.

50. Quoted in Lovejoy, *Great Chain*, p. 230.

51. Quoted in François Jacob, *The Logic of Life*, B. E. Spillman, trans., New York: Vintage, 1973, p. 67.

52. Quoted in Jacob, *Logic*, p. 80.

53. Quoted in Mayr, *Growth*, p. 261.

54. Mayr, *Growth*, p. 261.

55. Mayr, *Growth*, p. 261.

56. Jacob, *Logic*, p. 285.

57. Stubbe, *History*, p. 89.

58. B. E. Raikov, "Caspar Friedrich Wolff," *Zoologische Jahrbücher Abteilung für Systematik*, 91 (1964), p. 555; cited in Stubbe, *History*, pp. 89–90.

59. See Andreas-Holger Maehle, "Literary Responses to Animal Experimentation in Seventeenth-Century and Eighteenth-Century Britain," *Medical History*, 34 (1990), pp. 27–51.

60. Alexander Pope, "Against Barbarity to Animals," *Guardian*, May 21, 1713; cited in Maehle, "Literary," p. 36.

61. Immanuel Kant, *Critique of Judgement*, J. H. Bernard, trans., New York: Hafner, 1966, p. 222.

62. Timothy Lenoir, *The Strategy of Life*, Chicago: University of Chicago Press, 1989 (orig. pub. 1982).

63. Quoted in Lenoir, *Strategy*, p. 161.

64. Quoted in Mayr, *Growth*, p. 184.

65. Quoted in Jacob, *Logic*, p. 105.

66. J. B. Lamarck, *Zoological Philosophy*, H. Elliot., trans., Chicago: University of Chicago Press, 1984.

67. Richard W. Burkhardt, Jr., "The Zoological Philosophy of J. B. Lamarck" (introduction to Lamarck, *Zoological*, pp. xv–xxxix).

68. See Jacob, *Logic*, pp. 100–108; E. S. Russell, *Form and Function*, London: Murray, 1916; pp. 52–78.

69. Quoted in Mayr, *Growth*, p. 362.

70. Quoted in Stephen J. Gould, "The Gift of New Questions," *Natural History*, August 1993, pp. 4–13.

71. Mayr, *Growth*, pp. 104–105.

72. John F. Cornell, "God's Magnificent Law: The Bad Influence of Theistic Metaphysics on Darwin's Estimation of Natural Selection," *Journal of the History of Biology*, 20 (1987), pp. 381–412.

73. Robert J. Richards, *The Meaning of Evolution*, Chicago: University of Chicago Press, 1992.

74. Charles Darwin, *The Origin of Species by Charles Darwin: A Variorum Text*, M. Peckham, ed., Philadelphia: University of Pennsylvania Press, 1959, p. 345.

75. Stuart A. Newman, "Generic Physical Mechanisms of Tissue Morphogenesis: A Common Basis for Development and Evolution," *Journal of Evolutionary Biology*, 7 (1994), pp. 467–488.

76. Thomas Nagel, *The View from Nowhere*, London: Oxford University Press, 1986, p. 78.

77. Ronald L. Numbers, "George Frederick Wright: From Christian Darwinist to Fundamentalist," *Isis*, 79 (1988), pp. 624–645.

78. Julian S. Huxley, *New Bottles for New Wine*, New York: Harper, 1957, p. 293.

79. For critical discussions of this concept, see Ruth Hubbard, "The Theory and Practice of Genetic Reductionism—from Mendel's Laws to Genetic Engineering," in *Towards a Liberatory Biology*, S. Rose., ed., London: Allison and Busby, 1982, pp. 62–78; Susan Oyama, *The Ontogeny of Information*, Cambridge: Cambridge University Press, 1985; Stuart

A. Newman, "Idealist Biology," *Perspectives in Biology and Medicine*, 31 (1988), p. 353; and H. F. Nijhout, "Metaphors and the Role of Genes in Development," *BioEssays*, 12 (1990), pp. 441–446.

80. Robert Olby, *Origins of Mendelism*, 2nd ed., Chicago: University of Chicago Press, 1985; Marcel Blanc, "Gregor Mendel: La Légende du génie méconnu," *Recherche*, 15 (1984), pp. 46–59.

81. Peter J. Bowler, *The Eclipse of Darwinism*, Baltimore: Johns Hopkins University Press, 1983.

82. Charles Darwin, *The Variation of Animals and Plants under Domestication*, London: Murray, pp. 394–395.

83. H. Kalmus, "The Scholastic Origin of Mendel's Concepts," *History of Science*, 21 (1983), pp. 61–83.

84. R. A. Fisher, "Has Mendel's Work Been Rediscovered?" *Annals of Science*, 1 (1936), pp. 115–137.

85. L. A. Callender, "Gregor Mendel: An Opponent of Descent with Modification," *History of Science*, 26 (1988), 41–75.

86. Quoted in L. C. Dunn, *A Short History of Genetics*, New York: McGraw-Hill, 1965, p. 41.

87. Dunn, *Short*, p. 43.

88. Quoted in Mayr, *Growth*, p. 717.

89. Quoted in Dunn, *Short*, p. 93.

90. Quoted in Elof A. Carlson, *The Gene: A Critical History*, Philadelphia: Saunders, 1966, p. 37.

91. A. Mozzarelli, J. Hofrichter, and W. A. Eaton, "Delay Time of Hemoglobin-S Polymerization Prevents Most Cells from Sickling *in vivo*," *Science*, 237 (1987), pp. 500–506.

92. Newman, "Generic."

93. E. Schrödinger, *What is Life?* Cambridge: Cambridge University Press, 1945, p. 21.

94. Max Delbrück, "How Aristotle Discovered DNA," in *Physics and Our World: A Symposium in Honor of Victor F. Weisskopf*, K. Huang, ed., New York: American Institute of Physics, 1976, pp. 123–130.

95. A. Rich and S. H. Kim, "The Three-Dimensional Structure of Transfer RNA," *Scientific American*, January 1978, pp. 52–62.

96. Freeman Dyson, *Origins of Life*, Cambridge: Cambridge University Press, 1985, p. 6.

97. Richard Dawkins, *The Blind Watchmaker*, New York: Norton, 1986, p. 112.

98. Mary Midgley, *Science as Salvation*, London: Routledge, 1992, pp. 162, 147, 159.

99. Freeman Dyson, "Time without End: Physics and Biology in an Open Universe," *Reviews of Modern Physics*, 51, p. 449; quoted in Midgley, *Science*, p. 150.

100. Midgley, *Science*, p. 147.

101. Newman, "Generic."

102. Jonas, *Phenomenon*, p. 51.

103. Edward O. Wilson, *The Diversity of Life*, Cambridge, Mass.: Harvard University Press, 1992, pp. 73–74.

104. Edward O. Wilson, *Biophilia*, Cambridge, Mass.: Harvard University Press, 1984.

105. Stephen J. Gould, "Prophet for the Earth" (review of *The Diversity of Life* by E. O. Wilson), *Nature*, 361 (1993), p. 311.

106. Howard L. Kaye, *The Social Meaning of Modern Biology*, New Haven, Conn.: Yale University Press, 1986, pp. 95–135.

107. Wilson, *Diversity*, p. 345.

108. Wilson, *Diversity*, p. 302.

109. Freeman Dyson, *Disturbing the Universe*, New York: Harper and Row, 1979, p. 230.

110. Alun Anderson, "Oncomouse Released," *Nature*, 336 (1988), p. 300.

111. Leslie Roberts, "Genetic Engineers Build a Better Tomato," *Science*, 241 (1988), p. 1290.

112. Phillip E. Canuto, "Engineering a Rebirth: Genetics May Be Ohio's Hope," *Akron Beacon Journal*, January 19, 1988, pp. A1, A8–9.

113. Daniel E. Koshland, Jr., "The Future of Biological Research: What Is Possible and What is Ethical?" *MBL Science*, 3 (1988), pp. 11–15.

114. Nelson A. Wivel and LeRoy Walters, "Germ-line Gene Modification and Disease Prevention: Some Medical and Ethical Perspectives," *Science*, 262, pp. 533–538.

115. U.S. Office of Technology Assessment, *Patenting Life*, 1989, p. 14.

116. Indeed, the function of the identical gene in two different individuals of the same species may be entirely different. See Ruth Hubbard and Elijah Wald, *Exploding the Gene Myth*, Boston: Beacon, 1993, and Richard C. Strohman, "Ancient Genomes, Wise Bodies, Unhealthy People: Limits of a Genetic Paradigm in Biology and Medicine," *Perspectives in Biology and Medicine*, 37 (1993), p. 112.

12 | The Nazi Treatment of Animals and People

Arnold Arluke and Boria Sax

IT IS WELL known that the Nazis treated human beings with extreme cruelty.[1] Grisly "medical" experiments on humans have been carefully documented and analyzed,[2] as has the cold, calculated extermination of millions of people in the Holocaust.[3] Less well known are the extensive measures taken by Nazis to ensure humane care and protection of animals. Of course other societies have also exhibited a disdain for humans while also showing marked concern for animals, but the extent to which humans were brutalized and animals were idolized in Nazi Germany makes other cases pale by comparison. In short, Nazi Germany presents a particularly marked inversion of conventional morality in modern Western societies.

How could the Nazis have been so concerned about cruelty to animals while they treated people so inhumanely? While it would be easy to dismiss the apparently benevolent Nazi attitude toward animals as hypocrisy, this would be a facile way of evading an examination of the psychological, social, and cultural dynamics underlying Nazi thinking and behavior. Rather than questioning the authenticity of the motivations behind Nazi animal protection—a question that is unanswerable—it may be more useful to ask how such thinking was possible and what significance it had.

We offer three explanations for this contradiction. First, at a personal or psychological level, this behavior may not seem so contradictory because anecdotal reports and psychological assessments of many prominent Nazi political and military leaders suggest that they felt affection and regard for animals but enmity and distance toward humans. While love of animals is considered an admirable quality, under the Nazis it may have obscured brutality toward human beings on both the personal and the political level, whatever its roots were. Second, at a political level, animal protection measures, whether sincere or not, may have been a legal veil to level an attack on the Jews. In making this attack, the Nazis allied themselves with animals, since both were portrayed as victims of "oppressors" such as Jews. And third, at a cosmological level, the Nazis abolished moral distinctions between animals and people by viewing people as animals. The result was that animals could be considered "higher" than some people. All three of these explanations argue for a culture where it was possible to increase the moral status of animals and decrease the moral status of some humans by blurring the boundaries between humans and animals, making it pos-sible for National Socialists to rationalize their behavior and to disenfranchise large groups of humans.

While we assume a position of analytic detachment, this stance should not be read as excusing Nazi behavior. Our analysis of the Nazi movement has far-reaching ethical implications, but these are largely beyond the scope of this chapter, or indeed this book. We believe, in this instance, that moral concern is best channeled into understanding; indeed, a highly moralistic discussion might obscure the dynamics of the National Socialist movement.

Nazi Animal Protection

Around the end of the nineteenth century, kosher butchering and vivisec-tion were the foremost concerns of the animal protection movement in Ger-many.[4] These interests continued during the Third Reich and became formal-ized as laws. Before taking power, the Nazis had begun to prepare laws to address these issues. In 1927, a Nazi representative to the Reichstag called for measures against cruelty to animals and against kosher butchering.[5] In 1932, a ban on vivisection was proposed by the Nazi party,[6] and at the start of 1933, the Nazi representatives to the Prussian parliament met to enact this ban.[7] On April 21, 1933, almost immediately after the Nazis came to power, they passed a set of laws regulating the slaughter of animals. In August 1933, Hermann Göring announced an end to the "unbearable torture and suffering in animal experi-ments" and threatened to "commit to concentration camps those who still think they can continue to treat animals as inanimate property."[8] Göring decried the "cruel" experiments of unfeeling scientists whose animals were operated on, burned, or frozen without anesthetics. A ban on vivisection was enacted in Bavaria as well as Prussia,[9] although the Nazis then partially retreated from a full ban, which provoked some criticism within the Nazi movement.[10] The Nazi animal protection laws of November 1933 permitted experiments on animals in some circumstances but subject to a set of eight conditions and only with the explicit permission of the minister of the interior, supported by the recommen-dation of local authorities. The conditions were designed to eliminate pain and prevent unnecessary experiments. Horses, dogs, cats, and apes were singled out for special protection. Permission to experiment on animals was given not to individuals but to institutions.[11]

Inconspicuously buried in the animal protection laws of November 1933 (point four, section two) was a provision for the "mercy killing" of animals. The law not only allowed but actually required that domesticated animals which were old, sick, and worn out, for which "life has become a torment," be "pain-lessly" put to death. The wording of the provision was ambiguous; it was not entirely clear whether a family would be required to kill, say, an old dog which did nothing but sit by the fire. One binding commentary, passed immediately after the laws themselves, mandated that an expert should decide whether further life for an animal was a "torment" in unclear cases.[12]

In addition to the laws against vivisection and kosher slaughter, other legal documents regulating the treatment of animals were enacted from 1933 through 1943, probably several times the number in the previous half century.[13] These documents covered in excruciating detail a vast array of concerns, from the shoeing of horses to the use of anesthesia. One law passed in 1936 showed "particular solicitude" about the suffering of lobsters and crabs,[14] stipulating that restaurants were to kill crabs, lobsters, and other crustaceans by throwing them one at a time into rapidly boiling water.[15] Several "high officials" had debated the question of the most humane death for lobsters before this regulation was passed, and two officials in the Interior Ministry had prepared a scholarly treatise on the subject.[16]

The Nazis also sought to protect wildlife. In 1934 and 1935, the focus of Nazi legislation on animals shifted from farm animals and pets to creatures of the wild. The preface to the hunting laws of March 27, 1935, announced a eugenic purpose behind the legislation, stating, "The duty of a true hunter is not only to hunt but also to nurture and protect wild animals, in order that a more varied, stronger and healthier breed shall emerge and be preserved."[17] Nazi veterinary journals often featured reports on endangered species.[18] Göring in particular was concerned about the near extinction in Germany of the bear, bison, and wild horse and sought to establish conservation and breeding programs for dwindling species and to pass new and more uniform hunting laws and taxes.[19] His game laws are still operative today.

A uniform national hunting association was created to regulate the sport, restock lakes, tend forests, and protect dying species. Taxes levied on hunters would be used for upkeep of forests and game parks. Göring also established three nature reserves, introduced elk, and began a bison sanctuary with two pure bulls and seven hybrid cows on one of the reserves.[20] He eventually succeeded in rearing forty-seven local bison and created a game research laboratory where he reintroduced night owl, wood grouse, heathcock, gray goose, raven, beaver, and otter. Albert Speer referred to this reserve as "Göring's animal paradise."[21] Göring viewed forests almost in religious terms, calling them "God's cathedrals," and culling of game populations to prevent starvation or epidemics was conducted as a "pseudo-religious duty."[22]

It is doubtful whether there was a single, major driving force behind these laws, since the motives of the Nazis, like those of other people, were often contradictory and confusing. For some, there may have been a genuine concern for animals. After all, the objective of the laws was to minimize pain, according to one doctoral dissertation on animal protection, written in Germany primarily during the Nazi period.[23]

For others, the Nazi animal protection laws, formulated with considerable medical and legal sophistication, may have reflected bureaucratic thoroughness and an impulse toward centralization of power that would regulate all relationships. This certainly occurred in relation to animal protection groups. Thus a decree of August 11, 1938, provided for the merger of various organizations devoted to animal protection, and they were combined in a single bureaucracy under the ministry of the interior. All local organizations which resisted the merger were abolished. Membership in the umbrella organization, called Reichs-Tierschutzverein (Association for the Protection of Animals of the Reich), was confined to persons of "Germanic or related blood," and the organization was empowered to monitor and censure activities which could, in the view of the leadership, "place animal protection in jeopardy."[24]

But this attempt to control relationships seems also to have extended to human behavior toward animals. In their desire to achieve conformity and to control everything, the Nazis may have sought to require humane attitudes.[25] The purpose of the law for the protection of animals, as noted in its introduction, was "to awaken and strengthen compassion as one of the highest moral values of the German people."[26]

Another possible force behind the animal protection measures may have been changing attitudes toward crime and guilt that called for the protection of things deemed of value to the Nazi state. That animals were to be protected for their own sakes rather than for their relationship to humanity was a new legal concept.[27] By the same token, since Jews were of no value to Germany, there was no reason to have laws to protect them.[28]

According to Meyer, two additional forces may have accounted for the rapid passage of the animal protection laws.[29] Accidental circumstances, such as Hitler's personal contact with the veterinarian Dr. F. Weber, an opponent of kosher butchering, may have been a factor. A more important force, Meyer contends, was the need for the Nazis in 1933 to secure and consolidate their power by obtaining acclaim from broad segments of the population. Demonstrating their humanitarianism through the passage of animal protection laws could have been a device to improve the image of the Nazis because these laws would elicit sympathy from Germans as well as foreigners.

In many respects, the laws of November 1933 did not go far beyond the laws protecting animals in Britain, then considered the most comprehensive in the world. The severity of the punishments mandated by the German laws was, however, virtually unprecedented in modern times. "Rough mistreatment" of an animal could lead to two years in prison plus a fine.[30] It is not clear, however, how vigorously or conscientiously the animal protection laws were enforced, particularly outside of Prussia. Like virtually all legal documents, these laws contained ambiguities and possible loopholes. Barnard maintains that several experiments on animals were conducted secretly by Nazi doctors.[31] Hilberg also describes several Nazi medical experiments on animals that preceded those on human beings.[32] At any rate, Nazi Germany gradually became a state where petty theft could result in death while violent crimes might go unpunished; punishment did not fit the crime in any traditional sense. The new government retained the entire legal apparatus of the Weimar Republic but used it in the service of a different concept. In accordance with declarations by Hitler, for example, the laws of July 2, 1934, on measures for protection of the state provided that punishment was to be determined not by the crime itself but by the "fundamental idea" behind the crime.[33] Mistreatment of animals, then, might be taken by courts as evidence of a fundamentally antisocial mentality or even of Jewish blood.

The preoccupation with animal protection in Nazi Germany continued almost until the end of World War II. In 1934, the new government hosted an international conference on animal protection in Berlin. Over the speakers' podium, surrounded by enormous swastikas, were the words "entire epochs of love will be needed to repay animals for their value and service."[34] In 1936, the German Society for Animal Psychology was founded, and in 1938 animal protection was accepted as a subject to be studied in German public schools and universities. In 1943 an academic program in animal psychology was inaugurated at the Hannover School of Veterinary Medicine.[35]

The Ideological and Historical Context

Though it appeared politically monolithic, the Nazi movement contained a surprisingly wide range of intellectual opinions. The leaders, in general, showed little interest in abstract theory, and only Alfred Rosenberg even attempted to synthesize Nazism into a cohesive set of doctrines. One cannot, therefore, understand the movement as though it were centered on an abstract philosophy, searching for more formal kinds of logic and coherence. Nazism was far more a cluster of loosely associated concerns. Even leading National Socialists avoided committing themselves on the subject of ideology, emphasizing that in its totality, National Socialism was indefinable.[36]

Nevertheless, the National Socialists attempted to actualize a racial ideology and, in so doing, create a new Germanic identity.[37] The search for German national character certainly did not start during the Third Reich. The enormous anxiety and preoccupation of the Nazis over national identity and differentiation from other human groups was only a heightened version of Germany's long obsession with its identity and with its boundaries from other human groups and relationship with animals. Essential to this construction of national identity were certain themes regarding the connections of humans to nature and animal life that were articulated in German romantic poetry, music, and social thought. These ideas shaped Nazi thinking and served as intellectual resources that were drawn upon and distorted when expedient.

Man as Beast

One influential theme, particularly evident in the work of Friedrich Nietzsche, was the rejection of intellectual culture and the praising of animal instinct in humans. This view celebrated the earth and animals in mythical ways and glorified the blond-haired beast who purified the race by copulation. Possessing qualities of vitality, unscrupulousness, and obedience,[38] Nietzche's "blond beast" or "Raubtier" (beast of prey) played up the animal origin and character of humans.[39]

Nietzsche was one of several heroes under Nazism whose work was distorted to become more brutal and aggressive, particularly his conception of the "blond beast." Glaser calls this element of National Socialism "man as predator." Glaser writes: "The domestic animal who had been domesticated on the surface only was in the end superior to and more honest than man; in the predator one could 'rediscover his instincts and with that his honesty.' "[40] Animal instinct came to represent rebellion against culture and intellectualism. Returning to the animal nature within humans, communing with nature, and elevating animal life to the level of cult worship were seen as alternatives to modernity, technology, and urbanization, according to Glaser. Acceptance of this view, it was thought, would lead to spiritual and ideological changes necessary and desirable for a new national self-identity to emerge in German cultural life.[41]

Indeed, many Nazi leaders called for a return to the pre-Christian tribal mentality of barbarian hunters who worshiped nature and held animals in awe. The barbarian, according to advocates of this view such as Hermann Göring, was much closer to the stag, elk, and wild boar than to the financier or the teacher.[42] As Göring proclaimed, "Yes we are barbarians, and we think with our blood."

As part of the rejection of culture, the new German, according to National Socialist ideology, was to disavow humanitarian behavior toward fellow humans as insincere. One element of this totalitarian system was the principle of contempt for certain human beings. Himmler, for example, called for renouncing "softness."[43] "False" comradeship and compassion were derogated. Instead of encouraging compassion, Hitler emphasized that the new German should emulate certain animal behaviors such as the obedience and faithfulness of pets and the strength, fearlessness, aggressiveness, even cruelty supposedly found in beasts of prey, qualities that were among the movement's most stringent principles.[44]

The training of SS personnel illustrated the importance of these animal qualities, even if it ironically meant killing animals. According to Radde, an oral historian of Nazi Germany, after twelve weeks of working closely with a German shepherd, each SS soldier had to break his dog's neck in front of an officer in order to earn his stripes.[45] Doing so, it was thought, would instill teamwork, discipline, and obedience to the Führer—qualities that were deemed more important than feelings for anything, including animals.

Hitler himself pleaded for these qualities in German youth: "I want violent, imperious, fearless, cruel young people. . . . The free, magnificent beast of prey must once again flash from their eyes. I want youth strong and beautiful . . . , and athletic youth. . . . In this way I shall blot out thousands of years of human domestication. I shall have the pure, noble stuff of nature."[46] In another instance, Hitler called for German youth to be as "swift as whippets."[47] The new Germans were to be part animal while renouncing a side of their humanity. The compassion normally reserved for humans was to be redirected toward animals, and the supposed cold aggressiveness of animal instinct was to become the model German. Animals were to be identified with, and compassion toward animals rather than humans was to be encouraged, if not required. This was, in fact, part of the intent of the animal protection laws.

Animals as Moral Beings

A second theme was that animals were to be regarded as moral if not sacred beings. For example, the German zoologist Ernst Haeckel, writing at the turn of the century, attacked religion, primarily Christianity, for putting humans above animals and nature and for isolating humans from nature and creating contempt for animals. He believed that humans and animals had the same natural as well as moral status and that much of human morality stemmed from animals, claiming that Christian moral principles such as "do unto others as you would have them do unto you" were "derived from our animal ancestors."[48]

In Haeckel's view, animals were to be learned from, using the laws of nature as a way to reform human society. The function of human societies, like animal societies, was survival, and biological fitness was essential to both. Not surprisingly, Haeckel supported "racial hygiene" through euthanasia. He deduced the ideal state from his observations of animals and nature, maintaining that the most efficient organization to ensure survival among animals was to be highly centralized and hierarchical, like the brain and nervous system, and therefore human society should adopt it too. In his analyses, Haeckel stressed "duty" as essential to the success of an ideal society; duty, he claimed, was a biological impulse shared with all animals in that they were bound to care for family and the larger collectivity because both were necessary for survival.[49]

Haeckel's views informed the Nazi outlook toward animals as moral beings. Religion was blamed for the attitude that saw the lower moral status of animals as justifying their use by humans. Thus Germans had a humanitarian

duty to animals because there was little difference between humans and animals, as noted in the following Nazi propaganda statement:

> Most Germans were brought up with a notion that God created animals to
> benefit humans. We know only a few clergymen who represent this attitude.
> In general, they merely intend to make the difference between humans and
> the soulless animal appear as large as possible. Every friend of animals knows
> the extent of mutual understanding between humans and animals and the
> feeling of community that can develop.[50]

The Nazis called for redressing wrongs to animals; humans were to have a
regard for nature as a moral duty. Goebbels commented in his diaries:

> Man should not feel so superior to animals. He has no reason to. Man believes
> that he alone has intelligence, a soul, and the power of speech. Has not the
> animal these things? Just because we, with our dull senses, cannot recognize
> them, it does not prove that they are not there.[51]

The moral status of animals was to be changed in the coming German empire;
they were to be sentient beings accorded love and respect as a sacred and essential element in man's relationship with nature. For example, toward the end of
the war, the editors of a book on legal protection of animals proclaimed, "Animals are not, as before [the Nazi period], objects of personal property or unprotected creatures, with which a man may do as he pleases, but pieces of living
nature which demand respect and compassion." Looking to the future, they
quoted the words of Göring that "for the protection of animals, the education
of humanity is more important than laws."[52]

Society was not to violate animals by killing them, either for sport or for
food. The vision of the future included a world where animals would not be
unnecessarily harmed, holding out as role models various groups that were
seen as respectful toward animal life. Hunting became a symbol of the civilization left behind; meat eating became a symbol of the decay of other civilizations;
and vegetarianism became a symbol of the new, pure civilization that was to be
Germany's future. Hunting was seen as appropriate to an earlier stage of humanity when killing animals involved some "risk" to the hunter. Now, only
"sick" animals and those needed for food should be killed. When animals were
to be killed for food, they were given a "sacred" status and their death was seen
as a form of "sacrifice." This spiritual attitude toward animals, even when they
must be killed, can be seen in Nazi farm propaganda:

> The Nordic peoples accord the pig the highest possible honor . . . in the cult of
> the Germans the pig occupies the first place and is the first among the domes-
> tic animals. . . . The predominance of the pig, the sacred animal destined to
> sacrifices among the Nordic peoples, has drawn its originality from the great
> trees of the German forest. The Semites do not understand the pig, they do not

accept the pig, they reject the pig, whereas this animal occupies the first place in the cult of the Nordic peoples.[53]

Such a claim is rather ironic, since anti-Semitic propaganda from the late Middle Ages had depicted Jews in positions of intimacy with pigs, sometimes suckling from a sow.[54] The Hebraic people were sometimes referred to as "pig-Jews," or "Saujuden." On one occasion, the Nazis publicly humiliated a prostitute by forcing her to wear a sign with the words "I am the biggest pig in the neighborhood and only associate with Jews."[55]

Unity of Nature and Humans

A third theme, particularly expressed by thinkers such as the composer Richard Wagner, exalted synthesis against analysis, unity and wholeness against disintegration and atomism, and *Volk* legend against scientific truth.[56] Life, according to this view, had an organic unity and connectedness that should not be destroyed by theoretical analysis or physical dissection. Science, and the Jews behind it, were portrayed as a destructive intellectualism that treated nature and animals mechanically by dissolving the whole into parts, thereby losing the invisible force which makes the whole more than the sum of its parts.

Particularly influential was Wagner's thinking. Wagner had urged smashing laboratories and removing scientists and "vivisectors." The vivisector, to Wagner, came to represent both the scientists' "torture" of animals and the capitalists' torture of the proletariat. Wagner also portrayed the vivisector as both evil and Jewish, but he was not alone in this. In *Gemma, oder Tugend und Laster*, a sentimental novel of the 1870s that did much to arouse public sentiment against animal experimentation, the author portrayed the vivisectionists as cultists who, under the pretense of practicing science, ritualistically cut up living animals in orgiastic rites.[57] The author may not have intended to identify the vivisectionists in the novel with the Jews (membership in the cult of vivisectionists was a matter of volition rather than heredity), but the representation of vivisectionists in the book was so close to the popular stereotype of Jews engaged in kosher butchering that it was inevitable that many people would make the connection.

Such views helped to shape the Third Reich's criticisms of scientific thinking and practices such as vivisection. The path of Western civilization had taken an incorrect turn, according to National Socialism. Mechanistic, exploitative technology, attributed to the Jews and to "science," was seen as cutting humans off from their connections with nature and ultimately with their own spirit. It became important to portray German leaders as close to nature and having values compatible with a simple agricultural way of life; the soil was seen as the source of life and inspiration. Old Germans, Himmler argued, were nature worshipers, and so too should be new Germans, whom he tried to sell on the nobility and virtues of farm life.[58] Companionship with dogs, in particular, provided a link between the soil and humanity. A great deal was written about Hitler's fondness for dogs during the 1930s and 1940s, and many pictures were taken to prove it was so as part of a propaganda compaign to demonstrate Hitler's "modesty and simplicity," which according to Langer were key values behind rural glorification.[59] One example of such a propaganda photo shows Hitler and "two friends" (two dogs),[60] and another shows Hitler relaxing with a dog.[61]

Biological Purity

A fourth theme involved Nordic racism and the biological purity of Aryans. The human race, it was argued, had become contaminated and impure through a mixing of the races and the eating of animal flesh. "Regeneration of the human race" was linked to animal protection and vegetarianism.[62] Wagner's principal concern was with the notion of biological purification of Germany and its political future. He wrote that "present-day socialism must combine in true and hearty fellowship with the vegetarians, the protectors of animals, and the friends of temperance" to save mankind from Jewish aggression.[63] Viereck refers to this "fellowship" as Wagner's "united front of purifiers" who could oppose the antivegetarian stance of Jews.[64] According to Wagner, "the Jewish God found Abel's fatted lamb more savoury than Cain's offer" of a vegetable.

In an essay first published in 1881 entitled "Heldentum und Christenheit" (Heroism and Christianity), Wagner articulated an anti-Semitic theory of history which linked vegetarianism to Germany's future.[65] It drew on the racial theories of Arthur Gobineau, the philosophy of Schopenhauer, and Wagner's own idiosyncratic brand of Catholicism. In abandoning their original vegetarian diet, Wagner believed, people had become corrupted by the blood of slaughtered animals. This degeneration was then spread by the mixing of races. Interbreeding eventually spread through the entire Roman Empire, until only the "noble" Germanic race remained pure. After their conquest of Rome, the Germans, however, finally succumbed by mating with the subject peoples. "Regeneration" could be achieved, even by highly corrupted races such as the Jews, through a return to natural foods, provided this was accompanied by partaking of the Eucharist. Wagner also believed that one could not live without "animal food" in the northern climates, so he suggested that in the future there would be a German migration to warmer climates where it would not be necessary to eat animals, thereby permitting Europe to return to pristine jungle and wild beasts.[66]

Racial contamination, some argued, had mixed biologically inferior human stock with Aryan blood, thereby threatening the purity of the highest species. The physician Ludwig Woltmann, for example, described the Germans as the highest species because of their perfect physical proportions and their height-ened spirituality.[67] He argued that life was a constant struggle against the bio-logical decay of this highest species. This biological struggle was waged against the subhuman, a notion that can be linked to an intellectual undercurrent in the German movement known as the neo-Manichaean gnosis, a third-century cos-mology given a secular form by a defrocked Viennese monk at the beginning of the twentieth century.

The former monk, Adolf Lanz, who adopted the name Baron von Lieben-fels, published a book called *Theozoology*, which claimed that in the beginning there were two races, the Aryans and the Apes, whom Lanz called the "animal people." The Aryans were pure and good, while the animal people represented darkness and sought to sexually defile Aryans. Because of such interbreeding, the original Aryans and animal people no longer existed, but Lanz claimed that one could still distinguish and rank races according to the proportion of Aryan or ape blood they possessed. Thus Nordic people were close to pure Aryan and their racial rank was the highest, while Jews ranked the lowest because they were close to pure ape.[68]

Lanz published a pulp magazine entitled *Ostara, Briefbücher der Blonden Männesrechtler* (Newsletters of the Blond Fighters for Male Rights), in which all of history was explained in lurid terms as a struggle between the noble Aryan and the hairy subhuman, as the latter continually sought to contaminate the "higher" race by leading blond beauties astray. He founded an organization called the New Templars, which adopted the swastika as its banner. As a young man, Hitler read *Ostara* and may have met several times with Lanz, who called the future dictator "one of our pupils."[69] Similar ideas appeared in the writings of Wagner, who maintained in "Heldentum" that the Semitic races had always viewed themselves as descended from the apes, while the Aryan races traced their descent "from the Gods."[70]

The Nazis, in many ways, departed from the anthropocentric under-standing of the cosmos that has dominated Occidental civilization since at least the late Middle Ages. Their world was not so much centered on humans, at least as presently constituted, as on the process of evolution, conceived as a process of perpetual improvement through "survival of the fittest." This process, how-ever, was not viewed as a spontaneous process but as something that, in the con-temporary world, sometimes required assistance.[71] In other words, it became a project to biologically perfect what it meant to be German—a task not unlike that taken with German shepherd dogs, who were deliberately bred to represent and embody the spirit of National Socialism. Van Stephanitz, the creator of this breed, sought national status for a local population of coyotelike dogs in the 1920s that were to be regarded as racially better dogs, analogous to better bred humans, and whose only reason to exist was to go to war on the day hostilities began.[72]

One influential theorist of evolutionary progress in the Nazi party was Konrad Lorenz, who viewed the human race as having reached a turning point between extinction and development to a "higher" organism. In an essay published in 1940, Lorenz wrote that "the great decision of today depends on whether we learn to fight the degeneration, in our people and in the entire human race, resulting from a lack of natural selection, before it is too late. Precisely in this decision of survival or extinction we Germans are very far ahead of all other cultural people."[73] Today Lorenz is generally remembered as the benign, grandfatherly founder of ethology. Some of the Nazi legacy, passed on through him, probably survives in the work of contemporary students of animal behavior, who often interpret interaction among animals in very strict hierarchical terms.

Central to National Socialist ideology was the quest for racial purity by creating a "super race" and eliminating "inferior races." Indeed, laws passed under the Third Reich to improve the eugenic stock of animals anticipated the way in which Germans and non-Aryans were treated eugenically. Germans were to be treated as farm animals, bred for the most desirable Aryan traits while ridding themselves of weaker and less desirable specimens. Such remodeling of civilization was not to flout the "natural order," meaning that distinctions between humans, animals, and the larger "natural" world were not to make up the basic structure of life. Rather, the fundamental distinction made during the Third Reich was between that which was regarded as "racially" pure and that which was polluting and dangerous. The former was embodied in the Aryan people and nature, the latter in other humans who were synonymous with "lower" animals.

According to Hitler's own fanciful anthropology, non-Aryans were subhuman and should be considered lower than domestic animals. He stated in *Mein Kampf* that slavery came before the domestication of animals. The Aryans supposedly subjugated the "lower races": "First the vanquished drew the plough, only later the horse."[74] This, in Hitler's imagination, was the "paradise" which the Aryans eventually lost through the "original sin" of mating with the conquered people. The notion of race, as used by the Nazis, in many ways assumed the symbolic significance usually associated with species; the new phylogenetic hierarchy could locate certain "races" below animals. The Germans were the highest "species," above all other life; most "higher" animals, however, were above other "races" or "subhumans."

Understanding the Contradiction

Psychological Explanations

In trying to understand the pairing of apparent concern for animals with cruelty toward humans, we would be remiss to ignore the possibility that these attitudes can coexist in the same person. Nazis who advocated humane measures for animals and inhumane acts against humans may have had genuine interest in and affection for the former while simultaneously feeling indifference and hatred for the latter.

On the one hand, this explanation can be questioned because reports of Nazi compassion for animals are based on historical anecdotes or personal diaries and memoranda that may have been circulated or written to create a sympathetic image of Nazi leaders as warm and humane people or as having values consistent with the National Socialist movement. Bromberg and Small, for instance, contend that Hitler's compassion for animals was no more sincere than his interest in children; both were mere propaganda ploys. He supposedly once shot and killed a dog without reason.[75] Indeed, the following portrayal of Hitler, which appeared in the magazine *Neugeist/Die Weisse*, certainly smacks of propaganda:

> Do you know that your Führer is a vegetarian, and that he does not eat meat because of his general attitude toward life and his love for the world of animals? Do you know that your Führer is an exemplary friend of animals, and even as a chancellor, he is not separated from the animals he has kept for years? . . . The Führer is an ardent opponent of any torture of animals, in particular vivisection, and has declared to terminate those conditions . . . thus fulfulling his role as the savior of animals from continuous and nameless torments and pain.[76]

In addition, Meyer argues that prominent Nazis cared about animals when they were personally close to them but not when they were anonymous masses.[77] As an example, he points to Hitler's order to kill 30,000 horses at Krim to prevent them from falling into Russian hands.[78] On the other hand, there are reasons why these accounts should be taken seriously. Many of these reports were made before the Nazi rise to power in the 1930s when they were less likely to have served as propaganda, and some were made after the war by psychiatrists conducting assessments of prisoners of war. In addition, these sympathetic attitudes toward animals were consistent with the larger cultural and historical context of German thinking discussed above.

Numerous anecdotal reports suggest that Hitler enjoyed the company of dogs as companion animals.[79] Dogs, throughout much of his life, were Hitler's closest attachments.[80] For example, Hitler's landlady observed that a large dog named Wolf was his constant companion. Toland claims that Hitler "had a need for the faithfulness he found in dogs, and had a unique understanding of them," commenting once that some dogs "are so intelligent that it's agonizing."[81] According to Padfield, Hitler frequently praised his wolfhound Blondi's wholehearted devotion to him, while expressing doubts about the complete loyalty of his staff.[82] At the end of the war, Hitler came to depend on the companionship not only of Eva Braun but also of Blondi and her pups.[83] During his final days in the bomb shelter, Hitler permitted no one but himself to touch or feed Blondi's pup Wolf,[84] and he risked his life every day by taking Blondi for a walk outside his bunker.[85] When it came time for Hitler and others to commit suicide, he could not bring himself to give Blondi the poison or watch her die.[86]

Hitler was not the only prominent Nazi to keep pets. Göring was unusually fond of and dedicated to several pet adult lions kept at his estate. According to Irving, the chief forester, Ulrich Scherping, claimed that those who saw Göring with his lions could sense the fondness that they had for each other.[87] Goebbels, Hess, Höss, and several other elite Nazis had pet dogs. A typical case of such affection was that of Klaus Dönitz, admiral of the German navy, who was known to have a deep love for dogs. When he would return home, his first greeting was always for the family dog.[88] Later he had another pet dog named Wolf, whom "he loved dearly." He remarked more than once that "there is nothing in the world more faithful than a dog. He believes his master unconditionally. What he does is right."[89] Dönitz also expressed concern for the protection of stray dogs: "I think I shall start a kindergarten when I get out, a mixed one for puppies as well as children."[90] He did not, however, ever create such an orphanage. Padfield suggests that Dönitz may have simply fallen under the influence of Hitler, who emphasized the virtues of obedience in animals, or conversely had doubts about the correctness of the path he was following. Or perhaps he, like Hitler, had doubts about the complete loyalty of his staff.

When it came to hunting, the only advocate was Göring, and even in this case, as Bookbinder notes, "The stag that Göring could hunt, who would face him in an honest pitting of wits, strength, and skill in a contest Göring knew he would win, was a creature that Göring preferred to his fellow human beings."[91] Other Nazis, including Hitler, showed little interest in or staunchly opposed hunting. Hitler was known to have a strong distaste for it. Toland recounts that once, when dinner conversation turned to hunting, Hitler commented: "I can't see what there is in shooting, you go out armed with a highly perfected modern weapon and without risk to yourself kill a defenseless animal."[92] Hitler frequently criticized hunting as a "remnant" of a "dead feudal world": "How can a person be excited about such a thing? Killing animals, if it must be done, is the butcher's business. . . . I understand, of course, that there must be professional hunters to shoot sick animals. If only there were still some danger connected with hunting, as in the days when men used spears for killing game."[93] Himmler also opposed hunting. He had a "postively hysterical opposition to hunting"[94] and viewed it as "pure murder" of the "innocent."[95]

Vegetarianism was practiced by many leading Nazis. Hitler hired a vegetarian cook[96] and became very critical of others who were not vegetarian, sometimes referring to beef broth as "corpse tea" and sausage as "cadavers."[97] The vegetarianism of other Germans was a fad spawned by Hitler's preferences.[98] Some became even more fanatic about it than Hitler. Rudolf Hess was so worried about the food he ate with Hitler in the Chancellery that he would bring his own vegetarian food in containers, defending his practice by saying that his food had to contain "biologically dynamic ingredients."[99]

Hitler, following Wagner, viewed meat eating as contaminating because of the mixture of animal and Aryan blood and attributed much of the decay of civilization to meat eating.[100] Among the many ideas that the dictator adopted from the composer was a belief that civilization could be regenerated through vegetarianism.[101] Several entries in Goebbels's diaries underscore the notion that vegetarianism symbolized to Nazis a higher state of humanity to which they aspired. In one entry, Goebbels observed that "animals that live on plants have much greater powers of resistance than those that feed on meat."[102] In another entry, Goebbels noted: "Meat-eating is a perversion of our human nature. When we reach a higher level of civilisation, we shall doubtless overcome it."[103]

Identification with animals by Nazis was often paired with their contempt for humanity. The most common argument is that intimate relationships with other humans were difficult for these individuals to sustain because they were socially marginal or alienated. Not surprisingly, such individuals may have found it easier to develop relationships with animals.[104] Lifton suggests that caring for animals may have had more complex psychological roots by serving as a coping device that allowed Nazis to "double," seeing themselves as humane while behaving insensitively or cruelly toward humans.[105] For whatever motivation, key members of the German general staff may have personally identified with animals while having contempt for humanity.

Most notably, Hitler is often depicted as having contempt and fear of humans but compassion and warmth for animals. Toland notes that it became known in the Third Reich that Hitler had a deep affection "for all dumb creatures," but very little for men and women. "It was as though since the Viennese days he had turned away from the human race, which had failed to live up to his expectations and was therefore damned. At the heart of the mystery of Hitler was his fear and contempt of people."[106] Similarly, Payne observes that Hitler felt closer to and more compassion for certain animals than people, when it came to their suffering.[107] Payne reports that a German pilot recalled that "Hitler saw films given to him by a friendly Maharaja. During the scenes showing men savagely torn to pieces by animals, he remained calm and alert. When the films showed animals being hunted, he would cover his eyes with his hands and asked to be told when it was all over. Whenever he saw a wounded animal, he wept."[108] He hated people who engaged in blood sports, and several times he said it would give him the greatest pleasure to murder anyone who killed an animal.

Similarly, while Goebbels's attitude toward humans was contemptuous, his expressed attitude toward his pet dog was loving. His diary entries, especially those written in the mid-1920s, were explicit about this split in feelings. Goebbels revealed:

> As soon as I am with a person for three days, I don't like him any longer; and if I am with him for a whole week, I hate him like the plague. . . . I have learned to despise the human being from the bottom of my soul. He makes me sick in my stomach. Phoeey! Much dirt [gossip] and many intrigues. The human being is a canaille [riff raff, but also pack of dogs]. . . . The only real friend one has in the end is the dog. . . . The more I get to know the human species, the more I care for my Benno [his pet dog].[109]

Hitler and Goebbels were not the only members of the German Nazi elite to identify with animals, express compassion for them, and praise traits in them such as obedience and aggressiveness while simultaneously showing contempt for humanity. Rudolf Hess, for instance, had a pet wolfhound named Hasso and was quite pleased when Hasso gave birth to three puppies, named Nurmi, Hedda, and Nickl.[110] Höss, the commander of Auschwitz, was a "great lover" of animals, particularly horses. After a hard day of work at the camp, he "found relief walking through the stables at night."[111] Eduard Wirth, a prominent physician at Auschwitz, had three pet dogs at one point. When two became ill, he referred to one of his rooms as their "sick ward." When his favorite dog died, he wrote sadly to his wife of its death, noting that the dog "suffered a lot so I gave him morphine. . . . It is good that he dies; he was in the end blind in both eyes."[112]

Psychological assessments of the personalities of a number of leading Nazi political figures also show evidence of distancing from humans and interest in animals. In one study, Rorschach tests were administered to Nazi prisoners of war.[113] Results indicated several departures from "normal" test findings, with subjects seeing themselves as animals or subhuman in the Rorschach more often than controls did. Half the subjects depicted themselves, or aspects of themselves, as animals (typically unevolved, low-level bugs, beetles, or insects); six of the subjects also offered self-portraits of themselves as subhuman or inhuman figues such as gremlins. Miale and Selzer contend that the respondents' animal responses had a "lack of vitality" indicating that this group was "cut off from their vital impulses and were unable to be free and spontaneous. Their anti-social attitudes were not expressions of normal impulses, but rather of the repression and distortion of these impulses.[114] In short, the findings suggested that, on the whole, these men had an "incapacity to feel human feelings."[115] Dicks's research also found those Nazis studied to be "affectionless and lacking deep positive relations to human figures."[116]

Political Explanations

National Socialist propaganda often portrayed Germany as a woman at one with nature but exploited and oppressed by demonic Bolsheviks, capitalists, and Jews.[117] These victimizers were seen as endangering the purity of the German "blood" and "spirit." Animals, too, were being victimized by these oppressors, whether by slaughtering them according to kosher law or by using them as subjects in scientific experiments. Metaphorically, only a subtle difference separated the animal from the German victim in this struggle. The hated "vivisectors" became synonymous with the Jews, enemies of both animals and Germans.

Animal protection measures may have served as a legal vehicle to express these anti-Semitic feelings.[118] This is not to argue that all of these measures were part of a well-defined and orchestrated master plan. But as German law was recast to embody Nazi values, it is reasonable to think that some animal protection measures might have expressed animosity toward certain human groups. And as a totality, these animal protection efforts, at the least, provided an opportunity for attacking various segments of society.[119]

Clearly, the banning of kosher slaughter was a deliberate action to isolate the "Jewish disease" through legal means.[120] Laws passed by the Nazis on April 21, 1933, to regulate butchering constituted a barely concealed attack on the Jews, whose "ritualistic slaughter" was characterized as "torment of animals." The preamble to the laws stated:

> The animal protection movement, strongly promoted by the National Socialist government, has long demanded that animals be given anesthesia before being killed. The overwhelming majority of the German people have long condemned killing without anesthesia, a practice universal among Jews though not confined to them, . . . as against the cultivated sensitivities of our society.[121]

The discussion that followed contained further references to the horrors allegedly found in kosher butcher shops.[122]

The German movement against animal experimentation was also, from its inception, strongly associated with anti-Semitism.[123] In a decree issued on August 17, 1933, Göring, then chairman of the Prussian ministry, proclaimed that people "foreign" or "alien" to Germany viewed the animal as "a dead thing under the law." He declared:

> I . . . will commit to concentration camps those who still think they can continue to treat animals as inanimate property. . . . The fairy tales and sagas of the Nordic people, especially the German people, show the spirit of close contact, which all Aryan people possess, with the animals. It is the more incomprehensible, therefore, that justice, up to now, did not agree with the spirit of the people on this point as it did on many others. Under the influence of for-

eign [i.e., Jewish] conceptions of justice . . . the animal was considered a dead thing under the law.[124]

The statement is particularly noteworthy, since the very existence of concentration camps was generally not acknowledged at the time.

Nazi ideologues sought to link the history of Judaism to vivisection. The revelation of Abraham and Moses was understood as the dominant tradition of the Occident, which culminated in the industrial revolution and the human domination of nature.[125] The word *vivisection* (the same in German as in English) was often used broadly to refer to all dispassionate dissection and analysis. Judaism, in both actual and symbolic ways, was understood as the tradition of "vivisection."[126] Nazi racial theorists regularly contrasted the supposedly cold, analytic mentality of the Jew with that of "Nordic man," who, they claimed, understood things organically as part of the natural world.[127]

The anti-Semitism of the Nazis was a very radical form of an idea that is still familiar: that Jews and, by association, Christians had scorned the natural world. Some of the Nazis, such as chief ideologist Alfred Rosenberg, rejected Christianity as a sect of Judaism, while others tried to purify Christianity of its Jewish heritage.[128] As a result, the distinction between Christianity and paganism in Nazi Germany grew increasingly unclear.[129]

The link between animal protection and anti-Semitism is paradoxical, since the Old Testament celebrates animals with great passion and eloquence. Nevertheless, such an association may go back very far. In the fourteenth century, Geoffrey Chaucer satirized it in his *Canterbury Tales*. When the prioress is introduced, we are told how well she fed her hounds and how she would weep at the sight of a mouse caught in a trap. But this same prioress uses her tale for a furious attack on Jews, accusing them of ritual murder of children.[130] More recently, in the mid–nineteenth century, philosopher Arthur Schopenhauer held that Jewish traditions were responsible for a view of animals as objects.[131]

The key figure in promoting this association, however, was Wagner. Long after he died, his writing continued to have considerable impact on German thought. He dramatized his ideas respecting race and animal protection in the opera *Parzifal*, and his prose sometimes contained imagery featuring blood of a sort that was constantly used in the rhetoric of Hitler and his followers.[132] In a letter of August 1879 to Ernst von Weber, the founder of the Dresden Animal Protection Society and author of the influential *Die Folterkammern der Wissenschaft* (The Torture Chambers of Science), Wagner stated:

> One must begin by drawing people's attention to animals and reminding them of the Brahman's great saying "Tat twam asi" ["Thou art that"]—even though it will be difficult to make it acceptable to the modern world of Old Testament Judaization [the spread of Jewish blood and influence]. However, a start must be made here—since the commandment to love thy neighbor is be-

coming more and more questionable and difficult to observe—particularly in the face of our vivisectionist friends.[133]

Like Göring and others who would come later, Wagner identified vivisectionists with Jews.

A much expanded version of the above letter was published under the title "Offenes Schreiben an Ernst von Weber" (Open Letter to Ernst von Weber) and dated October 1879. The revision was even more emotional in tone. Wagner supported breaking into laboratories where experiments on animals were conducted, as well as physical attacks on vivisectionists. He closed with the melodramatic declaration that should the campaign against vivisection prove unsuccessful, he would gladly depart from a world in which "no dog would any longer wish to live . . . even if no 'Requiem for Germany' is played after us."[134] With Wagner's public and financial support and von Weber's skillful leadership, the Dresden Animal Protection Society soon became the center of the German antivivisection movement.[135]

As illustrated by the quotation from Wagner's original letter, in the anti-Semitic rhetoric in Germany persecution of Jews was sometimes put forward as revenge on behalf of aggrieved animals. Jews were identified as enemies of animals and implicitly of Germans. In Wagner's outrage against the use of frogs in experiments, he explicitly identified "vivisectors" as "enemies." Vivisection of frogs was "the curse of our civilization." Wagner urged the *Volk* to rid itself of scientists and rescue the frog martyrs. Viereck maintains that Wagner created "a sort of moral Armageddon" between those "who free trussed animals" and those "who truss them to torture them." Those who fail to untruss frogs were "enemies of the state."[136]

After the death of Wagner in 1883, his followers, such as the brothers Bernard and Paul Förstner, continued the anti-Semitic campaign against vivisection. The latter became editor of *Thier-und Menschenfreund* (Friend to Animals and Humans), the journal of the Dresden Animal Protection Society. Wagner's admirers in the twentieth century included such spokesmen for anti-Semitism as Houston Stewart Chamberlain, Alfred Rosenberg, and, most significantly, Adolf Hitler.[137]

Another close associate of von Weber who added prestige to the movement against vivisection was Friedrich Zöllner, a famous though controversial professor of astrophysics.[138] In a popular book entitled *Über den wissenschaftlichen Missbrauch der Vivisection* (On the Scientific Misuse of Vivisection), first published in 1880, Zöllner launched an attack against physiologists. For example, he attacked a Jewish zoologist named Semper, accusing him of showing gross insensitivity (a "thick skin" like that of an elephant) by hunting the birds that attacked his botanical gardens, with the following sarcastic remarks:

One would be justified in describing the anti-Semitic movement not as "persecution of Jews," but metaphorically as a "hunt for elephants." Because surely Professor Semper would recognize a right to hunt not only thrushes but also elephants if they broke into his garden, the German people have the same right to hunt overeducated, Semitic "elephants" as Semper does to hunt the thrushes.[139]

The reversal of roles between hunter and animal is an old motif that appears frequently in literature against misuse of animals.[140]

Although Zöllner did not unequivocally advocate physical attacks on Jews, this passage is an anticipation of the Nazi persecutions. Despite what the quotation suggests, Zöllner seems to have been far less a vicious man than a complaisant one. Confident that concern for animals proved his moral superiority, he could, elsewhere in his book, content himself with the most abstract expressions of compassion for the Jews. Many of his attitudes were later adopted by Nazi doctors who attempted to purify medicine of "Jewish" influence.[141]

Cosmological Explanations

While the Nazis tended to deride mathematics and physics as "Jewish" or "mechanistic," they granted a virtually unprecedented prestige to biology and medicine. These latter fields, in turn, acquired the sort of mystical overtones that are frequently associated with the most abstract disciplines. National Socialists looked to biology for ultimate answers, not only to questions of cause and effect but to ethical ones as well.

This medicalization of moral and aesthetic issues, far from being an aberration, had a basis in traditions going back for centuries and including some of the most illustrious names in German culture. It may be found, for example, in Goethe's famous remark to Eckermann that classicism is the art of health while romanticism is that of disease. The notion was made still more explicit in Schiller's essay "On Naive and Sentimental Poetry." Later thinkers such as Nietzsche and Thomas Mann broadened the concepts of "disease" and "health" until these terms appeared more metaphysical than scientific.

Guidelines for high school teachers during the Nazi period quoted the statement of Hans Schlemm that "National Socialism is politically applied biology."[142] Konrad Lorenz observed that "the most deeply committed and passionate National Socialists" were those who had most thoroughly internalized evolutionary theory.[143] Lifton has characterized Nazi Germany as a "biocracy," a state in which biological theory was elevated to the status of a religion.[144]

Biological distinctions among people assumed, in consequence, a nearly absolute importance. In accord with a philosophy which equated biology with destiny, institutions such as schools were strictly divided along lines of gender.[145] Homosexuality, since it suggested an ambiguous area between the two genders, was strictly forbidden by the Nazis. Nazi society, in other words, was intended to be strictly compartmentalized along biological lines. The actual practice, however, often contradicted this ideal, as official art featured homoerotic themes[146] and leaders such as Hitler appealed to values derided as "feminine."[147]

While stressing the biological distinctions among types of human beings, the Nazis saw human life as part of the larger biological order that they sought to create. As part of this order, all humans, including Germans, were treated as animals. Germans were regarded as livestock to breed the purest biological forms; non-Aryans were viewed as pests that could contaminate the racial purity so important to National Socialist aims. Such treatment of humans as animals was another reason why the combination of animal protection measures with cruelty toward humans may not have seemed paradoxical to Germans. By animalizing human life, moral distinctions between people and animals were obliterated, making it possible to treat animals as considerately as humans and humans as poorly as animals.

In *Mythos*, a book intended to have virtually scriptural authority within the Nazi movement, Alfred Rosenberg found it extremely ironic that more concern was shown about the racial pedigree of horses and donkeys than about that of human beings.[148] To correct this, the National Socialists treated Germans themselves, in the most literal sense, as animals. Just as the breeding stock of "less pure" animals had been improved, so too was the "pure blood" of Germans to be restored. According to Darré, "As we have restored our old Hanoverian horse from less pure male and female animals by selective breeding, we will also, in the course of generations, again selectively breed the pure type of the nordic German from the finest German bloodlines."[149]

Several leading Germans used their experience in farming, as well as their training in agriculture and veterinary medicine, to pursue this goal. For example, Martin Borman, often considered second in rank only to Hitler, had been an agricultural student and, in 1920, became the manager of a large farm.[150] The rector of the University of Berlin in the mid-1930s was by profession a veterinarian. He instituted twenty-five new courses in *Rassenkunde*—racial science—and by the time he finished rewriting the curriculum had instituted eighty-six veterinary courses applied to humans.[151] And for a period in the 1920s, Himmler was a chicken breeder.[152] Thus veterinary medicine and agricultural science became the means of teaching racial doctrine in German universities.[153] Indeed, National Socialism viewed Europe, including Germany, "as if it were a thoroughly neglected animal farm which urgently needed the elimination of racially poor and unhealthy stock, better breeding methods, etc. All of Europe and the East were finally to make biological sense."[154]

Much of Himmler's knowledge about animal breeding practices was directly applied to plans for human breeding to further Aryan traits.[155] Himmler was obsessed with *Lebensborn*, "Spring of Life," his program for breeding superior Nordic offspring,[156] although contrary to his racial beliefs he was willing to use blue-eyed, blond-haired children of murdered Jews to breed from.[157] Financial awards were made for giving birth if the child was of biological and racial value, and potential mothers of good Aryan stock who did not give birth were branded as "unwholesome, traitors and criminals."[158] Encouraging the propagation of good German blood was seen as so important that several Nazi leaders advocated free love in special recreation camps for girls with pure Aryan qualities. In one of Himmler's schemes, he argued that if 100 such camps were established for 1,000 girls, 10,000 "perfect" children would be born each year.[159]

Despite the criticism of the Reich minister of the interior who opposed the "idea of breeding Nordics" when it reached the point of "making a rabbit-breeding farm out of Germany,"[160] plans were developed for a series of state-run brothels where young women certified as genetically sound would be impregnated by Nazi men. The intent was to breed Aryans as if they were pedigree dogs.[161] But since, from a eugenic point of view, a weak animal will probably be of little use no matter how good the stock, young German women chosen to breed with specially selected "good" German male stock had their infants immediately taken away and put outside, unprotected, to see if the infant would survive.[162]

Other proposals and policies reflected a similar view of the German people as livestock to be improved through proper breeding. Laws passed to regulate marriage were based on "racial blood"; the goal was to prevent contamination of Germanic blood such that children born in Germany would be either purely Jewish or purely non-Jewish.[163] Even selection for membership in certain Nazi organizations, such as Himmler's SS, emphasized pure Aryan qualities, the object being to draw the sons of the best genetic families into Nazi ranks. Preference was given to those applicants having a certified family tree extending five or six generations, blond hair, blue eyes, and a height of at least six feet. They were to become the biological elite, the purest Germans.[164] One proposal suggested sending biologically unfit Germans into battle so that biologically superior individuals could be preserved for reproduction.[165]

Medical research under the Third Reich also approached Germans as livestock. For instance, those familiar with Josef Mengele's concentration camp experiments believed that his thoughtlessness for the suffering of his victims stemmed from his passion about creating a genetically pure super race "as though you were breeding horses."[166] The principal purpose of his experiments was to discover the secret of creating multiple births of babies with genetically engineered Aryan features and to improve the fertility of German women, as well as to find efficient and easy ways to mass sterilize "inferior races."[167]

While the German people themselves were dealt with as biological stock or farm animals, certain groups of people, considered contaminating or threatening to German blood and culture, were viewed as "lower animals" to be dispatched accordingly. When it came to discussing the goal of selecting out "inferior" races from the world's breeding stock, the language is full of references to contamination from contact with others considered dirty or polluting. Hitler referred to race "poisoning," while others used terms such as "race defilement" and "corruption," "decay," "rot," or "decomposition" of German "blood"[168] to refer to everything from casual acquaintanceships to sexual relations with Jews[169] and contact with their "harmful animal semen."[170] Even animals owned by Jews were seen as racially contaminating to other animals. Viereck cites the case of a German mayor who decreed that in order to further race purity, "cows and cattle which were bought from Jews, directly or indirectly, may not be bred with the community bull."[171]

Those peoples deemed genetically contaminating were thought of and treated as animals. Such labeling of people, typically emphasizing beastly or wild instincts, was not confined to Jews. "Foreign workers" were "pigs, dogs, they are creatures who are the counterfeits of human beings."[172] An SS propaganda booklet, *The Subhuman*, described all peoples of the "East" as "animalistic trash, to be exterminated."[173] Russian soldiers were a "conglomeration of animals,"[174] "unrestrained beasts" and "wild animals,"[175] and had "primitive animality."[176] Even the Romanian peasants, allies of the Germans, were described as "miserable pieces of cattle."[177]

When groups of people, most commonly Jews, were likened to specific animal species, it was usually "lower" animals or life forms, including rodents, reptiles, insects, and germs. Hitler, for instance, called the Jews a "pack of rats,"[178] and Himmler, in seeking to help some soldiers cope with having just shot one hundred Jews, told them that "bedbugs and rats have a life purpose . . . but this has never meant that man could not defend himself against vermin."[179] In one propaganda film, images of rats were superimposed over presumed "degenerate people" such as Jews, and the 1940 film *The Eternal Jew* portrayed Jews as lower than vermin, somewhat akin to the rat—filthy, corrupting, disease carrying, ugly, and group oriented.[180] Weinstein reports that because Jews were thought to be like chameleons—able to merge with their surroundings—they were made to wear the yellow Star of David so that innocent Aryans would not be contaminated by unwitting contact.[181] Jews were also likened to bacteria and "plagues" of insects.[182]

There was also the notion of *Untermenschen*, or subhumans, lower than animals. As described in one SS document:

> The subhuman—that creation of nature, which biologically is seemingly quite identical with the human, with hands, feet, and a kind of brain, with eyes and a mouth—is nevertheless a totally different and horrible creature, is merely an attempt at being man—but mentally and emotionally on a far lower level than any animal. In the inner life of that person there is a cruel chaos of wild un-

inhibited passions: a nameless urge to destroy, the most primitive lust, undisguised baseness. . . . But the subhuman lived, too. . . . He associated with his own kind. The beast called the beast. . . . And this underworld of subhumans found its leader: the eternal Jew![183]

The notion of the subhuman referred to an even lower and more distant (i.e., more dangerous in terms of pollution) form of life. This was the final twist on the Nazi phylogenetic inversion; Aryans and certain animals symbolized purity and were above human animals that were a contaminant involving impure "races" and "lower" animal species; the subhumans were below everything. Hitler, in fact, came to believe that Jews, as subhumans, were biologically demonic. He speculated that they descended from beings which "must have been veritable devils" and that it was only "in the course of centuries" that they had "taken on a human look" through interbreeding with Aryans. As the personification of the devil, Jews, to Hitler, were the main danger to the purity of the Aryan world.[184] Himmler, also buying into the notion of the subhuman, had studies made of the skulls of "Jewish-Bolshevik commissars" in order to arrive at a typological definition of the "subhuman."[185]

When coupled with a desire for racial purity, the conception of certain people as animal-like may have facilitated experimentation on concentration camp inmates as though they were as expendable as laboratory rats. At the Ravensbrück concentration camp for women, hundreds of Polish inmates—the "rabbit girls," as they were called—were given gas gangrene wounds while others were subjected to "experiments in bone grafting."[186] In some cases, concentration camp inmates were substituted for animals before trials on Germans themselves. For example, in 1941 Himmler approved the use of camp inmates in a sterilization study of a plant extract based on premature findings from rodent research, and in 1943 Himmler authorized the reversal of a research study on jaundice in which healthy animals had been injected with virus from jaundiced humans so that humans were injected with virus from diseased animals.[187] More typical were medical experiments on people that had not even been tried on animals. Experimenters such as Mengele referred to camp inmates as human "material" and their body parts as "war materials."[188] At Belsen, staff members viewed their work in terms of how many "pieces of prisoner per day" were handled, and letters from IG Farben's drug research section and Auschwitz camp authorities made reference to "loads" or "consignments" of human guinea pigs.[189]

Conceiving of certain people as animal-like also facilitated their killing. Those deemed "unfit" or "unworthy" of life were considered "degenerate"; if permitted to breed, they would only contaminate German stock and reduce its physical, mental, and moral purity.[190] Hence the need for "hygienic prophylaxis."[191] Jews in particular were viewed as "breeders of almost all evil."[192] The expectation was that those humans deemed polluting and dangerous to the race would be eliminated through a program of euthanasia. "Mercy killing," or *Gnadentod*, was for those with "lives not worth living,"[193] a notion that is strikingly similar to the 1933 animal protection regulation regarding euthanasia. The first to be given a "mercy death" were incurably insane persons or deformed infants under a 1939 plan which became known as the "euthanasia program."[194] The killing was then extended to older children. Ironically, Jewish children were at first excluded from the killing. According to the bizarre, dreamlike logic of the National Socialists, Jews did not deserve such an "act of mercy."[195]

The Holocaust victims included Jews, Gypsies, alcoholics, homosexuals, criminals, and almost anyone else the regime objected to. Extermination of humans considered to be contaminating extended beyond the killing of millions in concentration camps. By giving only limited medical and dental care and fostering abortions, the empire envisioned by the Nazis would not encourage native populations, such as those in southern Russia, to survive. It was a philosophy of utter contempt and revulsion for those thought of, in Himmler's terms, as "these human animals."[196] Speaking to his SS officers, Himmler commented: "We Germans, who are the only ones in the world who have a decent attitude toward animals, will also show a decent attitude toward these human animals, but it would be a crime against our own blood to worry about them."[197]

If the real Nazis were the comic-book figures of popular melodramas, their deeds would be no less horrible. The phenomenon we have examined, however, would be less profoundly disturbing. Our analysis raises a troubling and unsavory contradiction: that in Nazi Germany disregard for humans was coupled with concern for animals.

This paradox vanishes, however, if a broader context is placed around the treatment of humans and animals under the Third Reich. All cultures seek to order and classify human existence by grouping things in terms of shared qualities.[198] According to Mary Douglas, when something is ambiguous or hard to classify because it confuses or blurs socially constructed categories, it becomes a pollution and societal danger. The containment or elimination of polluting elements is society's effort to organize a conceptually "safe" environment by preserving the integrity of that deemed pure. Indeed, by containing the danger of pollution, people can further the illusion of their power as they seek to guard the ideal order of society against the dangers which threaten it.

At the core of this dichotomy of purity and danger is a design of society and a definition of what constitutes its boundaries and margins, and "laws of nature" are cited to strengthen the moral code and social rules that define these boundaries. In Nazi Germany, the conception of what it meant to be Aryan, or pure, relied heavily on seeing other groups of people as the societal danger. The

Nazis did this by blurring boundaries between humans and animals and by constructing a unique phylogenetic hierarchy that altered conventional Western human-animal distinctions and imperatives.

Although it may seem inconsistent, German identity was not contaminated by seeing itself closely related to animals in moral, if not biological, terms. As we saw, at times Germans saw in themselves "ideal" qualities embodied in animals, such as strength, loyalty, and fearlessness. To cope with their greatest threat, the "genetic pollution" of a pure, holistic, natural people, Germans were encouraged to fight for their survival with the same unfeeling determination as any "wild" species of life was said to do. We also saw this blurring in the alliance of Germans with animals against their "oppressors": Jews and others labeled as "vivisectors" and "torturers." In facing a common danger, Germans likened themselves, as "victims," to animals and distanced themselves from human "victimizers." Like animals, Germans were also "virtuous" and "innocent." Finally, we saw this blurring in the animalization of Germans themselves. For example, as part of the natural order, Germans of Aryan stock were to be bred like farm animals.

However, non-Aryans were seen as a polluted category precisely because they were animalized humans, or from Douglas's perspective, a freakish mix of the two categories—human and animal. That the animalization of Aryans was not seen as a pollution while the animalization of non-Aryans was so defined only points to the capriciousness of socially constructed categories. "Natural," taken-for-granted dichotomies such as human versus animal can assume various meanings and uses even in the same time and place.

Did Germans see these contradictions as we might? Ethnohistorians and ethnographers agree that what appears to the observer as a contradiction may not be experienced as a contradiction from the standpoint of the people who set up the distinctions. Nowhere was this point made clearer than in Gunnar Myrdal's *An American Dilemma*, which showed how Americans could maintain racist attitudes toward Americans of African descent while professing to believe in equality.[199] Did this paradox between racism and democratic ideals become invisible because of the historical and social psychological mechanisms that existed to justify the inconsistency? Sociologically, the answer is yes—to the extent that it was built into the culture in the same way that apple pie and motherhood are passed on to future generations as part of the American way of life.

Thus it was normative for Nazi Germans to behave cruelly and inhumanely toward polluted "lower" beings and affectionately and humanely toward morally elevated animals. Rather than seeing this as an inconsistency, in Nazi Germany it could exist as a consistency, given the consanguinity (in holistic, pure Nature) of certain "higher" humans and animals and the lack of consanguinity of certain humans and other humans. As part of the moral and intellectual context of Nazi Germany, this boundary blurring, while certainly not a causal factor, may have been one of the many contributing factors that made it possible for otherwise quite ordinary German individuals to implement the Holocaust.

Notes

1. An earlier version of this chapter appeared in *Anthrozoös* 5, 1992, pp. 6–31. We are particularly indebted to Andrew Rowan for his support and guidance at every stage of this project, as well as to those who commented on this article (see *Anthrozoös* 6, 1993, pp. 72–98).

2. R. Lifton, *The Nazi Doctors* (New York: Basic Books, 1986).

3. R. Hilberg, *The Destruction of the European Jews* (Chicago: Quadrangle, 1961).

4. H. Hoelscher, "Tierschutz und Strafrecht," doctoral dissertation presented at the University of Heidelberg, 1949; R. Neff, *Der Streit um den Wissenschaftlichen Tierversuch in der Schweiz des 19. Jahrhunderts* (Basel: Schwabe, 1989); U. Tröhler and A. Maehle, "Antivivisection in Nineteenth-Century Germany: Motives and Methods," in *Vivisection in Historical Perspective*, ed. N. Rupke (New York: Croom Helm, 1987, pp. 149–187).

5. Meyer, "Response to Arluke and Sax," *Anthrozoös* 6, 1993, pp. 88–90.

6. B. Schröder, "Das Tierschutzgesetz," dissertation, Berlin, 1933.

7. R. Proctor, *Racial Hygiene* (Cambridge: Harvard University Press, 1988).

8. H. Göring, *The Political Testament of Hermann Göring*, trans. H. W. Blood-Ryan (London: John Lang, 1939). Imagery of concentration camps is often linked with the issue of animal experimentation by both the opponents and the supporters of vivisection. The reasons for this association are complex, but the phenomenon should not be understood as either an unequivocal sanction or condemnation of such research. For a discussion of Holocaust imagery in the debate on animal experimentation and its implications, see B. Sax, "Holocaust Images and Other Powerful Ambiguities in the Debates on Animal Experimentation," *Anthrozoös* 6, 1993, pp. 108–114.

9. AMA, "New Regulations Concerning Vivisection," *Journal of the American Medical Association* 102(14), 1933, p. 1087.

10. C. Giese and Kahler, *Das Deutsche Tierschutzrecht: Bestimmungen Zum Schutze der Tiere* (Berlin: Duncker and Humbolt, 1944).

11. Ibid.

12. Ibid.

13. Ibid.

14. R. Waite, *The Psychopathic God* (New York: Basic Books, 1977).

15. Giese and Kahler, *Das Deutsche Tierschutzrecht*.

16. Waite, *The Psychopathic God*.

17. Giese and Kahler, *Das Deutsche Tierschutzrecht*.

18 Proctor, *Racial Hygiene*.

19. D. Irving, *Göring* (New York: Morrow, 1989, p. 181).

20. Ibid., p. 182.

21. A. Speer, *Inside the Third Reich* (New York: Macmillan, 1970, p. 555).

22. Irving, *Göring*, p. 182.

23. Hoelscher, "Tierschutz und Strafrecht."

24. Giese and Kahler, *Das Deutsche Tierschutzrecht*.

25. C. Bryant, "The Nazi Posture toward Animals: A Comment on Arluke and Sax," *Anthrozoös* 6, 1993, pp. 78–81.

26. Giese and Kahler, *Das Deutsche Tierschutzrecht*; Waite, *The Psychopathic God*, p. 41.

27. Giese and Kahler, *Das Deutsche Tierschutzrecht*; Hoelscher, "Tierschutz und Strafrecht;" Hans Meyer, *Der Mensch und das Tier: Anthropologische und Kultursoziologische Aspekte* (Munich: Heinz Moos, 1975).

28. P. Bookbinder, " 'Nazi Animal Protection': A Response," *Anthrozoös* 6, 1993, pp. 75–78.

29. Meyer, "Response to Arluke and Sax."

30. Giese and Kahler, *Das Deutsche Tierschutzrecht*.

31. N. Barnard, "The Nazi Experiments," *The Animal's Agenda*, April 1990, pp. 8–9.

32. Hilberg, *The Destruction of the European Jews*, pp. 600–604.

33. I. Staff, *Justiz im Dritten Reich: Eine Dokumentation*, Ilse Staff, Editor (Frankfort: Fischer Bücherei, 1964).

34. Meyer, *Der Mensch und das Tier*.

35. Giese and Kahler, *Das Deutsche Tierschutzrecht*.

36. J. Fest, *The Face of the Third Reich* (New York: Pantheon, 1970).

37. G. Mosse, *Nazi Culture: Intellectual, Cultural and Social Life in the Third Reich*, trans. Salvator Attanasio and others (New York: Grosset & Dunlap, 1966).

38. H. Glaser, *The Cultural Roots of National Socialism* (Austin: University of Texas Press, 1978, p. 138).

39. The conventional translation of *Raubtier* is "beast," but a more exact one would be "predator" or "carnivore." The Nazis, in identifying with predators celebrated in heraldry, were aligning themselves with warriors of old. While animal predation was praised in Germans, it became something to criticize in Jews. While visiting Munich in 1935, the head Gauleiter Julius Streicher offered "scientific evidence of the predatory nature of the Jews," at one point arguing insistently that, "if one were attentive while visiting zoos, one would note that the blond-haired German children always played happily in sandboxes while the swarthy Jewish children sat expectantly before the cages of beasts of prey, seeking vicarious satisfaction of their blood-tainted lusts"; G. Craig, *The Germans* (New York: New American Library, 1982).

40. Glaser, *The Cultural Roots of National Socialism*, p. 138.

41. D. Gasman, *The Scientific Origins of National Socialism: Social Darwinism in Ernst Haeckel and the German Monist League* (New York: American Elsevier, 1971).

42. Bookbinder, "Nazi Animal Protection."

43. Fest, *The Face of the Third Reich*, pp. 120, 293.

44. Ibid., p. 293.

45. G. Radde, personal communication, 1991.

46. H. Maltitz, *The Evolution of Hitler's Germany* (New York: McGraw-Hill, 1973, p. 62).

47. R. Grunberger, *A Social History of the Third Reich* (London: Weidenfeld and Nicolson, 1971, p. 136a).

48. A. Bramwell, *Ecology in the 20th Century: A History* (New Haven: Yale University Press, 1989, p. 49).

49. Ibid., pp. 49–50.

50. W. Wuttke-Groneberg, *Medizin im Nationalsozialismus* (Tübingen: Schwabische Verlaggesellschaft, 1980, p. 321).

51. F. Taylor, *The Goebbels Diaries, 1939–1941* (New York: Putnam, 1983, p. 77).

52. Giese and Kahler, *Das Deutsche Tierschutzrecht*.

53. R. Brady, *The Spirit and Structure of German Fascism* (New York: Howard Fertig, 1969, p. 53).

54. B. Sax, *The Frog King: On Legends, Fables, Fairy Tales and Anecdotes of Animals* (New York: Pace University Press, 1990, p. 82).

55. C. Koonz, *Mothers in the Fatherland: Women, the Family and Nazi Politics* (New York: St. Martin's Press, 1987, p. 220).

56. P. Viereck, *Metapolitics: The Roots of the Nazi Mind* (New York: Capricorn, 1965).

57. E. Melena, *Gemma, oder Tugend und Laster* (Munich: G. Franz, 1887).

58. W. Deuel, *People under Hitler* (New York: Harcourt Brace, 1942).

59. W. Langer, *The Mind of Adolf Hitler* (New York: Basic Books, 1972, p. 56).

60. J. Toland, *Adolf Hitler* (Garden City, N.Y.: Doubleday, 1976, p. 341).

61. Maltitz, *The Evolution of Hitler's Germany*, p. 232e.

62. Viereck, *Metapolitics*, p. 119.

63. Ibid.

64. Ibid.

65. R. Wagner, "Heldentum und Christenheit," in *Gesammelte Schriften und Dichtungen* (Leipzig: G. W. Fritsch, 1881, pp. 275–285).

66. Viereck, *Metapolitics*, p. 119.

67. L. Woltmann, *Politische Anthropologie. Woltmanns Werke*, vol. 1 (Leipzig: Dorner, 1936).

68. J. Rhodes, *The Hitler Movement* (Stanford: Hoover Institution, 1980, p. 107).

69. D. Sklar, *The Nazis and the Occult* (New York: Dorset, 1977, pp. 17–23).

70. Wagner, "Heldentum und Christenheit."

71. Proctor, *Racial Hygiene.*

72. Radde, personal communication.

73. A. Baumler-Schleinkofer, "Biologie unter dem Hakenkreuz: Biologie und Schule im Dritten Reich," *Universitas* 547, 1990, pp. 54–55.

74. A. Hitler, *Mein Kampf* (Munich: Franz Eher, 1937).

75. N. Bromberg and V. Small, *Hitler's Psychopathology* (New York: International Universities Press, 1983).

76. Wuttke-Groneberg, *Medizin im Nationalsozialismus*, p. 81.

77. Meyer, "Response to Arluke and Sax."

78. This incident tells us little about the extent or nature of Hitler's affinity for animals either as individuals in his personal life or as large and remote masses. Horses served as a major substitute for mechanized forces in World War II. When that war began, over 80 percent of the German army's motive power depended on horses. Eventually, two and a half million horses served on the eastern front, and an average of one thousand died each day (J. Lucas, *War on the Eastern Front* [New York: Basic Books, 1979]). Presumably, the killing of horses at Krim was a necessary strategic move, since Russia depended on them as much as Germany did.

79. Besides dogs, Hitler apparently felt some bond with other animals. In *Mein Kampf*, Hitler explained that deprivation had taught him to empathize with mice, so he shared his food with them. When living in Vienna, it was known that he would save bits of dried bread to feed the birds and squirrels when he read outdoors. He was particularly fond of birds, being drawn to ravens. He later gave special orders that ravens never be molested (Waite, *The Psychopathic God*, p. 41). Hitler, however, was most obsessed with wolves. In his earlier years, he used the nickname "Wolf" (Langer, *The Mind of Adolf Hitler*, p. 93); later, he chose for himself the cover name "Herr Wolf." He referred to Helena Bechstein, a mother figure to him, as "Mein Woelfchen" (O. Strasser, *Flight from Terror* [New York: R. M. McBride, 1943, p.301]), asked his sister Paula to change her name to Frau Wolf, and chose Johanna Wolf as his secretary. A number of prominent Nazis also had animal nicknames, although these names were not related to wolves. Martin Borman was known as the "bull" because of his short, thick neck; Klaus Barbie was known as "gorilla ears" in reference to the simian shape of his ears (B. Murphy, *The Butcher of Lyon* [New York: Empire, 1983, p.36]); and Goebbels was called "Mickey Mouse" (Grunberger, *A Social History of the Third Reich*, p. 335). Hitler's favorite dogs were Alsatians, or "Wolfhunde" in German, and these were the only ones with which he allowed himself to be photographed. In France he called his headquarters "Wolfschlucht" (Wolf's Gulch), in

the Ukraine, "Werwolf," and in East Prussia, "Wolfschanze" (Wolf's Lair)—saying to a servant there that "I am the wolf and this is my den." The only headquarters not named after wolves was still named after an animal. Hitler's other headquarters in 1940 was called the Eagle's Eyrie (Toland, *Adolf Hitler*, p. 832). And one of Hitler's favorite tunes was "Who's Afraid of the Big Bad Wolf?" (Langer, *The Mind of Adolf Hitler*, p. 246).

80. P. Padfield, *Dönitz: The Last Führer* (New York: Harper and Row, 1984, p. 475).

81. Toland, *Adolf Hitler*, p. 133.

82. Padfield, *Dönitz.*

83. N. Stone, *Hitler* (Boston: Little, Brown, 1980).

84. Waite, *The Psychopathic God*, p. 425.

85. J. Serpell, *In the Company of Animals: A Study of Human-Animal Relationships* (Oxford: Basil Blackwell, 1986).

86. R. Payne, *The Life and Death of Adolf Hitler* (New York: Praeger, 1960).

87. Irving, *Göring*, p. 180.

88. Padfield, *Dönitz*, p. 115.

89. Ibid., p. 331.

90. Ibid., p. 475.

91. Bookbinder, "Nazi Animal Protection," p. 78. Göring was clearly a driven hunter, a fact that bothered Hitler. Göring was so involved with his hunting expeditions that he kept extensive hunting diaries interspersed with notes of diplomatic and political meetings at hunts. He also considered being a good hunter necessary for promotion in the Luftwaffe.

92. Toland, *Adolf Hitler*, pp. 424–425.

93. Speer, *Inside the Third Reich*, pp. 115–116.

94. Fest, *The Face of the Third Reich*, p. 121.

95. A. Wykes, *Himmler* (New York: Ballantine, 1972).

96. Payne, *The Life and Death of Adolf Hitler*, p. 566.

97. Waite, *The Psychopathic God*, p. 19.

98. Stone, *Hitler*, p. 62.

99. R. Manvell and H. Fraenkel, *Hess* (London: George Allen and Unwin, 1971, p. 64).

100. Waite, *The Psychopathic God*, p. 64.

101. Although following Wagner's practices is the most persuasive and common explanation of Hitler's vegetarianism, several other attempts to explain this vegetarianism have been made. In at least one instance, Hitler's diet was attributed to his inability to tolerate the thought of animals being slaughtered for human consumption (P. Huss, *The Foe We Face* [Garden City, N.Y.: Doubleday Doran, 1942, p. 405]). For Langer (*The Mind of Adolf Hitler*, p. 56), vegetarianism was a propaganda tool to portray Hitler as kind and gentle. Langer (p. 191) also contends that Hitler became a vegetarian only after the death of his niece; in clinical practice, compulsive vegetarianism often occurs after the death of a loved one. Another writer maintains that his vegetarianism was due to chronic indigestion and the medical necessity to avoid meat (W. Bayles, *Caesars in Goosestep* [New York: Harper, 1940, p. 47]). Finally, Rauschning claims that Hitler's vegetarianism stemmed from his "absolute conviction" that decadence "had its origin in the abdomen—chronic constipation, poisoning of the juices, and the results of drinking to excess" (H. Rauschning, *The Voice of Destruction* [New York: Putnam, 1940]). Decay resulting from constipation was something which in his mind could be avoided by not eating anything resembling feces and by purging often.

102. L. Lochner, *The Goebbels Diaries, 1942–1943* (Westport, Conn: Greenwood, 1948).

103. Taylor, *The Goebbels Diaries*, p. 6.

104. Bryant ("The Nazi Posture toward Animals") astutely notes that this psychiatric profile can be seen in many infamous villains who also developed significant relation-

ships with animals. One example he cites is Robert Stroud, the "Birdman of Alcatraz," who fixated on canaries but was antisocial to other inmates in prison.

105. Lifton, *The Nazi Doctors*.
106. Toland, *Adolf Hitler*, p. 425.
107. Payne, *The Life and Death of Adolf Hitler*, p. 461.
108. Ibid., p. 461.
109. Lochner, *The Goebbels Diaries*, p. 8.
110. J. Leasor, *Rudolf Hess* (London: George Allen and Unwin, 1962).
111. Glaser, *The Cultural Roots of National Socialism*, p. 240.
112. Lifton, *The Nazi Doctors*, p. 399.
113. F. Miale and M. Selzer, *The Nuremberg Mind* (New York: Quadrangle, 1975).
114. Ibid., p. 276.
115. Ibid., p. 282.
116. H. Dicks, *Licensed Mass Murder* (New York: Basic Books, 1972).
117. J. Fest, *Hitler* (New York: Vintage, 1975); B. Lane and L. Rupp, *Nazi Ideology before 1933* (Austin: University of Texas Press, 1978).
118. Animal protection measures were not turned against Jews only. The 1941 German film *I Accuse* was released to test public response to legislation to kill persons with mental disorders that followed on the heels of legislation authorizing the euthanasia of animals. The film had jurors discussing the guilt of a physician who performed human euthanasia. One juror noted to his peers: "A few weeks ago, gentlemen, I gave my old dog the coup de grâce. He was blind and paralyzed . . . but he had served me well." Another juror responded: "But animals are different." To which the first speaker retorted: "Should people be treated worse than animals?"
119. Bryant, "The Nazi Posture toward Animals."
120. R. Lerner, *Final Solutions: Biology, Prejudice, and Genocide* (Univeristy Park: Pennsylvania State University Press, 1992).
121. Giese and Kahler, *Das Deutsche Tierschutzrecht*.
122. Accounts from this period of kosher butchering as a form of ritualistic torture resemble other slanders that have been used against the Jews, such as the kidnaping and murder of children and the killing of Christ. Cultural attitudes tend to find expression in common symbols, even when the views are never made explicit. The connection between the previously mentioned accusations against Jews and kosher butchering were likely to have been reinforced by Christian symbolism, where Christ is represented by the sacrificial lamb.
123. The association between anti-Semitism and vivisection was not confined to Germany. It was also strong in Switzerland and England. For example, a British magazine (Anonymous, "Review of Über den Wissenschaftlichen Missbrauch der Vivisection by Friedrich Zöllner," *Animal's Defender and Zoolphilist* 1, 1881, pp. 27–28) ran a highly favorable review of this author's book, offering the following summary: "Professor Zöllner has no difficulty in tracing many evils to the uncongenial influences of Judaism and Materialism. It would be wrong to say that vivisection is a Jewish pursuit, yet medicine is, in Germany at least, an eminently Jewish profession . . . " Although the anonymous British reviewer obviously shared Zöllner's anti-Semitic views, the latter sometimes expressed them in a particularly extreme manner, maintaining that Jews were by nature callous and bloodthirsty.
124. Göring, *The Political Testament of Hermann Göring*.
125. This is not to say that the Nazis were against technology. They took pride in feats of engineering such as the construction of the autobahn (A. Giesler, *Biotechnik* [Leipzig: Quelle and Meher, 1938]). In many ways, they carried technocratic control to a unique extreme. Hitler (*Mein Kampf*) often invoked the ideal of "progress." But the movement also exploited a longing for a simpler, preindustrial way of life. The Nazis wished

to take full credit for the advantages of technology while using Jews as scapegoats for the accompanying problems.

126. While the vivisectionist was explicitly identified with the Jew, vivisectionist imagery was also used to express the Romantic critique of society. For Wagner and others, animals were dynamic and sacred expressions of life that should not be destroyed politically by the atomistic state, mentally by analysis, or physically by vivisection. In at least one case, Wagner used vivisectionist imagery to attack the uninspired "dusty office desks" of government bureaucracies which he described as "modern torture-rooms . . . between files of documents and contracts, the hearts of live humanity are pressed like gathered leaves" (Viereck, *Metapolitics*, p. 109).

127. Giesler, *Biotechnik*; Proctor, *Racial Hygiene*.

128. Mosse, *Nazi Culture*.

129. Glaser, *The Cultural Roots of National Socialism*.

130. G. Chaucer, *Complete Works*, ed. W. Skeat (London: Oxford University Press, 1969).

131. A. Schopenhauer, *The Basis of Morality*, trans. A. B. Bullock (London: Swann, Sonnenschein, 1903).

132. Craig, *The Germans*.

133. R. Wagner, *Selected Letters of Richard Wagner*, trans. S. Spencer and B. Millington (New York: Norton, 1987).

134. R. Wagner, "Offenes Schreiben an Ernst von Weber," in *Gesammelte Schriften und Dichtungen* (Leipzig: G. W. Fritsch, 1888, pp. 195–210).

135. Tröhler and Maehle, *Antivivisection in Nineteenth-Century Germany*.

136. Viereck, *Metapolitics*, p. 108.

137. J. Katz, *The Darker Side of Genius: Richard Wagner's Anti-Semitism* (Hanover: Brandeis University Press, 1986).

138. H. Bretschneider, *Der Streit um die Vivisektion im 19. Jahrhundert* (Stuttgart: Gustav Fischer, 1962).

139. F. Zöllner, *Über den wissenschaftlichen Missbrauch der Vivisection* (Leipzig: Gustav Fock. 1885).

140. Sax, *The Frog King*.

141. Proctor, *Racial Hygiene*.

142. Baumer-Schleinkofer, "Biologie unter dem Hakenkreuz," pp. 57–58.

143. Ibid.

144. Lifton, *The Nazi Doctors*, p. 17.

145. Koonz, *Mothers in the Fatherland*.

146. G. Mosse, "National Socialism, Nudity and the Male Body," *Culturefront* 3(1), 1944, pp. 89–96.

147. Koonz, *Mothers in the Fatherland*.

148. A. Rosenberg, *Der Mythos des 20. Jahrhunderts* (Munich: Hoheneichen, 1935).

149. Glaser, *The Cultural Roots of National Socialism*, p. 154.

150. J. McGovern, *Martin Bormann* (N.Y.: William Morrow, 1968, pp. 11–12).

151. W. Shirer, *The Rise and Fall of the Third Reich* (New York: Simon and Schuster, 1960, p. 250).

152. Fest, *The Face of the Third Reich*, p. 116.

153. J. Bendersky, *A History of Nazi Germany* (Chicago: Nelson-Hall, 1985, p. 156).

154. Maltitz, *The Evolution of Hitler's Germany*, p. 289.

155. P. Bookbinder, personal communication, 1989.

156. Shirer, *The Rise and Fall of the Third Reich*, p. 984.

157. S. Gittleman, "Comments on Arluke/Sax Article," *Anthrozoös* 6, 1993, pp. 81–82.

158. Deuel, *People under Hitler*, pp. 164–165.

159. Ibid., p. 165.

160. Ibid., p. 203.

161. Glaser, *The Cultural Roots of National Socialism.*

162. C. Gailey, personal communication, 1990.

163. Deuel, *People under Hitler,* p. 217.

164. Bayles, *Caesars in Goosestep,* p. 155.

165. Gasman, *The Scientific Origins of National Socialism.*

166. G. Posner and J. Ware, *Mengele* (New York: McGraw-Hill, 1986, p. 31).

167. Ibid., p. 31.

168. Weinstein, *The Dynamics of Nazism,* p. 136.

169. Deuel, *People under Hitler,* pp. 210–211.

170. Brady, *The Spirit and Structure of German Fascism,* p. 53.

171. Viereck, *Metapolitics,* p. 254.

172. Grunberger, *A Social History of the Third Reich,* p. 166.

173. R. Herzstein, *The War That Hitler Won* (New York: Putnam, 1978, p. 365).

174. Lochner, *The Goebbels Diaries,* p. 206.

175. Maltitz, *The Evolution of Hitler's Germany,* p. 61.

176. Herzstein, *The War That Hitler Won,* p. 357.

177. Maltitz, *The Evolution of Hitler's Germany,* p. 61.

178. Hitler, *Mein Kampf.*

179. Hilberg, *The Destruction of the European Jews,* p. 219.

180. Herzstein, *The War That Hitler Won,* p. 309.

181. F. Weinstein, *The Dynamics of Nazism* (New York: Academic Press, 1980, p. 141).

182. Herzstein, *The War That Hitler Won,* p. 354.

183. Maltitz, *The Evolution of Hitler's Germany,* pp. 61–62.

184. H. Staudinger, *The Inner Nazi* (Baton Rouge: Louisiana State University Press, 1981).

185. Fest, *The Face of the Third Reich,* p. 113.

186. Shirer, *The Rise and Fall of the Third Reich,* p. 979.

187. Hilberg, *The Destruction of the European Jews,* pp. 601–602, 604.

188. Posner and Ware, *Mengele,* pp. 17, 39.

189. Grunberger, *A Social History of the Third Reich,* p. 330.

190. Deuel, *People under Hitler,* pp. 221, 225.

191. Herzstein, *The War That Hitler Won,* p. 66.

192. Shirer, *The Rise and Fall of the Third Reich,* p. 250.

193. Lifton, *The Nazi Doctors;* Proctor, *Racial Hygiene.*

194. Hilberg, *The Destruction of the European Jews,* p. 561; D. Peukert, *Inside Nazi Germany: Conformity, Opposition, and Racism in Everyday Life,* trans. R. Deveson (New Haven: Yale University Press, 1987).

195. Proctor, *Racial Hygiene.*

196. Maltitz, *The Evolution of Hitler's Germany,* pp. 288–289.

197. Ibid., p. 41.

198. M. Douglas, *Purity and Danger: An Analysis of Concepts of Pollution and Taboo* (Baltimore: Penguin, 1966).

199. G. Myrdal, *An American Dilemma* (New York: Harper, 1944).

13 | Working across the Human-Other Divide

Emily Martin

WHAT DIFFERENCE DOES it make how scientists think about and treat nonhuman beings? For example, if scientists thought of animals used in experimental procedures as distant kin rather than alien species, would it affect the procedures themselves? What if scientists recognized a different kind of kinship in the machines surrounding them, considering the implications of seeing them as having memory, intelligence, skill, needs, personalities, or tempers? In this chapter, I will explore some implications of several recent approaches to blurring the boundaries between humans and other creatures.

Historically, many ways have been found to breach the boundaries between humans and other sentient forms, whether to assert commonality with other forms or, in an opposite direction, to create an unbridgeable gap between them. For example, in the direction of establishing commonality between humans and other animals, in Victorian England the finely graded individual differences among the prize dogs favored by the middle classes "offered a vision of a stable, hierarchical society, where rank was secure and individual merit, rather than just inherited position, appreciated."[1] Owners of pedigree dogs were metaphorically equated with their elite pets and metonymically linked to the upper classes.[2]

The upper classes, meanwhile, were appalled at this abrogation of privileges they once monopolized. The historical prototypes of dog breeds that had been kept by the elite bore no resemblance to the dogs being bred by their new middle-class owners.[3] Demonstrations of affection and effusions of emotions toward pets that had been the preserve of the upper classes until the early nineteenth century were now accessible to other classes.

Also during the Victorian period, and working in the opposite direction to express distance from humans, other animals, considered exotic, were housed in zoos to "serve not just as a popular symbol of human domination, but also as a more precise and elaborate figuration of England's imperial enterprise."[4] This whole complex of activities surrounding the association of social class with animals in England has led Marilyn Strathern to speculate that because the English are culturally more comfortable naming those different from them by social class than Americans are, more secure, so to speak, about the reality of class differences among humans, they are more able to explore a variety of boundary crossings with nonhumans than Americans are.[5] To this difference she attributes what she describes as the puzzling amazement Americans sometimes have upon realizing that animals have been used in science as templates for all sorts of differences among humans. Reading the male dominance and aggression and female dependence and nurturance characteristic of some human societies some of the time into the nature of chimpanzee or monkey societies is one among many examples of this. I will explore some of the implications of this idea later on. Here I want to consider understanding what is at stake in moves to blur the human-other divide by exploring how some recent interpretations of scientific practices accord interesting new modes of being to nonhuman beings. I begin by examining a very influential recent theory, the "actor-network theory" promulgated by Bruno Latour and Michel Callon.[6] I will then proceed to examine some ways that various anthropologists have begun to describe how people experience the fact that the boundaries between humans and other life forms are being sundered and breached.

How Latour and Callon Work across the Human-Other Divide

Latour and Callon argue that scientists try to make their discoveries become accepted as "facts" by enrolling allies (often including other scientists, but not only them) in the new view of reality their discoveries have revealed. Scientists are agents active in the world, vigorously and energetically taking whatever steps they can to make others take account of or use the thing or concept they have discovered or invented. They write reports, publish articles, speak at conferences, visit patent offices, hire lawyers, spend time at granting agencies, lecture to diverse audiences, encourage manufacturers, and undertake a host of other concrete activities. Only insofar as scientists succeed in making other people take account of their findings, referring to them in print or the spoken word, using them in their own labs, or buying them from a distributor, do their discoveries or inventions come to be "facts" in the world.[7]

A paradigmatic case for the Latour-Callon notion might be the invention of the diesel engine. At the point in Latour's book where we meet its inventor, Rudolf Diesel, he has already succeeded in some fact building, but it has been done within the laboratory. He has a plan for "a perfect engine working according to Carnot's thermodynamic principles, an engine where ignition could occur without an increase in temperature. He wrote a book and took out a patent. This is a paper world similar to the paper world of the laboratory." To get this far, Diesel had to have garnered support inside science, obtaining the money to run his lab, support himself, and so forth. To publish and apply for a patent he had to have gained the credibility of many other people. He has some allies in the enterprise of bringing his engine to life.

But what has happened so far is simple compared to what lies ahead. "At this point," Latour writes, "he needs others to transform the two-dimensional project and patent into the form of a three-dimensional working prototype."[8] For almost fifteen years, various firms and engineers built and modified prototypes but were constantly plagued by the engine's unreliability and frequent breakdowns. At one point, disillusioned firms returned the prototypes and asked for their money back: Diesel went bankrupt and had a nervous breakdown. Just as Diesel's health was in jeopardy, so was the health of his invention: the reality of the engine, its very existence in the world, was receding.

Finally, the few engineers who had continued to work on the engine managed to get it to work reliably, just as the patent fell into the public domain. From this point on the reality of the engine became more and more solid. It was too late to save Diesel, who committed suicide by jumping off a transatlantic ship, but the work he and others had done to gain allies in believing his engine could work, in building more and stronger sets of associations with his engine, recruiting people away from other interests and to his, negotiating revisions of the design of the engine, were exactly what made the engine gain reality in the world, and the principles behind its construction become accepted as fact.

One way Latour describes the making of facts like the diesel engine is to say they become "black boxes." By that he means an entity which performs a function so reliably that it ceases to be necessary for its users to understand how it works. Black boxes in Latour's sense can become blacker and denser the more elements they attract to themselves. They can become more and more powerful the more people regard them as "obligatory passage points" in the doing of science or the carrying out of any activity.[9] So, for example, when the mechanics of a working windmill for grinding grain were mastered, "the whole windmill will act as *one piece*, resisting dissociation in spite of/because of the increasing number of pieces it is now made of."[10] It becomes a "black box." Concomitantly, as for the people around the mill:

> No matter what they want, no matter how good they were at handling the pestle, they now have to pass through the mill. Thus they are kept in line *just as much* as the wind is. If the wind had toppled the mill, then they could have abandoned the miller and gone their usual ways. Now that the top of the mill revolves, thanks to a complicated assembly of nuts and bolts, they cannot compete with it. It is a clever machination, isn't it, and *because* of it the mill has become an obligatory passage point for the people, for the corn and for the wind.[11]

Whereas black boxes in Latour's account are notable for their inertness, lack of agency, and automaticity, virtually all the other elements in a Latour-Callon network are described as agents. The preeminent agents are the scientists who recruit, engage, persuade, entice, convert, and convince through their writing, visual evidence, and prior credibility and status. Perhaps most of the converts they attempt to recruit are other humans, but many are members of nonhuman species or inanimate objects.

In an extended case study of an attempt to raise scallops in St. Brieuc Bay, France, Callon tells us that a group of scientific researchers are trying to enlist other entities (fishermen, colleagues, scallops) in their enterprise, which is to get scallops to attach themselves to artificial net bags and so increase their population in the overdredged bay. This process is called "interessement": "the group of actions by which an entity . . . attempts to impose and stabilize the identity of the other actors it defines through its problematization."[12] The group of researchers, in joining forces with the scallops, fishermen, and scientific colleagues to attain their goal, "define the identity, the goals, or the inclinations of their allies."[13] In so doing, the scientists must "build devices which can be placed between them and all other entities who want to define their identities otherwise." "Competitive" pulls on the identities of potential allies must be "cut" or "weakened" so that only links to the researchers still operate. Allies from fishermen to scallops must be "cornered,"[14] be it by physical violence, seduction, transaction, or consent without discussion.[15]

In this process the scallops are described as having every bit as much agency as the researchers or the fishermen. These scallops (*Pecten maximus*) "anchor" (attach to something) or "refuse" to anchor.[16] They are negotiated with; they can be enrolled.[17] If all sorts of transactions with the scallops (the distance anchorages are placed from shore and from the bay bottom, the material out of which the anchoring lines are made, protection from currents, etc.) are conducted successfully and the scallops do anchor, then Callon says that the negotiation is successful.[18] Their anchoring is called "voting": "the anchorage is equivalent to a vote and the counting of anchored larvae corresponds to the tallying of ballots."[19] The scientific researchers count the votes, convert them into tables or graphs, and make them part of an article or paper. As the results are discussed at conferences, if they are judged significant, the "researchers are authorized to speak legitimately for the scallops of St. Brieuc Bay."[20] The scallops themselves cannot speak, but the researchers are their "spokesmen."

The network of entities who can come together over the resolution of any scientific problem includes nonhuman organisms such as scallops, but also many nonliving objects. In one of Latour's examples, early Portuguese sailing expeditions went forth in heavy carracks that did not disintegrate in storms or long journeys. The carracks were "mobile and versatile tools, able to extract compliance from the waves, the winds, the crew, the guns and the natives, but not yet from the reefs and the coastline. These were always more powerful than the carracks since they appeared unexpectedly, wrecking the ships one after the other. How to localise in advance all the rocks instead of being, so to speak, *localised* by them without warning?"[21] The solution, which came out of a small sailing commission convened by King John II of Portugal in 1484, came to be to use the sun and stars, together with a sextant, tables, and records of bearings.

> Before this commission, capes, reefs and shoals were stronger than all the ships, but after this, the carracks plus the commission, plus the quadrants, plus the sun, had tipped the balance of forces in favour of the Portuguese carracks: the dangerous coastline could not rear up treacherously and interrupt the movement of the ship.[22]

It would be a mistake to castigate Latour and Callon for thinking in some simple-minded way that scallops or coastlines are just like people. There is a much more compelling and important reason why they adopt a way of speaking that gives agency to scallops and coastlines. This is to preserve what they call the principle of symmetry. The idea behind this is that any division between what counts as "nature" and what counts as "society" must be seen as a result of historical processes, not as a given with which an account starts.[23] Our task as scholars of science is to understand how such allocations come to be made, not to start with them, and use them as explanations. "Symmetry" refers to both sides, nature and society, playing an equal role in our description of any outcome. As Latour states, "We cannot use Society to explain how and why a controversy has been settled. We should consider *symmetrically* the efforts to enrol and control human and non-human resources."[24] To describe events this way, we have to position ourselves at the "median point" from which we can follow how people attribute both human and nonhuman properties to other beings and things.[25]

But Latour insists on symmetry only at the *start* of a series of negotiations. He tells us in many ways how intensely scientists (and others, such as lawyers) "are fighting endlessly to *create* an asymmetry between claims, an asymmetry no one can reverse easily."[26]

In the case of the early Portuguese explorers, voyages out and back are part of a "whole cycle of accumulation: how to bring things back to a place for someone to see it for the first time so that others might be sent again to bring other things back."[27]

> At every run of the accumulation cycle more elements are gathered in the centre . . . ; at every run the asymmetry . . . between the foreigners and the natives grows, ending today in something that indeed looks like a Great Divide, or at least like a disproportionate relation between those equipped with satellites who localise the "locals" on their computer maps without even leaving their air-conditioned room in Houston, and the helpless natives who do not even see the satellites passing over their heads.[28]

Claiming to preserve symmetry between the sides in what Latour calls a "trial of strength" ignores what is obvious from other places in the account: only one side *desires* to accumulate elements; only one side is driven to enroll, enlist, or convince others in an energetic entrepreneurial effort to garner as many allies as possible. One commentator on Latour's account summarizes it as treating "research as a kind of war whose only objective is domination. Winning the 'proof race' consists of establishing networks consisting of a large number of allies whose behavior one can control so as to 'make dissent impossible.' "[29]

Given this, one can only represent the account to be symmetrical by imagining that the locals desired to accumulate in the same way but carried it off less well. But surely the burden of a great deal of anthropological research is that some societies cohere in ways that would make such accumulation nonsensical. Malinowski reported that the Trobriand Islanders treated the pearls in the oysters they dove for as toys and threw them to their children. When Western merchants tried to pay the islanders to dive for more pearls, they refused beyond the point where they could use the payment in tobacco to satisfy their and their relatives' daily needs. Accumulation beyond this point would have been nonsensical.[30]

Latour and Callon could reply to this train of thought by saying that describing scallops and coastlines (not to mention locals on the other side of the world) as if they had a desire to accumulate is only a rhetorical device. Nothing less would serve to jolt us out of our habitual way of explaining that "science" is the way it is because something in "society" causes it to be that way (social studies of science), or that "society" is the way it is because of laws derived from "science" (social science). Both these moves assume from the start the division between science and society that, in fact, needs to be explained.

We might want to ponder, though, whether modes of speech can be so innocent. Marx, for one, worried about the tendency of people living in capitalist societies to speak of commodities, especially money, as if they were alive and could act on their own in the world. Saying that an investment "made" our capital "grow," that our money "earned" interest, that prices "fell," makes it seem that these inanimate objects are things mysteriously possessed of life, like "fetishes" that play a part in ritual cults in some societies. The problem Marx saw in this was that the locutions diverted attention away from the human social activities that make up the reality of capital "growing" or prices "falling": employers squeeze labor for more profit; businesses create or abolish jobs. In this topsy-turvy world, "the productions of the human brain appear as independent beings endowed with life, and entering into relation with one another and the human race."[31] Thus "fetishism," in which "the productions of the human brain appear as independent beings endowed with life," makes invisible the social relations that in reality give life to such things as profit and prices.

Parallel to what Marx called the "fetishism of commodities," have Latour-Callon created a "fetishism of the nonhuman"? If so, what does such a move conceal from us? I think the main cost would be that an active role in the production of facts is offered to many nonhuman entities only on the condition they carry out that role exactly like a Latour-Callon entrepreneurial accumulator. That is, the scallops are offered a role in the scenario that involves possible construction of the fact "the scallops have anchored" only on the assumption that they, like the scientists, are intensely trying to accumulate resources.

If everyone, every being, and every thing becomes an accumulating entrepreneur, then the conditions of capitalism, especially late capitalism, have truly spread to the ends of the earth. This is a strange kind of symmetry indeed: I will offer you an equal role in the game as long as you play *my* game!

What such a view conceals is that the scallops (not to mention the "locals") may not see the world as the putative Latourian scientist does: we do not get a chance to imagine how the scallops might perceive the activities of those strangely shaped creatures who are forever dropping ropes and nets into the water; we do not need to face squarely the fact that describing the scallops as if they had the choice to anchor conceals the manipulation of their lifeways for the greater profit of certain men.

In part Latour wants to ensnare everything that exists within his networks of accumulation for a valid reason: such a snare would catch up both the activities of Western science and the activities of "premodern" societies in the same field and make them both amenable to the same kind of account. As Latour writes: "The collectives are all similar, except for their size, like the successive helixes of a single spiral. . . . [It is] possible to respect the differences (the dimensions of the helixes do vary) while at the same time respecting the similarities (all collectives mix human and nonhuman entities together in the same way)."[32] In Latour's view, this is the only way toward a comparative anthropology. While I would agree that a crude casting of "us" versus "them" in which "our" science is used to explain "them" is misguided, I would question whether, in countering this, it is necessary to say that all creatures—human, nonhuman animal, nonhuman inanimate—necessarily operate according to the same calculus and that all collectives mix human and nonhuman "in the same way."

In addition to the profound differences which can be perceived among societies when one gives up the presumption that the accumulating, aggressive individual under capitalism is forming "his" networks and gathering "his" allies everywhere, there may be as yet poorly perceived differences between humans and other animals. For example, perhaps, as Barbara Noske states, "it is not just human subjects who socially and collectively construct their world but that animal subjects may do so too."[33] She continues: "Not many people have seriously tried to imagine what it must be like to perceive and conceive the world in terms of 'olfactory images' (such as dogs must do) or 'tactile images' (as horses do to a large extent) or 'acoustic pictures' (as dolphins and whales must do)."[34]

Beyond this, we may wonder whether all scientists have, or ever had, such an aggressive, individualistic competitive approach to the world.[35] In particular we might wonder whether some women scientists (being, for a host of reasons, more excluded from playing these games than men) may have lived their scientific lives very differently and garnered a measure of success in spite of it.[36] We might wonder whether the growth of a science, such as high energy physics, where research is conducted by large collaborating groups with cooperative links to other such groups, would mitigate any tendency toward individual competition.[37]

How Haraway and Some Anthropologists Work across the Human-Other Divide

Other scholars engaged in the cultural study of science are also grappling with the problem of crossing boundaries between the human and the other. Donna Haraway has criticized the ferocious resistance of some radical science critics to genetic engineering and to high technology generally.[38] She finds their abhorrence of the "mixing" of "natural kinds" problematic, and dependent on an impossible "imagined organic body."[39] I take this "imagined organic body" to refer to the notion that if we leave bodies alone in their given organic state, we will avoid the deleterious consequences of technological processes such as genetic engineering. She cautions that as feminists we need to tolerate exchange of substances across "kin" as widely as possible. This means being attentive to new kinds of linkages with all kinds of "kin": other animals, engineered creatures, or inanimate creatures.[40] Being attentive to these linkages does not mean accepting them uncritically, but it does mean looking into their implications in depth and with an open mind. In Haraway's view, hybrids (creatures that combine two species) or cyborgs (creatures that combine human or animal and machine) can be "potent myths for resistance" in a world where "people are not afraid of their joint kinship with animals and machines."[41] Haraway urges us to move forward, imagining new relationships across the human-other divide, in part because whether anyone likes it or not, everyone on the globe is in the process of crossing the divide in one way or another.[42]

The ethnographic evidence is overwhelming that borders are being breached on all sides: boundaries between humans and other species by patents on genetically altered creatures such as a pig with a human immune system;[43] boundaries between unique individuals and the group by cloning, which produces a group of identical individuals;[44] boundaries between humans and machines by the enormous proliferation of human-machine cyborgs in fantasy (Robocop, Terminator 2) and reality ("intelligent" machines, prosthethic devices).[45] Bodies of individuals are increasingly being disassembled and reassembled, broken down into their components, which can then be reassembled or reused in someone else. Organ transplantation, egg and fetal harvesting, and genetic engineering are only a few of the ways body parts can be taken apart and put back together in new ways.[46]

Anthropologists are finding that in the present day "human subjects and subjectivity are crucially as much a function of machines, machine relations, and information transfers as they are machine producers and operators. . . . Machines and other technologies should be understood as agents in the construction of subjectivities and bounded realms of knowledge."[47] In other words, how we think of ourselves as persons is partly a function of our interaction with machines. The writer who declares his memory is mostly "in" his personal computer, the construction worker who cooperates with a "smart" machine to do her job, the lawyer who uses computerized searches to find relevant case materials: in all these cases, the person's identity, his or her self, may include "social" interactions with machines.

As Arturo Escobar explains these processes, "Technoscience is motivating a blurring and implosion of categories at various levels, particularly the modern categories that have defined the natural, the organic, the technical and the textual."[48] Hybrid forms that cross these older categories have become commonplace, such as a nuclear reactor, in which the "natural" (radioactive material), the "organic" (humans), the "technical" (machinery), and the "textual" (computerized information) come together to form one intricately interrelated (hopefully) functioning system.

Yet to gain imaginative ways of conceptualizing relations among humans, machines, and other entities, it may not be necessary to dwell solely on the new. It is true that the sense of self need no longer be limited to a single human body, but perhaps it has never been.[49] Relations with ancestors in another world, with plants, with the divine, with animals, with machines, could in the past partake of many of the same features we now associate with strange new hybrids, ambiguous, boundary-transgressing creatures like cyborgs. And relations with nonhuman animals who act as one's hands, voice, eyes, legs, or ears have long been known from the testimonies of the handicapped.[50] Perhaps it is the weight of human-centered accounts that has prevented us from seeing that human consciousness has very frequently stretched beyond the border of the skin. Perhaps it was the dominance of the ideology of individualism in the West that shut our ears to ways our identities never were single.[51]

In the present, there seems to be some intensification of old preoccupations, such as human relationships with pets, perhaps influenced by the extent of boundary crossing all around us. Books expressing great appreciation of two-way interspecies communication are now on the best-seller lists.[52] It is striking that in some of this literature, the nonhuman animal is described as having its own "form of life" that he or she may invite or attempt to compel the human to join. Elizabeth Marshall Thomas (author of the anthropological classic *The Harmless People*) was frequently invited by a husky to participate in his nightly journeys into the city streets. When she accepted the invitation, she learned many "dog secrets."[53] Thomas encourages us to rethink our own assumptions about human dominance and consider the possibility of dog- or cat-human collectives in which we adjust to each other's lifeways and acknowledge each other's agency. However successful one judges her effort, she has a point: who in a household with a dog or cat has not been persistently ushered toward the empty food bowl?[54]

Although Latour intends the following remark to lend weight to his attempt to spread the entrepreneurial network universally, it can be read (somewhat out of context!) as expressing the richness that can come from giving up the separation between human and other:

> If the human does not possess a stable form, it is not formless for all that. . . . The expression "anthropomorphic" considerably underestimates our humanity. *We should be talking about morphism.*[55]

Morphism here, as in the computer software term *morphing*, for making one thing appear to turn into another, refers to shape changing that carries along with it a change in identity. The morphing that has produced advertisements in which a car seems to literally transform itself into a tiger and movies in which the villain transforms himself into a pool of liquid metal entails a profound change in form with continuity of identity. Latour is saying that humanity is not limited to beings in the shape of humans. Humans can morph into human-machine hybrids, for example, and retain their humanity.

Perhaps what is new about morphism is not the existence of instances of crossing the border between the human and the other, but the *normalizing* of these transitions in a general way in the United States. Perhaps we could speculate that the profusion of cyborg images and the heightened interest in interspecies communication are two arms of the same phenomenon: a generalized and cautiously eager expectation of the continuous crossing of the human-other divide.

I would like to end with two points about possible implications of this development. First, recalling the class-specific character of the English involvement with pets, are there likely to be class dimensions to these processes? We might speculate that an ease about one's security in the social hierarchy might produce a certain ease about boundary crossings of new kinds, between human and animal or human and machine. Certainly material resources necessary to access cyberspace are selectively available to those in the first world, men, and those in the privileged classes.[56] Cyberspace (the term was originally used in *Neuromancer*, a novel by William Gibson published in 1984) now usually refers to the electronic "space" one travels through by means of the Internet, electronic bulletin boards, on-line services such as Prodigy and America Online, and so on. Despite the frequently iterated goal of making access to cyberspace free and open to all, the reality falls far short of that ideal. Equipment (computers, software, telephone), knowledge, even access itself are limited in economically stratified ways.

Such stratification recalls Strathern's observations about Anglo-American differences with respect to recognizing class. It is possible that those in the elite strata in the United States who have access to the necessary technologies are becoming more "English" in the sense that they feel both a certain security of class position and simultaneously experience a willingness to explore across divides into the nonhuman.

Second, to return to my beginning questions, what might be the impact of these processes on the practice of science? Presumably, if the phenomena I have glossed as crossing the human-other divide are being widely experienced in U.S. society, they would affect scientists themselves as well as members of the general public who may have had their eggs harvested or organs transplanted, or who may think of their memories as partly residing in a personal computer.

By what route would these processes enter science, however? Traditionally wedded to a series of sharp distinctions—subject versus object, Man versus Nature, rationality versus irrationality—the blurrings I have described might be anathema to the practicing scientist.[57]

Furthermore, even the very scientists who are developing the techniques by which the body can be broken down, reassembled, and combined with machines or nonhuman parts may not have taken in the implications of these developments. In my research with immunologists, it was the rare practicing immunologist who had thought deeply about the image of the body as maintained by an immune system (a complex, internal, ready-response system) for daily life. The opposite was true for the many nonscientists in my research. Outside the confines of the laboratory, people in different social, ethnic, and economic walks of life wove rich accounts of what it means to be a body which is a scintillating, ever-changing system, connected complexly to an endless series of other systems: other people, animals, plants, society, climate, weather, and so on indefinitely.[58]

It would be risky to predict the future of processes still under way, but the "crossings" I have explored in this chapter can at least demonstrate how labile the uses of these processes can be. Making scallops into active agents can extend the reach of the capitalist entrepreneurial spirit; making dogs strange coinhabitants of the world but equally as complex as humans can challenge our human-centered assumptions about other sentient beings.

The Latour-Callon account offers participation in the human collective to other animals and to machines only on condition they too become accumulating entrepreneurs. Although this move breaks down the divide between Man and Nature, it allows room for only one kind of being, who resembles all too closely a Western businessman. It universalizes dominance of a certain kind of human activity. Other accounts I have discussed permit us to consider a collective, including humans and nonhumans, in which all participate but human dominance is not assumed. These accounts would, provided they could find their way into the practice of science, pose the greatest challenge to the boundaries between categories on which science as we know it is based.

Notes

1. Harriet Ritvo, *The Animal Estate: The English and Other Creatures in the Victorian Age* (Cambridge, Mass: Harvard University Press, 1987), p. 84.

2. Ritvo, p. 93.

3. Ritvo, p. 83.

4. Ritvo, p. 206. For a wide-ranging treatment of the depiction of animals in culture, see Steve Baker, *Picturing the Beast: Animals, Identity and Representation* (Manchester: Manchester University Press, 1993).

5. Marilyn Strathern, "Primate Visionary," *Science as Culture*, 2, no. 11 (1991): 282–295.

6. One measure of the influence of this theory is the number of positive reviews received by one of Latour's recent books, *Science in Action*; for example: Wiebe E. Bijker, "Review of Bruno Latour's *Science in Action*," *Technology and Culture*, 29 (October 1988): 982–983; Susan Leigh Star, "Review of Bruno Latour's *Science in Action*," *Sociological Review*, 36 (May 1988): 385–388; Trevor Pinch, "Review of Bruno Latour, *Science in Action*," *Sociology*, 21 (August 1987): 984–985; Henry Etzkowitz, "The Process of Science," *Science*, 238 (October 30, 1987): 695–696; Steven Yearley, "The Two Faces of Science," *Nature*, 326 (April 23, 1987): 754. Another is the citation of the theory in feminist accounts of science, without sustained critical attention. See, for example, Donna Haraway, "A Cyborg Manifesto: Science, Technology, and Socialist-Feminism in the Late Twentieth Century," *Simians, Cyborgs, and Women: The Reinvention of Nature* (New York: Routledge, 1991), pp. 149–181, or Sandra Harding, *The Science Question in Feminism* (Ithaca: Cornell University Press, 1986), esp. p. 198. Donna Haraway, "The Promises of Monsters: A Regenerative Politics for Inappropriate/d Others," in *Cultural Studies*, ed. Lawrence Grossberg, Cary Nelson, and Paula A. Treichler (New York: Routledge, 1992), pp. 295–337, has a very interesting critique of Latour, but it is buried in the footnotes (pp. 331–332). By grouping Latour and Callon, I do not mean to imply that they agree on every point, only that on the issues I discuss, they do. On this point, see Langdon Winner, "Upon Opening the Black Box and Finding It Empty: Social Constructivism and the Philosophy of Technology," *Science, Technology, and Human Values*, 18, no. 3 (1993): 362–378. Nor do I mean to imply that other scholars do not also share these views. For reviews of varying positions in science studies, see Andrew Pickering, "From Science as Knowledge to Science as Practice," in *Science as Practice and Culture*, ed. Andrew Pickering (Chicago: University of Chicago Press, 1992), pp. 1–26, and David J. Hess, "Introduction: The New Ethnography and the Anthropology of Science and Technology," in *Knowledge and Society: The Anthropology of Science and Technology*, ed. David J. Hess and Linda L. Layne (Greenwich, Conn: JAI Press, 1992), pp. 1–26.

7. Bruno Latour, *Science in Action* (Cambridge, Mass: Harvard University Press, 1987), p. 103.

8. Latour, *Science in Action*, p. 103.

9. Latour, *Science in Action*, p. 131.

10. Latour, *Science in Action*, p. 129.

11. Latour, *Science in Action*, p. 129.

12. Michel Callon, "Some Elements of a Sociology of Translation: Domestication of the Scallops and the Fishermen of St. Brieuc Bay," in *Power, Action and Belief*, ed. John Law (London: Routledge & Kegan Paul, 1986), pp. 207–208.

13. Callon, "Some Elements," p. 208.

14. Callon, "Some Elements," p. 211.

15. Callon, "Some Elements," p. 214.

16. Callon, "Some Elements," p. 209.

17. Callon, "Some Elements," p. 211.

18. Callon, "Some Elements," p. 212.

19. Callon, "Some Elements," p. 215.

20. Callon, "Some Elements," p. 216.

21. Latour, *Science in Action*, p. 221.

22. Latour, *Science in Action*, p. 222.

23. In Bruno Latour, *We Have Never Been Modern* (Cambridge, Mass: Harvard University Press, 1993), there is some elaboration of what Latour thinks these historical processes are. For lack of space, I do not consider these views here.

24. Latour, *Science in Action*, p. 144.

25. Latour, *We Have Never Been Modern*, p. 96.

26. Latour, *Science in Action*, p. 196.

27. Latour, *Science in Action*, p. 220.

28. Latour, *Science in Action*, p. 221.

29. Olga Amsterdamska, "Surely You Are Joking, Monsieur Latour!" *Science, Technology, and Human Values* 15, no. 4 (1990): 449.

30. Bronislaw Malinowski, *Coral Gardens and Their Magic* (London: George Allen and Unwin, 1935), p. 20.

31. Karl Marx, *Capital*, ed. Frederick Engels (New York: International, 1967), p. 72.

32. Latour, *We Have Never Been Modern*, p. 108.

33. Barbara Noske, *Humans and Other Animals: Beyond the Boundaries of Anthropology* (London: Pluto, 1989), p. 157–58.

34. Noske, *Humans*, p. 158. For other approaches to describing the subject worlds of animals, see Tim Ingold, Introduction to *What Is an Animal?* ed. Tim Ingold (London: Unwin Hyman, 1988), pp. 1–16; Thomas A. Sebeok, " 'Animal' in Biological and Semiotic Perspective," in *What Is an Animal?* ed. Ingold, pp. 63–76.

35. Lynda Birke, *Women, Feminism, and Biology: The Feminist Challenge* (New York: Methuen, 1986).

36. Evelyn Fox Keller, *A Feeling for the Organism: The Life and Work of Barbara McClintock* (New York: Freeman, 1983). Langdon Winner, "Upon Opening the Black Box and Finding It Empty: Social Constructivism and the Philosophy of Technology," *Science, Technology, and Human Values* 18, no. 3 (1993): 362–378, takes Latour and other social constructionists to task for ignoring groups that are "consistently excluded from power." The result is "an account that attends to the needs and machinations of the powerful as if they were all that mattered." For an account of the development of technology by Winner that does not ignore the powerless, see his "Do Artifacts Have Politics?" in *The Social Shaping of Technology: How the Refrigerator Got Its Hum*, ed. Donald MacKenzie and Judy Wajcman (Milton Keynes, England: Open University Press, 1985).

37. See Sharon Traweek, *Beamtimes and Lifetimes: The World of High Energy Physics* (Cambridge, Mass: Harvard University Press, 1988), pp. 149, 153, and Karin Knorr-Cetina, "The Couch, the Cathedral, and the Laboratory: On the Relationship Between Experiment and Laboratory in Science," in *Science as Practice and Culture*, ed. Andrew Pickering (Chicago: University of Chicago Press, 1992), pp. 113–138.

38. Haraway, "A Cyborg Manifesto," p. 154.

39. Haraway, "A Cyborg Manifesto," p. 154.

40. Remarks made by Haraway at the conference on Cyborg Anthropology at the School for American Studies, Santa Fe, NM, Oct. 1993.

41. Haraway, "A Cyborg Manifesto," p. 154.

42. Haraway, "A Cyborg Manifesto," p. 42.

43. Edmund L. Andrews, "U.S. Resumes Granting Patents on Genetically Altered Animals," *New York Times*, Feb. 3, 1993, p. A1.

44. For examples, see Sarah Franklin, "Fetal Fascinations: New Dimensions to the Medical-Scientific Construction of Fetal Personhood," in *Off-Centre: Feminism and Cultural Studies*, ed. Sarah Franklin, Celia Lury, and Jackie Stacey (London: HarperCollins, 1991), pp. 190–205; Gina Kolata, "Cloning Human Embryos," *New York Times*, 1993 p. A1; Michael Waldholz, "Scientists Halt Research to Duplicate Human Embryos After Furor Erupts," *Wall Street Journal*, (Oct. 26, 1993, pp. B6–7.

45. For a selection, see Scott Bukatman, *Terminal Identity: The Virtual Subject in Postmodern Science Fiction* (Durham, N. C.: Duke University Press, 1993); Constance Penley and Andrew Ross, eds., *Technoculture* (Minneapolis: University of Minnesota Press, 1991); Jonathan Crary and Sanford Kwinter, eds. *Incorporations* (New York: Urzone, 1992).

46. Andrew Kimbrell, *The Human Body Shop: The Engineering and Marketing of Life* (San Francisco: HarperSanFrancisco, 1993). Ruth Hubbard and Elijah Wald, *Exploding the Gene Myth: How Genetic Information Is Produced and Manipulated by Scientists, Physicians, Employers, Insurance Companies, Educators, and Law Enforcers* (Boston: Beacon Press, 1993), explore the disturbing social implications of increasing reliance on genetically based accounts of human biology and behavior.

47. Gary Lee Downey, Joseph Dumit, and Sarah Williams, "Granting Membership to the Cyborg Image," paper presented at the 91st Annual Meeting of the American Anthropological Association, San Francisco, 1992.

48. Arturo Escobar, "Welcome to Cyberia," *Current Anthropology* 35, no. 3 (1994).

49. Henrietta Moore made this point at the 1993 decennial meetings of the Association of Social Anthropologists of the Commonwealth.

50. A newspaper reported that a woman sued a restaurant, claiming she and her dog were denied service. Her dog "is trained to perform such tasks as opening doors and picking up dropped items for Mrs. Stein," who has multiple sclerosis; Katherine Richards, "Disabled Woman Charges Inn with Discrimination," *Baltimore Sun* Nov. 23, 1993, pp. B1–2

51. Franklin, "Fetal Fascinations," pp. 190–205, esp. 202.

52. Elizabeth Marshall Thomas, *The Hidden Life of Dogs* (Boston: Houghton Mifflin, 1993) is on the best-seller list as I write this, but it is one of many currently popular books that convey the way the world looks through the eyes of another species, from dogs and cats to cougars and snakes. See Michael J. Rosen, ed., *The Company of Cats* (New York: Doubleday, 1992), and Michael J. Rosen, ed., *The Company of Animals* (New York: Doubleday, 1993).

53. Thomas, *Hidden Life*.

54. Reviewers of Marshall's book have sometimes found it disappointing, claiming that her descriptions turn dogs into rather appealing people, rather than helping us see through dogs' eyes. See, for example, Harriet Ritvo, "The Hidden Life of Dogs," *New York Review of Books* 41, nos. 1–2, Jan. 13, 1994, pp. 3–5. In my opinion, parts of the book do succeed in helping us "enter into the consciousness of a nonhuman creature" (p. 121). In particular, her descriptions of time she spent with her dogs after they had acclimatized to each other and to a large, wooded, fenced area in the country evoke for me something of what it might be like to be a member of another species: "When dogs feel serene and pleased with life, they do nothing. So there on the hillside, in the warm autumn afternoons, nothing was what we did. . . . To sit idly, not doing, merely experiencing, comes

hard to a primate, yet for once I wasn't among primates. . . . Primates feel pure, flat immobility as boredom, but dogs feel it as peace" (pp. 120–121).

55. Latour, *We Have Never Been Modern*, 137.

56. For discussions pertinent to this point, see Donna Haraway, "The Politics of Postmodern Bodies: Constitutions of Self in Immune System Discourse," in *Simians, Cyborgs, and Women* (New York: Routledge, 1991), pp. 203–230; Sandy Stone, "The Empire Strikes Back: A Posttranssexual Manifesto," in *Body Guards: The Cultural Politics of Gender Ambiguity*, ed. Julia Epstein and Kristina Straub (New York: Routledge, 1991); Escobar, "Welcome to Cyberia."

57. See Evelyn Fox Keller and Christine R. Grontkowski, "The Mind's Eye," in *Discovering Reality*, ed. Sandra Harding and Merrill B. Hintikka (Dordrecht: Reidel, 1983), pp. 207–224; Evelyn Fox Keller, *Secrets of Life, Secrets of Death: Essays on Language, Gender and Science* (New York: Routledge, 1992).

58. More about this is in Emily Martin, *Flexible Bodies: Tracking Immunity in America from the Days of Polio to the Age of AIDS* (Boston: Beacon Press, 1994).

Envoi

Ruth Hubbard and Lynda Birke

THE VARIOUS TOPICS and viewpoints the contributors to this collection address meet on the central issue of respect for the organisms around which biologists are constructing their research questions and activities. Such respect, or its lack, is expressed in the ways we theorize about our experimental subjects as well as in the way we treat them. Its implications for human beings are discussed most explicitly by Karen Messing and Donna Mergler, Arnold Arluke and Boria Sax, and Vandana Shiva. Its implications for work with other organisms are the subject of most of the other contributions, while Emily Martin concentrates on experiments in which inanimate surrogates are "recruited" into supposed partnerships with animals and humans.

Clearly, biologists and biomedical scientists express their respect differently in these different situations. However, even if the scientists acknowledge the subjectivity of the organisms with whom they work, in every instance intrinsic power imbalances arise from the fact that they set, and to a large extent control, the terms of the transactions between themselves and their experimental subjects. Scientists tend to believe that they are uncovering truths about nature and therefore only have to frame the initial questions and set the experimental conditions. However, they are of course constantly making decisions that guide both the experiments and their interpretations.

Scientists always insist on remaining in control of nature, even when we acknowledge that we are part of it. Although the experiment or theory sometimes "leads" us in unanticipated directions, it can take us only as far afield as we allow ourselves to go. Even in the most open-ended, "pure" research, we cannot afford to follow every lead or else we would go off in too many directions at once, the work would get messy, and we could not "explain" what was happening. So we constantly close doors, leave loose ends to be tidied up later, and forget about most of them in order to pursue our main direction. If a set of experiments "leads nowhere," we may then turn back and explore some of the leads we neglected before. But even though the experiment proceeds by guesswork, by trial and error, we are always the ones who guide the process. The organisms are our tools, even when we try our best to imagine them to be our companions in the search.

Granted that imbalances always exist between experimenter and subject, when the subject is another person it becomes possible, at least in theory, to engage in consultations and collaborate in the formulation of the experimental agenda and methodologies. Comparable levels of collaboration cannot be achieved with nonhuman organisms. Listen as we will to what they are "trying to tell us," we are the ones who put the words into their mouths. Feminist and other politically aware anthropologists have long wrestled with problems of interpretation and translation. The most modest demand to make of biologists is that we at least acknowledge that the problems exist.

Unfortunately, to date, biological research is so untouched by the critical thinking of the last two decades and so mired in the myth of objectivity that only the rare biologist even knows that these problems exist. And, as Emily Martin points out here and Hilary Rose has noted elsewhere,[1] even critical sociologists of science, such as Bruno Latour, do not take proper account of their own role in the process by which their research questions get framed.

But disrespect for research subjects abounds also at much less subtle levels. Both human and nonhuman research subjects are routinely used (and abused) without proper regard for their individuality and needs. The current revelations about abuses that have occurred in radiation experiments on children, pregnant women, and members of the U.S. military undoubtedly are only the tip of the iceberg.[2] Messing and Mergler's discussion suggests that workers are routinely injured and exposed to toxic chemicals simply because it would cut into profits to provide better protection for them.

When it comes to animals, Hilary Rose raises the important question to what extent it is necessary, or even useful, to let animals stand in as surrogates for humans in medical research. This question is becoming more poignant and pressing because the ability to move genes around among widely different types of organisms is already resulting in the increased use of animals to model human diseases. This practice will not only increase the "recruitment" of animals for medical research; it will probably also lower the already meager respect they have hitherto been granted. As organisms increasingly get looked on as products of human invention, they come ever closer to embodying the Cartesian metaphor of animated machines or devices that merit no more respect than the inanimate products of human artifice do—a respect linked primarily to their market value. The "Harvard oncomouse," the "geep," Herman the transgenic bull, and the other genetically engineered animals and plants discussed by Vandana Shiva are valuable not as organisms in their own right but because they are expensive tools, designed to bring prestige and profit to their manufacturers.

Whatever awe biologists and the public may be able to muster for organisms as the bearers of a long history of evolutionary transformations is being replaced by a celebration of yet further evidence of human ingenuity and will. The artificial construction of novel organisms at the same time elevates and degrades these animated inventions. A similarly ambiguous status extends to humans: to the extent that worn-out or malfunctioning human parts become exchangeable for newer and better ones and "good" human genes get substituted for "faulty" ones, the healing of ill people inspires awe for medical ingenuity more than for human resilience.

But to return to more traditional laboratory practices: actually, the increase in attention to animal welfare during the last decades is probably a consequence of two different types of concerns. Keeping animals in squalid conditions wastes money because it yields questionable results. It also poses threats to human health. The fact that most of us are now infected with the presumably benign SV40 virus, unexpectedly transmitted to humans through contact with monkeys imported for medical experiments, probably has a lot to do with the fact that animal rooms are now cleaner and cages less crowded than they were when the two of us were students.

Of course, whatever the reasons for it may be, more humane treatment of animals is all to the good. But to return to the question to what extent the use of animals is necessary for medical research, the evidence has not been explored and debated enough to have a good answer. It is certainly true that to be approved for use on humans, all procedures and drugs eventually must be tested in people. So to what extent does preliminary animal research benefit us? Animal rights activists say it doesn't. Medical researchers say stopping or curtailing research on animals would destroy medical science. Is there any middle ground?

And what about the pursuit of knowledge for its own sake? This question also is hard to answer. It is difficult to judge how much current biology is done simply to know more or to advance one's career rather than with an eye to the possible applications of the knowledge. For one thing, "basic research" is usually justified by arguing that we can never foresee what research will eventually have practical utility. For another, current funding practices by government agencies, supported by taxes, have taught biologists to justify whatever experiments they want to do in terms of their potential benefits to the taxpayers.

However, a further issue that is not usually considered when scientists explain why they must use animals in order to know more about them is what gets defined as knowledge. Few biologists take account of the vast amounts of biological knowledge held by traditional societies and also by members of our own society, whose lives and work involve daily contact with plants and animals. The ethnobotanist Richard Schultes, who has spent his life cataloguing and describing the plants of the Amazon Basin, has pointed out that the native plant and animal species are being destroyed more rapidly than they can be recorded by Western scientists. In addition, sophisticated, detailed traditional knowledge about them is disappearing as local populations get drawn into modern ways of living and lose interest in acquiring the kinds of knowledge that have long been transmitted from one generation to the next.[3] And not just in faraway places but in our midst, farmers, beekeepers, animal trainers, gardeners, amateur naturalists, and other "nonscientists" have a great deal to teach biologists about the lives and habits of organisms. This means that vast amounts of already existing knowledge could be brought to our consciousness if we just listened. There is a lot that biologists could learn even without doing further experiments.

As long as biological research continues to be done, however, it is proper to insist that all uses of organisms must first be explained and justified, that as few organisms as possible must be used, and that they must be properly cared for and kept as humanely as possible. This must happen out of concern for their needs as individuals and not just because they are expensive research tools or because poor animal hygiene is a threat to human health.

Above and beyond such modest demands, reinventing biology would require that biologists, at a gut level, recognize our membership in the natural world and identify with it. At the same time, we must acknowledge the responsibilities that arise by virtue of the power we have as scientists to name, define, and categorize nature. We must learn to respect all organisms, take their needs as seriously as we do our own, and cherish the ecological relationships among us.

Biologists, as a professional group, will not come to such reevaluations unless the culture that shapes us and our practices insists on more empathy and makes our current, increasingly manipulative approaches unacceptable. This is a lot to ask at a time of great cruelty, on a global scale, of people toward each other as well as toward nonhuman organisms. But perhaps we should take courage from the fact that in recent years books that try to imagine how animals feel or conceptualize the world have become quite popular. To some extent, this interest may express a sentimental pursuit of the "exotic," but it also suggests a growing desire to learn about ways of knowing about the world other than our own.

Indeed, this interest may be related to the current popularity, noted by the authors of several chapters in this collection, of fantasies of crossing barriers of space, time, and species—say, by constructing animal parks in which people can pet dinosaurs (when the dinosaurs do not get out of hand and begin to eat the people). The creation of these fantasies is reminiscent of the setting up of zoos and safari parks in nineteenth-century Europe to underline European domination of "wild" animals and of the "primitive" lands and people from which they were removed. Like the older exemplars, the modern fantasies further the illusion that there is nothing Western Man cannot accomplish, and this presumably includes reversing the harm being done to the biosphere—in a word, nothing technology destroys cannot be fixed by more technology.

Given these diverse and contradictory trends in contemporary biology and in the culture at large—some of which imply increasing respect for organisms, others of which imply utter disdain—it is hard to know what aspects of the "re-inventions" suggested in this collection have a chance of gaining a foothold. Yet we hope that the very diversity of our contributions and approaches will stimulate people who read this book to explore out of their own backgrounds not only what aspects of biology need to be turned around but also how to do it.

Notes

1. Hilary Rose, *Love, Power and Knowledge* (Cambridge: Polity Press, 1994).

2. Brian McGrory and Sean P. Murphy, "Inmates Used in '60s drug test," *Boston Globe*, January 1, 1994, pp. 1, 16; David Armstrong, "Fernald School Chief Says Experimentation Was Routine," *Boston Globe*, January 1, 1994, p. 15; Dolores Kong, "1,800 Tested in Radiation Experiments: Vulnerable Subjects Twice US Estimate," *Boston Sunday Globe*, February 20, 1994, pp. 1, 22; Charles C. Mann, "Radiation: Balancing the Record," *Science*, 263 (1994), pp. 470–473.

3. Richard Evans Schultes, "Burning the Library of Amazonia," *The Sciences*, March/April 1994, pp. 24–30.

About the Authors

Arnold Arluke is Professor of Sociology and Anthropology at Northeastern University. His research has focused on the contradictory nature of human treatment of animals in modern societies.

Lynda Birke is a biologist in the Centre for the Study of Women and Gender at the University of Warwick in Coventry. After fifteen years of research in animal behavior, she now does research in the social studies of science. Much of her work has centered on feminist critiques of science. Her publications include *Women, Feminism and Biology: The Feminist Challenge* and *Feminism, Science and Animals: The Naming of the Shrew.*

Anne Fausto-Sterling is Professor of Medical Science in the Division of Biology and Medicine at Brown University. She has done biological research in developmental genetics and has examined the role of race and gender in the construction of scientific theories about these signifiers. She has written numerous articles in biology and in women's studies and is author of *Myths of Gender: Biological Theories about Women and Men.* She strongly believes that it is important for feminist students and scholars to understand science and that understanding feminist insights into science is essential for science students and researchers.

Ruth Hubbard is Professor Emerita of Biology at Harvard University, having retired from teaching in 1990. A founding member of the Council for Responsible Genetics, she has written and lectured extensively on scientific representations of women and on women and health. In recent years, she has focused especially on the impact of reproductive and genetic technologies on women and society. Her most recent books are *The Politics of Women's Biology, Exploding the Gene Myth* (coauthored with Elijah Wald), and *Profitable Promises: Essays on Women, Science and Health.*

Emily Martin is Professor of Anthropology at Princeton University. Her work on ideology and power in Chinese society was published in *The Cult of the Dead in a Chinese Village.* Beginning with *The Woman in the Body: A Cultural Analysis of Reproduction,* she has been working on the anthropology of science and reproduction in the United States. Her latest research is described in *Flexible Bodies: Immunity in America from the Days of Polio to the Age of AIDS*.

Judith C. Masters is Research Assistant Professor in the Department of Anthropology at the State University of New York at Stony Brook. For the past fifteen years, she has studied evolutionary theory—in particular, the biology of species formation and of species. She is currently investigating patterns of speciation among the prosimian primates of mainland Africa and Madagascar and is teaching primatology and primate evolution at Stony Brook.

Donna Mergler is a neurophysiologist and professor in the Department of Biological Sciences at the Université du Québec à Montréal and a researcher at the Centre pour l'étude des interactions biologiques entre la santé et l'environnement (CINBIOSE). She has published numerous articles on the effects of exposure to toxic substances and on women's occupational health.

Karen Messing is Professor of Biology and Director of the Center for the Study of Biological Interactions between Health and the Environment (CINBIOSE) at the Université du Québec à Montréal. Her research and publications center on women's occupational health and she is currently studying the physical workload of hospital cleaners. The research team she and Donna Mergler direct won the 1990 Muriel Duckworth Prize for action-oriented research given by the Canadian Institute for Research and Study of Women.

Stuart A. Newman is Professor of Cell Biology and Anatomy at New York Medical College and a founding member of the Council for Responsible Genetics. His research interests include the nature and evolution of mechanisms of embryonic development, an area in which he has published numerous scientific papers. He has also written articles on social and philosophical issues in the biological sciences.

Lesley J. Rogers is Professor and Head of the Department of Physiology at the University of New England in Armidale, New South Wales (Australia), where she runs an active research group. Her main research interests are in brain lateralization and the factors that influence its development. She has published widely in scientific journals and for general audiences. She recently co-authored, with John Bradshaw, a book on the evolution of brain asymmetry, language, tool use, and intellect.

Hilary Rose is Professor of Social Policy at the University of Bradford and an Honorary Research Fellow at the Institute of Education, London. She has published extensively in the politics and sociology of science. Her most recent book is *Love, Power and Knowledge: Towards a Feminist Transformation of the Sciences*.

Boria Sax is author of many books, including *The Frog King, The Parliament of*

Animals, and *The Serpent and the Swan* (forthcoming). He is founder and director of the organization Nature in Legend and Story (NILAS).

Vandana Shiva is Director of the Research Foundation for Science, Technology and Natural Resource Policy in New Delhi. She is also visiting professor at the University of Oslo and offers courses at the Schumacher College in the United Kingdom. She has written numerous books, including *Staying Alive, Ecology and the Politics of Survival, Violence of the Green Revolution,* and *Monocultures of the Mind.*

Marianne van den Wijngaard is a Researcher at the Institute on Women and Health in the Faculty of Psychology at the University of Amsterdam. Trained as a biologist, she wrote her dissertation about the way cultural ideas about femininity and masculinity have affected research on sex differences in the brain, which attracted much attention in the Dutch media. She is at work on a book entitled *Reinventing the Sexes: Feminism and the Biomedical Construction of Femininity and Masculinity, 1959–1985.*

Betty J. Wall spent the first twenty years of her career studying the physiology of nerves and muscles of insects as a researcher at a number of universities in the United States and Europe and at the Marine Biological Laboratory in Woods Hole, Massachusetts. In the 1970s, she changed her career to become a health practitioner. She has published numerous scientific papers and articles in both areas.

Index